TRANSVERSAL THEORY

This is Volume 75 in
MATHEMATICS IN SCIENCE AND ENGINEERING
A series of monographs and textbooks
Editor RICHARD BELLMAN, *University of Southern California*

A complete list of the books in this series appears at the end of this volume.

TRANSVERSAL THEORY

An account of some aspects of combinatorial mathematics

L. Mirsky

1971

ACADEMIC PRESS New York and London

ACADEMIC PRESS, INC.
111 Fifth Avenue, New York, New York 10003

United Kingdom Edition published by
ACADEMIC PRESS, INC. (LONDON) LTD.
Berkeley Square House, London, W1X 6BA

LIBRARY OF CONGRESS CATALOG CARD NUMBER: 71-142083
AMS 1970 SUBJECT CLASSIFICATIONS 05-00,05A05

PRINTED IN THE UNITED STATES OF AMERICA

PREFACE

Transversal theory, the study of combinatorial questions of which Philip Hall's classical theorem on 'distinct representatives' is the fount and origin, has only recently emerged as a coherent body of knowledge. The pages that follow represent a first attempt to provide a codification of this new subject and, in particular, to place it firmly in the context of the theory of abstract independence. I have sought to make the exposition leisurely, systematic, and as nearly self-contained as possible; but since the length of the book had to be kept within conventional bounds, it has been necessary to exclude certain topics even though they impinge on my central theme. Thus I say nothing about the subject of 'flows in networks' initiated by Ford and Fulkerson; I pass in silence over the exciting possibilities of establishing combinatorial theorems by the method of linear programming; and I refer only occasionally to the theory of graphs. I hope that as a result my presentation has gained in care and clarity what it has undoubtedly lost in breadth of treatment.

The account offered here is intended primarily for three classes of readers. It aims to serve as a detailed introduction to the methods of transversal theory for postgraduate students who wish to specialize in combinatorial mathematics. It will, perhaps, provide a convenient work of reference for experts in the field. And finally, it is a repository of combinatorial results which those engaged in the application of mathematical techniques to practical problems may find occasion to invoke.

The stock of knowledge requisite for the study of the book is modest, although a few of the arguments presuppose some degree of mathematical sophistication. The reader needs to be conversant with a small number of results from the theory of sets, including Zorn's lemma, and with some concepts in general topology: as much of this as is necessary is summarized in the first chapter. Beyond this, I assume some familiarity with the theory of vector spaces and, here and there, a nodding acquaintance with other basic structures of elementary algebra.

An early ancestor of the book is the survey article 'Systems of Representatives' contributed by Dr Hazel Perfect and myself to Volume 15 of the *Journal of Mathematical Analysis and Applications*. I have made entirely free with material from this source and I have to thank the editor and publishers (Academic Press, Inc.) for permission to do so.

I am eager to record my very deep sense of gratitude to a number of friends. My indebtedness to Dr Perfect will be plain to anyone who compares the survey with the present account. Indeed, Dr Perfect's influence has been pervasive, for she and I discussed at length almost every topic treated here, and she has put me under yet a further obligation by scrutinizing the entire manuscript. Dr J. S. Pym has read and commented on several chapters and has saved me from many blunders. Further, he extracted from his own investigations the proof of the difficult Theorem 10.4.4 presented below. I must add that it was only his and Dr Perfect's active encouragement which enabled me to complete the project. I have had many discussions with Professor R. A. Brualdi and with Dr D. J. A. Welsh, and I have benefitted greatly from their insight into combinatorial problems. My debt to Professor Richard Rado is very extensive. I owe to him the pleasure and stimulus of countless mathematical conversations and the use of much unpublished material that he most generously placed at my disposal. Above all, his contributions to transversal theory have had a decisive influence on the growth of the subject and, consequently, on the shape of this book.

I am grateful to the editor, Professor R. Bellman, for inviting me to write a volume for his series 'Mathematics in Science and Engineering.' Finally, I should like to express my appreciation of the helpfulness and impressive efficiency of Academic Press, Inc. and of the excellence of their printing.

University of Sheffield L. Mirsky
August 1970

CONTENTS

5 Variants, Refinements, and Applications of Hall's Theorem

6 Independent Transversals

7 Independence Structures and Linear Structures

8 The Rank Formula of Nash-Williams

9 Links of Two Finite Families

10 Links of Two Arbitrary Families

11 Combinatorial Properties of Matrices

12 Conclusion

TRANSVERSAL THEORY

1 Sets, Topological Spaces, Graphs

In this introductory chapter, we shall pass in brief and somewhat informal review a series of definitions and results from set theory, general topology, and the theory of graphs. This preliminary discussion will provide the necessary background for the study of transversal theory with which we shall come to grips in Chapter 2.

1.1 Sets and mappings

We shall assume that the reader is familiar with the elementary algebra of sets. Consequently, much of what is said here is intended to serve as no more than a reminder.

We regard the notion of a *set* as primitive and shall not attempt to subject it to further analysis. A set is specified by the *elements* (or *members*) which belong to or are contained in it. Two sets are therefore called *equal* if they contain the same elements. A particular set we often have to consider is the *empty set*, denoted by the symbol $\emptyset$, which contains no elements at all. If an element x belongs to a set X, we write $x \in \mathrm{X}$; in the contrary case, we write $x \notin \mathrm{X}$. We shall, whenever possible, use lower case italic letters for elements and capital Roman letters for sets. Complete notational consistency is, however, impossible since the terms 'element' and 'set' are only relative: thus given sets can themselves be elements of other sets. To avoid verbal contortions such as 'set of sets' we shall occasionally use the term *collection* as a synonym for a set. A set is called *finite* if it contains only a finite number of elements; otherwise it is called *infinite*.

We can often specify a set by listing its elements. Thus $\{1, 2, ..., k\}$ denotes the set whose elements are the first k natural numbers. It is an immediate consequence of the notion of a set that if, in such a catalogue, an element is mentioned more than once, then all but one of its occurrences can be ignored. Again, if the order in which the elements are listed is changed, the set is not affected. Thus, for example,

$$\{1, 1, 2\} = \{1, 2, 1\} = \{2, 1, 1\} = \{1, 2\} = \{2, 1\}; \tag{1}$$

and each of the above expressions stands for the set whose elements are the integers 1 and 2. Order and repetition of elements are, then, irrelevant in a set.

We shall use the symbol $\{x_1, \ldots, x_k\}_{\neq}$ to denote the set consisting of the elements $x_1, \ldots, x_k$ and at the same time express the fact that these elements are distinct. If the suffix '$\neq$' is not appended to the curly bracket, then no assumption is made about the distinctness of the elements listed.

If x is an element of some set, then, by the convention just laid down, $\{x\}$ denotes the set containing x as its only element. A set of this type is called a *singleton*. The objects x and $\{x\}$ are logically quite distinct. Thus, for example, $\{\emptyset\}$ is a set containing *one* element, namely $\emptyset$.

When the elements present in a set cannot be catalogued, we may still be able to give a 'descriptive' definition of the set by specifying some distinguishing property of its elements. Let X be a given set and let $\mathfrak{S}(x)$ be a statement about the element x of the set X. Then either of the expressions

$$\{x \in \mathrm{X} : \mathfrak{S}(x)\}, \qquad \{x : x \in \mathrm{X},\ \mathfrak{S}(x)\}$$

denotes the set of all those elements x in X for which the statement $\mathfrak{S}(x)$ is valid. More concisely, if less explicitly, we can also write $\{x : \mathfrak{S}(x)\}$ for this set if it is clear from the context that we are concerned with the elements of X.

Next, let X, Y be sets. We write $\mathrm{X} \subseteq \mathrm{Y}$ (or, equivalently, $\mathrm{Y} \supseteq \mathrm{X}$) to indicate that every element of X is also an element of Y. We then say that X is *contained in*, or is a *subset* of, Y. (This relation does not preclude the possibility of X and Y being equal, i.e. $\mathrm{X} = \mathrm{Y}$). The empty set is consequently a subset of every set. If X is not contained in Y, we write $\mathrm{X} \nsubseteq \mathrm{Y}$. If $\mathrm{X} \subseteq \mathrm{Y}$ but $\mathrm{X} \neq \mathrm{Y}$, we write $\mathrm{X} \subset \mathrm{Y}$ (or $\mathrm{Y} \supset \mathrm{X}$) and we say that X is *properly* or *strictly contained* in Y, or that it is a *proper subset* of Y. The relation $\subseteq$ is called the relation of *inclusion*, $\subset$ that of *strict* (or *proper*) *inclusion*. We shall write $\mathrm{X} \subset\subset \mathrm{Y}$ to indicate that X is a finite subset of Y.

The basic operations by means of which sets can be combined are the formations of unions, intersections, and differences. The *union* of X and Y, denoted by $\mathrm{X} \cup \mathrm{Y}$, is the set consisting of all elements which belong to at least one of X, Y. The *intersection* of X and Y, denoted by $\mathrm{X} \cap \mathrm{Y}$, is the set consisting of all elements which belong to both X and Y. Analogous definitions and obvious notational modifications apply in the case of more than two sets. The *difference* of X and Y, denoted by $\mathrm{X} \setminus \mathrm{Y}$, is the set of all elements which belong to X but not to Y.

We say that X and Y are *disjoint* if they have an empty intersection, i.e. $\mathrm{X} \cap \mathrm{Y} = \emptyset$. We say that X and Y *intersect* if they have a non-empty intersection, i.e. $\mathrm{X} \cap \mathrm{Y} \neq \emptyset$.

Let X be a subset of a set E. Then by the *complement* of X (relative to E) we mean the set $\mathrm{E} \setminus \mathrm{X}$, and we denote this set by $\mathscr{C}_{\mathrm{E}}(\mathrm{X})$ or, more briefly, by $\mathscr{C}(\mathrm{X})$ or $\mathscr{C}\mathrm{X}$.

The collection of all subsets of X, including the empty set and X itself, is called the *power set* of X and is denoted by $\mathscr{P}(X)$. If X is a (finite) set of n elements, then $\mathscr{P}(X)$ has 2^n elements.

Our next step is to introduce the fundamental notion of a mapping. Let X, Y be sets and suppose that, with each element x of X, is associated a definite element of Y, which we denote by $\phi(x)$. We then say that ϕ is a *mapping of* X *into* Y, and we express the situation symbolically by writing $\phi\colon X \to Y$. The element $\phi(x)$ is called the *image* of x (under ϕ). The sets X and Y need not be different: if $Y = X$, then ϕ is called a mapping of X *into itself*. The set X on which ϕ is defined is called its *domain*. When $A \subseteq X$, the subset $\phi(A)$ of Y is defined by the equation

$$\phi(A) = \{\phi(x) : x \in A\}$$

(so that, in particular, $\phi(\varnothing) = \varnothing$). The set $\phi(X)$ ($\subseteq Y$) is called the *range* of ϕ. In other words, the range of ϕ is the subset of Y consisting of those elements which are images, under ϕ, of elements in X. We shall normally use Greek letters for mappings.

Let $\phi\colon X \to Y$ be a mapping, suppose that $Y \subseteq Y'$, and let the mapping $\phi'\colon X \to Y'$ be defined by the equation $\phi'(x) = \phi(x)$ $(x \in X)$. Strictly speaking, ϕ and ϕ' are different objects but we shall, in practice, not distinguish between them and shall even designate them by the same symbol. Again, let $X' \subseteq X$ and let the mapping $\psi\colon X' \to Y$ be defined by the equation $\psi(x) = \phi(x)$ $(x \in X')$. We then call ψ the *restriction of* ϕ *to* X′, and write

$$\psi = \phi|X'.$$

It is useful to specify certain types of mappings. The mapping $\phi\colon X \to Y$ is called *injective* (or an *injection*) if distinct elements have distinct images, i.e. if $\phi(x_1) \neq \phi(x_2)$ whenever $x_1, x_2 \in X$ and $x_1 \neq x_2$. It is called *surjective* (or a *surjection*) if every element of Y is the image of at least one element in X, i.e. if the range of ϕ is Y. A mapping which is both injective and surjective is called *bijective* (or a *bijection*). Thus, a bijection is what, in traditional mathematical language, is known as a 'one-one correspondence.'† A bijection of a set X into itself is called a *permutation* of X. Further, if $\phi\colon X \to Y$ is an injection, then $\phi\colon X \to \phi(X)$ is a bijection. In some mathematical literature, the terms 'one-one' and 'onto' are used where we employ 'injective' and 'surjective'. Happily, this base coinage is likely before long to be withdrawn from currency.

Let $\phi\colon X \to Y$ be a mapping. It is then easily shown that

$$\phi(A \setminus B) = \phi(A) \setminus \phi(B)$$

for all $A, B \subseteq X$ if and only if ϕ is an injection.

† More precisely, a one-one correspondence is a bijection together with its 'inverse'. This last term is defined a few lines below.

Let $\phi: X \to Y$ be a bijection. We then define its *inverse* (*mapping*) $\phi^{-1}: Y \to X$ by the specification that, for each $y \in Y$, $\phi^{-1}(y)$ is the unique element x of X such that $\phi(x) = y$. Clearly ϕ^{-1} is again a bijection and $(\phi^{-1})^{-1} = \phi$. If $\phi: X \to Y$ is merely an injection, we can still define its inverse since $\phi: X \to \phi(X)$ is a bijection. In that case, ϕ^{-1} is a mapping of $\phi(X)$ into X.

Next, let $\phi: X \to Y$, $\psi: Z \to W$ be two mappings such that the range of the first is contained in the domain of the second. Then the *product* (or *composition*) $\psi\phi$ is defined as the mapping of X into W given by the equation

$$(\psi\phi)(x) = \psi\{\phi(x)\} \qquad (x \in X).$$

The mappings $\psi\phi$ and $\phi\psi$ are, in general, quite distinct: indeed, it may well happen that one of them is defined while the other is not. However, if ϕ, ψ, θ are mappings, then $\phi(\psi\theta) = (\phi\psi)\theta$ provided all the products are defined. It is therefore permissible to omit brackets and to write $\phi\psi\theta$ without risk of ambiguity—a remark which applies equally to the product of any number of mappings. If ϕ is a mapping of a set into itself, we define its powers inductively by the relations

$$\phi^1 = \phi, \qquad \phi^n = \phi\phi^{n-1} \qquad (n = 2, 3, \ldots).$$

Let $X_1, \ldots, X_n$ be pairwise disjoint sets and let $\sigma_k: X_k \to Y$ $(1 \leqslant k \leqslant n)$ be mappings. Let the mapping $\sigma: X_1 \cup \ldots \cup X_n \to Y$ be specified by the requirement that $\sigma(x) = \sigma_k(x)$ when $x \in X_k$. We then call σ the *direct sum* of the mappings $\sigma_1, \ldots, \sigma_n$.

Exercises 1.1

1. Show that two subsets A, B of a set E are unequal if and only if
$$(A \cap \mathscr{C}B) \cup (\mathscr{C}A \cap B) \neq \emptyset.$$

2. Let A, B, C be sets. Show that $C = A \cup B$ if and only if both the following conditions are satisfied. (i) $A \subseteq C$, $B \subseteq C$. (ii) The relations $A \subseteq D$, $B \subseteq D$ imply $C \subseteq D$.

3. Let $\phi: X \to Y$ be a mapping. Show that, for any subsets A and B of X,
$$\phi(A \cap B) \subseteq \phi(A) \cap \phi(B).$$
Establish the equivalence of the following statements.
   (i) ϕ is injective.
   (ii) $\phi(A \cap B) = \phi(A) \cap \phi(B)$ whenever $A, B \subseteq X$.
   (iii) $\phi(A \setminus B) = \phi(A) \setminus \phi(B)$ whenever $A, B \subseteq X$.

4. Show that the number of surjective mappings of a set of m elements into a set of n elements is equal to
$$(-1)^n \sum_{k=0}^{n} (-1)^k \binom{n}{k} k^m.$$

5. Let $A \triangle B$, the *symmetric difference* of A and B, be defined by the equation

$$A \triangle B = (A \cup B) \setminus (A \cap B).$$

Prove that $\triangle$ is an associative operation and verify that, for $n \geqslant 2$,

$$A_1 \triangle A_2 \triangle \ldots \triangle A_n$$

is the set of all those elements which belong to precisely an odd number of A's.

6. Let S be arbitrary set consisting of at least 2 elements and let $\mathscr{P}(S)$ be the power set of S. Show that, with respect to the operation $\triangle$ as addition and $\cap$ as multiplication, $\mathscr{P}(S)$ is a commutative ring with identity. Determine all divisors of zero in this ring.

7. A is a set and $\mathfrak{F}$ is a collection of subsets of A with $A \in \mathfrak{F}$ which is closed under arbitrary intersections. Prove that, for each subset X of A, there exists one and only one subset X^* of A such that (a) $X \subseteq X^*$; (b) $X^* \in \mathfrak{F}$; (c) if $X \subseteq Y$ and $Y \in \mathfrak{F}$, then $X^* \subseteq Y$. Also show that $(X^*)^* = X$.

1.2 Families

In a set, the order in which the elements occur and the frequency of their occurrence play no part. Since we need, at times, to consider totalities in which these features are present, we shall now introduce the notion of a 'family'. Strictly speaking, this is not a new object at all: we have at hand just what we need in the concept of a mapping.

Let, then, E and I be sets; let $\phi: I \to E$ be a mapping, and write $\phi(i) = x_i$ for all $i \in I$. We shall often find it useful to denote the mapping ϕ by the symbol $(x_i : i \in I)$ and to call it a *family* (or *system*) of elements of E *indexed* by I (or with *index set* I). A family is thus a mapping and not a set.† This distinction must be firmly maintained throughout our discussion: in the theory we shall develop, failure to do so is visited by calamity. To emphasize the distinction, we shall adopt the convention of using curly brackets for sets and round brackets for families. We note, in particular, that the domain of the family $(x_i : i \in I)$ is I and that its range is $\{x_i : i \in I\}$.

We shall normally use capital German letters for families. Let $\mathfrak{X} = (x_i : i \in I)$ be a family of elements of E. Unless the contrary is expressly stated we do not, of course, require that x_i, x_j should be distinct whenever $i \neq j$. Indeed, a given element x of E may occur infinitely often among the x_i. Thus the phenomenon of repetition is allowed for in the notion of a family. And equally, if the index set I possesses an 'ordinal structure',‡ say if it is the set of real numbers, then

† The distinction between sets and mappings is not as absolute as may appear from our remarks. If we were offering a systematic treatment of the theory of sets, we should take care to define mappings as certain kinds of sets (of 'ordered pairs'). This is, however, irrelevant in the present context, for we are now merely concerned to avoid confusion between $(x_i : i \in I)$ and $\{x_i : i \in I\}$.

‡ Cf. § 1.5 below.

an 'order of precedence' is set up among the elements of the family. Thus, if $i, j \in I$ and $i < j$, we may say that x_i precedes x_j.

The family $(x_i : 1 \leqslant i \leqslant n)$ will often be written in the alternative form $(x_1, x_2, \ldots, x_n)$. This notation conveys clearly that the index set (or domain) is $\{1, 2, \ldots, n\}$ and that the family is the mapping which, for $1 \leqslant i \leqslant n$, carries i into x_i. On the other hand, the meaning of a succession of n unindexed symbols such as

$$(a, b, \ldots, r) \tag{1}$$

is less immediately obvious. However, we shall agree to regard (1) as a legitimate notation for the family in which (unless the contrary is stated) the index set is understood to be $\{1, 2, \ldots, n\}$ and its elements $1, 2, \ldots, n$ are mapped into $a, b, \ldots, r$ respectively. We often call an object such as (1) an 'ordered set' or, more explicitly, an 'ordered n-tuple.' In particular, when $n = 2$, we speak of an 'ordered pair.'

With the conventions just laid down, (1, 1, 2) for example denotes the family $(x_i : 1 \leqslant i \leqslant 3)$, where $x_1 = 1, x_2 = 1, x_3 = 2$. We note that in contrast to the relation (1) in §1.1, the families

$$(1, 1, 2), \qquad (1, 2, 1), \qquad (2, 1, 1), \qquad (1, 2), \qquad (2, 1)$$

are all different (and that they do not even have the same index set) although all have the range $\{1, 2\}$. Again, a sequence $(x_n : n = 1, 2, \ldots)$ is simply a mapping of the set of positive integers into, say, the set of real numbers: here we have a further instance of a family. We shall frequently denote this family by the symbol $(x_1, x_2, x_3, \ldots)$.

For us the distinction between sets and families is fundamental. Nevertheless, it is often convenient to allow ourselves a certain latitude of language by using expressions about families which are, strictly speaking, only appropriate to sets. Indeed, we have already anticipated this convention by speaking about the 'elements of a family'. To be more explicit, we shall say that an element x of E belongs to, or is contained in, or is an element of the family $\mathfrak{X} = (x_i : i \in I)$ if $x = x_i$ for some $i \in I$. Again, let $I' \subseteq I$. Then the family $(x_i : i \in I')$ will be said to be a *subfamily* of $\mathfrak{X}$; we shall denote it by $\mathfrak{X}(I')$, and we shall write $\mathfrak{X}(I') \subseteq \mathfrak{X} = \mathfrak{X}(I)$. Further, when we speak of 'k x_i' s' in the family $\mathfrak{X} = (x_i : i \in I)$, we simply mean a subfamily $\mathfrak{X}(I')$ in which the index set I' consists of k elements. (The k x's in question need not, of course, be distinct.) A family will be called finite or infinite according as its index set is finite or infinite. If $\mathfrak{A} = (a_i : i \in I)$, $\mathfrak{B} = (b_j : j \in J)$ are two families with disjoint index sets, then the family $\mathfrak{A} + \mathfrak{B}$ is defined as $(c_k : k \in I \cup J)$, where $c_k = a_k$ or b_k according as $k \in I$ or $k \in J$.

The notions of union and intersection are easily extended to families of sets.

Thus, let $(A_i: i \in I)$ be a family of subsets of E. Then the union

$$\bigcup_{i \in I} A_i \tag{2}$$

denotes the set consisting of all those elements of E which belong to at least one A_i; and the intersection

$$\bigcap_{i \in I} A_i \tag{3}$$

is the set of those elements which belong to all A_i. It follows that, for $I = \emptyset$, the union (2) is $\emptyset$ while the intersection (3) is E.

A family $(A_i: i \in I)$ of subsets of E is called a *partition* of E if the A_i are pairwise disjoint and their union is E. We shall sometimes describe the same situation by saying that $E = \bigcup (A_i: i \in I)$ is a partition.

Again, let as before $(A_i: i \in I)$ be a family of subsets of E; let F be a set; and consider any mapping $\phi: E \to F$. Then

$$\phi\left(\bigcup_{i \in I} A_i\right) = \bigcup_{i \in I} \phi(A_i).$$

On the other hand, we have

$$\phi\left(\bigcap_{i \in I} A_i\right) \subseteq \bigcap_{i \in I} \phi(A_i);$$

but if ϕ is injective, then the relation of inclusion can be replaced by that of equality.

We next turn to the subject of cartesian products. Let (A, B) be a family of two sets (or, more precisely, a family of sets whose index set consists of two elements). We define the *cartesian product* $A \times B$ of this family as the set of all ordered pairs (x, y) with $x \in A$, $y \in B$. More generally, the cartesian product of the family $(A_1, \ldots, A_n)$ is defined as the set of all ordered n-tuples

$$(x_1, \ldots, x_n) \tag{4}$$

with $x_1 \in A_1, \ldots, x_n \in A_n$. This product is denoted by $A_1 \times \ldots \times A_n$ or, alternatively, by

$$\mathop{\times}_{i=1}^{n} A_i.$$

It is not immediately obvious how this definition is to be extended to the case of infinite families. To meet the difficulty, we shall make a fresh start. Let

$A_1, \ldots, A_n$ be subsets of E, and consider a mapping

$$\phi : \{1, \ldots, n\} \to E$$

with the property that

$$\phi(1) \in A_1, \ldots, \phi(n) \in A_n.$$

Thus ϕ is simply a mapping which 'chooses' one element from each of the sets $A_1, \ldots, A_n$; and we shall accordingly call it a 'choice function' of the family $(A_1, \ldots, A_n)$. We now recognize that the n-tuple (4) is just such a choice function and that the cartesian product $A_1 \times \ldots \times A_n$ is simply the set of all choice functions of the family $(A_1, \ldots, A_n)$.

When the definition of a cartesian product is framed in this manner, it can easily be extended to families with arbitrary index sets. Let, then, E, I be sets and let $\mathfrak{A} = (A_i : i \in I)$ be a family of subsets of E. Any mapping $\phi : I \to E$ such that $\phi(i) \in A_i$ for all $i \in I$ is called a *choice function* of $\mathfrak{A}$; and the set of all such choice functions is known as the cartesian product of $\mathfrak{A}$ and is denoted by

$$\bigtimes_{i \in I} A_i.$$

If A_i is empty for some $i \in I$, then the cartesian product is evidently also empty. Now there are many mathematical situations which call for the converse inference; and in the usual axiomatic presentations of the theory of sets (such as that associated with the names of Zermelo and Fraenkel) the legitimacy of this inference is guaranteed not by a theorem but by a separate axiom. This axiom, which is known as the *axiom of choice*, asserts that *every (non-empty) family of non-empty sets possesses at least one choice function*. Its status in the theory of sets is accounted for by the fact that it cannot be derived from the other axioms. The importance of the axiom of choice can hardly be overstated: modern mathematics would be a different and a poorer thing if a self-denying ordinance compelled us to relinquish its use.

Exercises 1.2

1. Let E, F be sets; let $\phi : E \to F$ be mapping; and let $(A_i : i \in I)$ be a family of subsets of E. Show that

$$\phi\left(\bigcup_{i \in I} A_i\right) = \bigcup_{i \in I} \phi(A_i), \qquad \phi\left(\bigcap_{i \in I} A_i\right) \subseteq \bigcap_{i \in I} \phi(A_i),$$

and that the relation of inclusion can be replaced by that of equality when ϕ is injective.

2. Let X, Y be sets and let $\psi : \mathscr{P}(X) \to \mathscr{P}(Y)$ be a mapping. Show that, if $\psi(A \cup B) = \psi(A) \cup \psi(B)$ for all $A, B \subseteq X$, then $\psi(A \cap B) \subseteq \psi(A) \cap \psi(B)$ for all $A, B \subseteq X$.

3. Let $(A_i\colon i \in I)$ be a family of subsets of E. Establish the 'De Morgan identities':

$$\mathscr{C}\Big(\bigcup_{i\in I} A_i\Big) = \bigcap_{i\in I} \mathscr{C}A_i, \qquad \mathscr{C}\Big(\bigcap_{i\in I} A_i\Big) = \bigcup_{i\in I} \mathscr{C}A_i.$$

4. Let $(X_i\colon i \in I)$ and $(Y_j\colon j \in J)$ be two families of subsets of a set E. Verify the relation

$$\bigcup_{i\in I} X_i \cap \bigcup_{j\in J} Y_j = \bigcup_{(i,j)\in I\times J} (X_i \cap Y_j).$$

5. If $(A_i\colon i \in I)$, $(B_i\colon i \in I)$ are two families of subsets of E with the same index set I, show that

$$\bigtimes_{i\in I} A_i \cap \bigtimes_{i\in I} B_i = \bigtimes_{i\in I} (A_i \cap B_i).$$

6. Let $(A_i\colon i \in I)$ be an infinite family of sets of which at least one is finite. Show that, if $\bigcap(A_i\colon i \in J) \neq \emptyset$ for every $J \subset\subset I$, then $\bigcap(A_i\colon i \in I) \neq \emptyset$. Show also that the qualification 'of which at least one is finite' cannot be omitted.

7. Let $(A_i\colon i \in I)$ be a family of sets. Can every subset of the cartesian product $\bigtimes(A_i\colon i \in I)$ be expressed in the form $\bigtimes(B_i\colon i \in I)$, where $B_i \subseteq A_i$ $(i \in I)$?

1.3 Mapping theorems and cardinal numbers

We shall next develop a number of results on pairs of mappings. These will be crucial to many of the arguments in combinatorial theory.

Our first result is concerned with mappings of power sets. A mapping $\Theta\colon \mathscr{P}(X) \to \mathscr{P}(Y)$ (which maps every subset of X into some subset of Y) is said to be *isotone* if the relation $X_1 \subseteq X_2$ implies $\Theta(X_1) \subseteq \Theta(X_2)$.

LEMMA 1.3.1. *Let* X, Y *be sets and* $\Theta\colon \mathscr{P}(X) \to \mathscr{P}(Y)$, $\Psi\colon \mathscr{P}(Y) \to \mathscr{P}(X)$ *isotone mappings. Then there exist partitions* $X = X_1^* \cup X_2^*$, $Y = Y_1^* \cup Y_2^*$ *such that* $\Theta(X_1^*) = Y_1^*$, $\Psi(Y_2^*) = X_2^*$.

To prove this, we observe that the collection

$$\mathscr{F} = \{F \subseteq X\colon X \setminus \Psi(Y \setminus \Theta(F)) \subseteq F\}$$

(of subsets of X) is non-empty since $X \in \mathscr{F}$. Let

$$X_1^* = \bigcap_{F\in\mathscr{F}} F,$$

so that

$$X_1^* \subseteq F \quad \text{for each } F \in \mathscr{F}. \tag{1}$$

Let $A \subseteq B \subseteq X$. Then, since Θ and Ψ are isotone, we have

$$X \setminus \Psi(Y \setminus \Theta(A)) \subseteq X \setminus \Psi(Y \setminus \Theta(B)). \qquad (2)$$

In view of (1), this implies that, for each $F \in \mathscr{F}$,

$$X \setminus \Psi(Y \setminus \Theta(X_1^*)) \subseteq X \setminus \Psi(Y \setminus \Theta(F)) \subseteq F.$$

Hence

$$X \setminus \Psi(Y \setminus \Theta(X_1^*)) \subseteq \bigcap_{F \in \mathscr{F}} F = X_1^*.$$

Denoting the expression on the left-hand side by X_1^{**}, we thus have $X_1^{**} \subseteq X_1^*$. Hence, by (2),

$$X \setminus \Psi(Y \setminus \Theta(X_1^{**})) \subseteq X \setminus \Psi(Y \setminus \Theta(X_1^*)) = X_1^{**},$$

i.e. $X_1^{**} \in \mathscr{F}$ and so, by (1), $X_1^* \subseteq X_1^{**}$. It follows that $X_1^{**} = X_1^*$, i.e.

$$X \setminus \Psi(Y \setminus \Theta(X_1^*)) = X_1^*.$$

Finally, put $X_2^* = X \setminus X_1^*$, $Y_1^* = \Theta(X_1^*)$, $Y_2^* = Y \setminus Y_1^*$. Then

$$\Psi(Y_2^*) = \Psi(Y \setminus \Theta(X_1^*)) = X \setminus X_1^* = X_2^*,$$

as required.

THEOREM 1.3.2. (Perfect & Pym) *Let* X, Y, X', Y' *be sets with* $X' \subseteq X$, $Y' \subseteq Y$, *and let* $\theta: X' \to Y$, $\psi: Y' \to X$ *be mappings. Then there exist sets* X_0, Y_0 *with* $X' \subseteq X_0 \subseteq X$, $Y' \subseteq Y_0 \subseteq Y$ *and partitions* $X_0 = X_1 \cup X_2$, $Y_0 = Y_1 \cup Y_2$ *such that* $X_1 \subseteq X'$, $Y_2 \subseteq Y'$, $\theta(X_1) = Y_1$, $\psi(Y_2) = X_2$.

We define isotone mappings $\Theta: \mathscr{P}(X) \to \mathscr{P}(Y)$ and $\Psi: \mathscr{P}(Y) \to \mathscr{P}(X)$ by means of the equations

$$\Theta(A) = \theta(A \cap X') \quad (A \subseteq X),$$
$$\Psi(B) = \psi(B \cap Y') \quad (B \subseteq Y).$$

Let X_1^*, X_2^*, Y_1^*, Y_2^* be sets with the properties specified in Lemma 1.3.1. Write

$$X_1 = X_1^* \cap X', \quad X_2 = X_2^*, \quad X_0 = X_1 \cup X_2,$$
$$Y_1 = Y_1^*, \quad Y_2 = Y_2^* \cap Y', \quad Y_0 = Y_1 \cup Y_2.$$

It is clear that these equations define partitions of X_0 and Y_0, and that

$X_1 \subseteq X'$, $Y_2 \subseteq Y'$. Furthermore

$$X_0 = (X_1{}^* \cap X') \cup X_2{}^* = (X_1{}^* \cup X_2{}^*) \cap (X' \cup X_2{}^*)$$
$$= X \cap (X' \cup X_2{}^*) = X' \cup X_2{}^* \supseteq X',$$

and similarly $Y_0 \supseteq Y'$. Finally

$$\theta(X_1) = \theta(X_1{}^* \cap X') = \Theta(X_1{}^*) = Y_1{}^* = Y_1,$$
$$\psi(Y_2) = \psi(Y_2{}^* \cap Y') = \Psi(Y_2{}^*) = X_2{}^* = X_2,$$

and the assertion is therefore established.

The case $X' = X$, $Y' = Y$ of the result just proved is of sufficient interest to be worth stating separately.

Corollary 1.3.3. (Banach) *Let* X, Y *be sets and let* $\theta: X \to Y$, $\psi: Y \to X$ *be mappings. Then there exist partitions* $X = X_1 \cup X_2$, $Y = Y_1 \cup Y_2$ *such that* $\theta(X_1) = Y_1, \psi(Y_2) = X_2$.

Another consequence of Theorem 1.3.2 runs as follows.

Theorem 1.3.4. (O. Ore) *Let* X, Y, X', Y' *be sets with* $X' \subseteq X$, $Y' \subseteq Y$, *and let* Δ *be a subset of the cartesian product* $X \times Y$. *Further, let* $\theta: X' \to Y$, $\psi: Y' \to X$ *be injective mappings and suppose that* $(x, \theta(x)) \in \Delta$ *for all* $x \in X'$ *and* $(\psi(y), y) \in \Delta$ *for all* $y \in Y'$. *Then there exist sets* X_0, Y_0 *with* $X' \subseteq X_0 \subseteq X$, $Y' \subseteq Y_0 \subseteq Y$ *and a bijection* $\sigma: X_0 \to Y_0$ *such that* $(x, \sigma(x)) \in \Delta$ *for all* $x \in X_0$.

Let $X_0, X_1, X_2, Y_0, Y_1, Y_2$ be sets with the properties specified in Theorem 1.3.2. Then

$$(x, \theta(x)) \in \Delta \quad \text{for all} \quad x \in X_1,$$
$$(\psi(y), y) \in \Delta \quad \text{for all} \quad y \in Y_2. \tag{3}$$

Now ψ is an injective mapping and $\psi(Y_2) = X_2$. Hence the restriction of ψ to Y_2 is a bijection of Y_2 into X_2. Thus ψ^{-1} exists, and (3) can be written in the form

$$(x, \psi^{-1}(x)) \in \Delta \quad \text{for all} \quad x \in X_2.$$

Defining the mapping $\sigma: X_0 \to Y_0$ by the equations

$$\sigma(x) = \begin{cases} \theta(x) & (x \in X_1) \\ \psi^{-1}(x) & (x \in X_2), \end{cases}$$

we arrive at the desired conclusion.

As for Theorem 1.3.2, we again formulate the special case $X' = X$, $Y' = Y$.

COROLLARY 1.3.5. *Let* X, Y *be sets and let* Δ *be a subset of* $X \times Y$. *Further, let* $\theta: X \to Y$, $\psi = Y \to X$ *be injective mappings and suppose that* $(x, \theta(x)) \in \Delta$ $(x \in X)$ *and* $(\psi(y), y) \in \Delta$ $(y \in Y)$.*Then there exists a bijection* $\sigma: X \to Y$ *such that* $(x, \sigma(x)) \in \Delta$ $(x \in X)$.

Much of the subsequent discussion of combinatorial theory is concerned with problems involving the 'size' of sets. To be able to handle such questions, we need to introduce some measure of size; and this is provided by the notion of a 'cardinal number'.

With each set X, there is associated a certain uniquely defined object called its *cardinal number* (or simply *cardinal*) and denoted by $|X|$. When X is finite, $|X|$ is defined as the number of elements in X. In the case of infinite sets, the definition of cardinal number presupposes an elaborate technical discussion, but fortunately we can dispense with it for our limited purpose. We shall not, in fact, need to define cardinal numbers: it will be sufficient to explain what constitutes equality and inequality between them.

Let, then X and Y be arbitrary sets. We say that their cardinal numbers are equal, and we write $|X| = |Y|$, if and only if there exists a bijective mapping of X into Y (or, equivalently, of Y into X). It is plain that, for finite sets, this definition is in conformity with the definition laid down a few lines earlier.

Inequalities between cardinal numbers can also be defined without difficulty. If X, Y are sets and there exists an injective mapping of X into Y, then we say that the cardinal of X is not greater than the cardinal of Y, and we write $|X| \leqslant |Y|$. If, in addition, $|X| \neq |Y|$, we write $|X| < |Y|$. We leave it to the reader to verify that this notation, too, is consistent with the definition of cardinal numbers for finite sets.

If a, b are real numbers, then the relations $a \leqslant b$ and $b \leqslant a$ imply the equality $a = b$. The analogous inference for cardinal numbers is not at all obvious. That it is nevertheless valid is guaranteed by the following important theorem.

THEOREM 1.3.6. (Schröder–Bernstein) *If* X *and* Y *are any sets, then the relations* $|X| \leqslant |Y|$ *and* $|Y| \leqslant |X|$ *imply that* $|X| = |Y|$.

By hypothesis, there exist injective mappings $\theta: X \to Y$, $\psi: Y \to X$. It follows by Corollary 1.3.5 (with $\Delta = X \times Y$) that there exists a bijection of X into Y, so that $|X| = |Y|$.

Another result of a similar kind will be needed subsequently.

LEMMA 1.3.7. *If* $X' \subseteq X$, $Y' \subseteq Y$, $|X'| \leqslant |Y'|$, $|X| = |Y|$, *then there exists a set* X^* *such that* $X' \subseteq X^* \subseteq X$ *and* $|X^*| = |Y'|$.

Since $|X'| \leqslant |Y'|$, there exists an injection $\theta: X' \to Y'$. Moreover, since $|Y'| \leqslant |Y| = |X|$, there exists an injection $\psi: Y' \to X$. Hence, by Theorem 1.3.4 (with $\Delta = X \times Y'$), there exists a set X^* with $X' \subseteq X^* \subseteq X$ and a bijection $\sigma: X^* \to Y'$. Hence $|X^*| = |Y'|$.

In the discussion of cardinal numbers, the set (say N) of natural numbers plays a special role since $|N|$ (universally denoted by $\aleph_0$) is the smallest infinite cardinal. This follows from the fact, the proof of which requires the axiom of choice, that every infinite set has a subset of cardinal $\aleph_0$.

Any set which can be mapped bijectively into N is called an (infinite) *denumerable* (or *countable*) set. The elements of such a set can be 'numbered off', i.e. the set can be written in the form $\{x_1, x_2, x_3, \ldots\}$. We recall a basic result proved in almost all books of analysis, namely that while the set of all rational numbers is denumerable, the set of all real numbers (or, indeed, the set of all real numbers in any non-degenerate interval) is not denumerable.

We sometimes speak of the cardinal number of a family, say $\mathfrak{A} = (A_i : i \in I)$. By this we mean the cardinal number of its index set I, and we denote this occasionally by $|\mathfrak{A}|$.

Exercises 1.3

1. Show that, for any finite sets $X_1, \ldots, X_n$,

$$|X_1 \cup \ldots \cup X_n| = \sum_{\emptyset \subset I \subseteq \{1, \ldots, n\}} (-1)^{|I|-1} \left| \bigcap_{i \in I} X_i \right|.$$

Also show that this identity remains valid if the symbols $\cup$ and $\cap$ are interchanged.

2. Establish the following implications. (i) If $X \subseteq Y$, then $|X| \leqslant |Y|$. (ii) If $|X| \leqslant |Y|$ and $|Y| \leqslant |Z|$, then $|X| \leqslant |Z|$.

3. Prove Cantor's theorem that, for any set X, $|X| < |\mathscr{P}(X)|$.

[Halmos (**1**), 93]

4. Let N be the set of all natural numbers. Show that the collection of all finite subsets of N is denumerable.

5. Let E be a set and ϕ an isotone mapping of the power set of E into itself. Show that there exists a non-empty subset E^* of E such that $\phi(E^*) = E^*$.

[Sierpiński (**1**), 145-7]

6. Denote by F the field of two elements (say 0 and 1). Let S be a set and let addition, multiplication, and multiplication by scalars in $\mathscr{P}(S)$ be defined by the equations

$$A + B = (A \cup B) \setminus (A \cap B),$$
$$A\,B = A \cap B,$$
$$1A = A, \qquad 0A = \emptyset.$$

Show that, with respect to these operations, $\mathscr{P}(S)$ is a commutative algebra of dimension $|S|$ over F. (Cf. Ex. 1.1.6.)

7. Prove that every infinite set possesses a denumerable subset.

1.4 Boolean atoms

The way in which the sets of a given family $(A_1, \ldots, A_n)$ overlap can be very complicated. To describe it effectively, it is sometimes convenient to consider certain associated sets known as 'boolean atoms'.

Write $N_0 = \{1, \ldots, n\}$ and, when $\emptyset \subset N \subseteq N_0$, put

$$A[N] = \bigcap_{i \in N} A_i \setminus \bigcup_{i \notin N} A_i.$$

Thus $A[N]$ is the set of all elements which belong to each A_i with $i \in N$ and to no A_i with $i \notin N$. The $2^n - 1$ sets $A[N]$ $(\emptyset \subset N \subseteq N_0)$ will be called the *boolean atoms generated by* $(A_1, \ldots, A_n)$.

LEMMA 1.4.1. *The family* $(A[N] : \emptyset \subset N \subseteq N_0)$ *of boolean atoms generated by* $(A_1, \ldots, A_n)$ *is a partition of* $A_1 \cup \ldots \cup A_n$. *In other words, the boolean atoms are pairwise disjoint and their union is* $A_1 \cup \ldots \cup A_n$.

Let $\emptyset \subset N_1, N_2 \subseteq N_0$ and $N_1 \neq N_2$. If $A[N_1] \cap A[N_2] \neq \emptyset$, let $x \in A[N_1] \cap A[N_2]$. Denote by i_1 an integer which belongs to one of N_1, N_2 but not to both, say $i_1 \in N_1$, $i_1 \notin N_2$. Now the relations $x \in A[N_1]$, $i_1 \in N_1$ imply $x \in A_{i_1}$; and $x \in A[N_2]$, $i_1 \notin N_2$ imply $x \notin A_{i_1}$. We thus arrive at a contradiction and conclude that the atoms are pairwise disjoint.

Again, whenever $\emptyset \subset N \subseteq N_0$, we have

$$A[N] \subseteq \bigcup_{i=1}^{n} A_i.$$

It follows that

$$\bigcup_{\emptyset \subset N \subseteq N_0} A[N] \subseteq \bigcup_{i=1}^{n} A_i.$$

On the other hand, let $x \in A_1 \cup \ldots \cup A_n$. Then there exists a set N' such that $\emptyset \subset N' \subseteq N_0$ and

$$x \in A_i \quad (i \in N'), \qquad x \notin A_i \quad (i \notin N').$$

Therefore

$$x \in A[N'] \subseteq \bigcup_{\emptyset \subset N \subseteq N_0} A[N]$$

and so

$$\bigcup_{i=1}^{n} A_i \subseteq \bigcup_{\emptyset \subset N \subseteq N_0} A[N].$$

Thus the union of the atoms is, indeed, equal to $A_1 \cup \ldots \cup A_n$.

Our next result shows how the original sets may be recovered from the boolean atoms.

LEMMA 1.4.2. *Let* $(A_1, \ldots, A_n)$ *be a family of sets. Then, for* $1 \leqslant k \leqslant n$, *we have*

$$A_k = \bigcup_{k \in N \subseteq N_0} A[N].$$

If $x \in \bigcup(A[N]: k \in N \subseteq N_0)$, then $x \in A[N]$ for some N with $k \in N \subseteq N_0$; and therefore $x \in A_k$. On the other hand, let $x \in A_k$. Then, by Lemma 1.4.1, x is contained in some atom $A[N']$ (and in no other). Assume, now, that $k \notin N'$. Then, since

$$x \in A[N'] = \bigcap_{i \in N'} A_i \setminus \bigcup_{i \notin N'} A_i$$

and so

$$x \notin \bigcup_{i \notin N'} A_i,$$

we infer that $x \notin A_k$. This contradicts our hypothesis so that, in fact, $k \in N'$. Hence

$$x \in A[N'] \subseteq \bigcup_{k \in N \subseteq N_0} A[N];$$

and the proof is complete.

Exercises 1.4

1. We call χ the 'characteristic function' of a set A if $\chi(x) = 1$ or 0 according as $x \in A$ or $x \notin A$.

Denote by μ an 'additive measure', i.e. a mapping of sets into real numbers such that $\mu(A \cup B) = \mu(A) \cup \mu(B)$ whenever $A \cap B = \emptyset$.

Let $A_1, \ldots, A_n$ be arbitrary sets, $\chi_1, \ldots, \chi_n$ their characteristic functions, and $k_1, \ldots, k_n$ real numbers. Show that, if

$$\sum_{r=1}^{n} k_r \chi_r(x) = 0$$

for all $x \in A_1 \cup \ldots \cup A_n$, then

$$\sum_{r=1}^{n} k_r \, \mu(A_r) = 0. \tag{1}$$

[R. Rado]

2. Let the notation be as in Ex. 1.4.1. An additive measure μ will be called 'special' if there exists an element x such that $\mu(A) = 1$ or 0 according as $x \in A$ or $x \notin A$. Show that, if the relation (1) holds for every special additive measure, then it holds for every additive measure.

[R. Rado]

3. With the same notation as in Ex. 1.4.1, establish the identities

$$\mu\Big(\bigcap_{i=1}^{n} A_i\Big) = \sum_{\varnothing \subset I \subseteq \{1,\ldots,n\}} (-1)^{|I|-1} \mu\Big(\bigcup_{i\in I} A_i\Big),$$

$$\mu\Big(\bigcup_{i=1}^{n} A_i\Big) = \sum_{\varnothing \subset I \subseteq \{1,\ldots,n\}} (-1)^{|I|-1} \mu\Big(\bigcap_{i\in I} A_i\Big).$$

[R. Rado]

4. Extend the definitions and results of § 1.4 to the case of infinite families.

5. Let $\mathfrak{A} = (A_i : i \in I)$ be an arbitrary family of arbitrary sets. A non-empty subset J of I will be called '$\mathfrak{A}$-distinguished' if

$$\bigcap_{i\in J} A_i \nsubseteq \bigcup_{i\notin J} A_i.$$

Show that there exists a set S such that $|S \cap A_i| = 1$ for all $i \in I$ if and only if I can be partitioned into $\mathfrak{A}$-distinguished subsets.

1.5 The lemmas of Zorn and Tukey

Let X be a set and let R be a subset of the cartesian product $X \times X$. If $x, y \in X$ and $(x, y) \in R$, then we shall write $x \leqslant y$ (or, equivalently, $y \geqslant x$) and we shall call '$\leqslant$' a *relation* defined on X. (It need not, of course, be connected in any way with the special relation of numerical inequality.) Suppose, further, that this relation satisfies the following axioms.

(i) $x \leqslant x$ for all $x \in X$.

(ii) If $x \leqslant y$ and $y \leqslant x$, then $x = y$.

(iii) If $x \leqslant y$ and $y \leqslant z$, then $x \leqslant z$.

Then '$\leqslant$' is said to be a *partial order* on X, and the pair $(X, \leqslant)$ is called a *partially ordered set*. If the nature of '$\leqslant$' is clear from the context, we may refer, rather loosely, to X as a partially ordered set.

Let $(X, \leqslant)$ be a partially ordered set and let $x, y \in X$. If $x \leqslant y$ but $x \neq y$, we shall write $x < y$. It is then easily verified that the relation '$<$' so defined has the following properties.

(i′) The statements $x < y$ and $y < x$ are incompatible.

(ii′) If $x < y$ and $y < z$, then $x < z$.

Conversely, if '$<$' is specified axiomatically by (i′) and (ii′) and if we write $x \leqslant y$ whenever $x < y$ or $x = y$, then conditions (i)–(iii) can be shown to hold for the relation '$\leqslant$'. Thus a partial order on X can be specified equally well by the relation '$\leqslant$' or by '$<$'. Consequently, we shall sometimes speak of a partially ordered set $(X, \leqslant)$ and sometimes of $(X, <)$; but in either case, we shall be referring to essentially the same entity.

Let $(X, \leqslant)$ be a partially ordered set, and let $x, y \in X$. If $x \leqslant y$ or $y \leqslant x$, then the elements x, y are said to be *comparable*; in the contrary case, they are said to be *incomparable*. A subset of X is called a *chain* if any two of its elements are comparable; it is called an *antichain* if no two (distinct) elements are comparable. The empty set is both a chain and an antichain; and so is every singleton.

A partially ordered set $(X, \leqslant)$ which is a chain (i.e. one in which any two elements are comparable) is called an *ordered set* or, for emphasis, a *totally ordered set*; and the relation '$\leqslant$' (or the associated relation '$<$') is then called a (*total*) *order*.

Let X be a set and let $\mathfrak{X}$ be any collection of subsets of X. Then $\mathfrak{X}$ is said to be *partially ordered by inclusion* if the statement $A \leqslant B$, where $A, B \in \mathfrak{X}$, is taken to mean $A \subseteq B$. This is the most natural partial order and, indeed, every partial order can be 'represented' as partial order by inclusion (cf. Ex. 1.5.6). Again, as can be inferred from the discussion in §1.3, the relation of inequality between cardinals is a partial order: it is also a total order, but this fact is much harder to prove. More obvious instances of total order come readily to mind. Thus any set of real numbers is totally ordered with respect to the relation of (ordinary) inequality.

If $(X, \leqslant)$ is a partially ordered set, then we can define another partial order, say $\leqslant^*$, on X by declaring that $x \leqslant^* y$ if and only if $y \leqslant x$. We shall say that $\leqslant^*$ is the *reciprocal order* of $\leqslant$.

Let $(X, \leqslant)$ be a partially ordered set and let A be a subset of X. If an element x of X (which need not be in A) has the property that $x \geqslant a$ for each $a \in A$, then x is said to be an *upper bound* of A. If $x \leqslant a$ for each $a \in A$, then x is said to be a *lower bound* of A.

Again, an element x of X is said to be *maximal* (in X) if, for each $y \in X$ which is comparable with x, we have $y \leqslant x$. Minimal elements in X are defined analogously. In particular, if $\mathfrak{X}$ is a collection of sets partially ordered by inclusion, then $A \in \mathfrak{X}$ is said to be maximal if it is not properly contained in any member of $\mathfrak{X}$.

That every finite partially ordered set possesses at least one maximal element is obvious. The corresponding statement for infinite sets is, of course, false. Thus, for example, the set of all natural numbers ordered by inequality has no maximal element. There is, however, a wide class of situations in which the existence of a maximal element can be asserted.

THEOREM 1.5.1. (Zorn's lemma) *If every chain in a (non-empty) partially ordered set possesses an upper bound, then the set has at least one maximal element.*

We shall not offer a proof of Zorn's lemma. It may, however, be of interest to recall that this result can be shown to be logically equivalent to the axiom of choice. The reason for introducing this particular version of the axiom of choice is that it is cast in a form which makes it extremely convenient for a great variety of applications.

By considering the reciprocal partial order in Zorn's lemma, we obtain an equivalent formulation: *if every chain in a (non-empty) partially ordered set possesses a lower bound, then the set has at least one minimal element.*

In the remainder of this section, we shall be concerned with collections of sets partially ordered by inclusion.

THEOREM 1.5.2. *Let* $\mathfrak{F}$ *be a (non-empty) collection of sets and suppose that, for each chain* $\mathscr{C}$ *in* $\mathfrak{F}$, *the union* $\bigcup(\mathrm{X} : \mathrm{X} \in \mathscr{C})$ *is an element of* $\mathfrak{F}$. *Then, for each* $\mathrm{A} \in \mathfrak{F}$, *there exists a maximal element* $\mathrm{M} \in \mathfrak{F}$ *such that* $\mathrm{A} \subseteq \mathrm{M}$.

Let $\mathfrak{F}^*$ denote the collection of all members of $\mathfrak{F}$ which contain A as a subset. Let $\mathscr{C}$ be any (non-empty) chain in $\mathfrak{F}^*$, and so in $\mathfrak{F}$. Then, by hypothesis, $\mathrm{C} = \bigcup(\mathrm{X} : \mathrm{X} \in \mathscr{C}) \in \mathfrak{F}$. Moreover, we clearly have $\mathrm{A} \subseteq \mathrm{C}$, so that $\mathrm{C} \in \mathfrak{F}^*$. Thus every chain in $\mathfrak{F}^*$ has an upper bound (in $\mathfrak{F}^*$) and so, by Zorn's lemma, $\mathfrak{F}^*$ has a maximal element M, i.e. there exists an $\mathrm{M} \in \mathfrak{F}$ with $\mathrm{A} \subseteq \mathrm{M}$ and such that M is not properly contained in any element of $\mathfrak{F}^*$, and so of $\mathfrak{F}$. Hence M is maximal in $\mathfrak{F}$.

Using the alternative form of Zorn's lemma relating to minimal elements, we obtain in the same way the following result. *Let* $\mathfrak{F}$ *be a (non-empty) collection of sets and suppose that, for each chain* $\mathscr{C}$ *in* $\mathfrak{F}$, $\bigcap(\mathrm{X} : \mathrm{X} \in \mathscr{C}) \in \mathfrak{F}$. *Then, for each* $\mathrm{A} \in \mathfrak{F}$, *there exists a minimal element* $\mathrm{N} \in \mathfrak{F}$ *such that* $\mathrm{N} \subseteq \mathrm{A}$.

Next, we introduce a concept which will play a fundamental role in much of the subsequent discussion. Suppose that a collection $\mathfrak{S}$ of sets is such that a set belongs to $\mathfrak{S}$ if and only if all its finite subsets belong to $\mathfrak{S}$. Then $\mathfrak{S}$ is said to have *finite character*. It is, of course, plain that $\mathfrak{S}$ has finite character if and only if it satisfies the following requirements. (i) Every subset of a member of $\mathfrak{S}$ is itself a member of $\mathfrak{S}$. (ii) If every finite subset of a set X is a member of $\mathfrak{S}$, then X is also a member of $\mathfrak{S}$.

THEOREM 1.5.3. (Tukey's lemma) *Let a collection* $\mathfrak{S}$ *of sets have finite character. Then every member of* $\mathfrak{S}$ *is a subset of some member maximal with respect to inclusion. In particular,* $\mathfrak{S}$ *possesses a maximal member.*

Let $\mathscr{C}$ be a chain in $\mathfrak{S}$ and put $\mathrm{C} = \bigcup(\mathrm{X} : \mathrm{X} \in \mathscr{C})$. Let $\mathrm{C}^* \subset\subset \mathrm{C}$ and write $\mathrm{C}^* = \{c_1, \dots, c_m\}$. Then each c_r belongs to some X and so all c's belong to the same (suitably chosen) X, say $\mathrm{C}^* \subseteq \mathrm{X}_0$, where $\mathrm{X}_0 \in \mathscr{C} \subseteq \mathfrak{S}$. It follows that $\mathrm{C}^* \in \mathfrak{S}$, i.e. every finite subset of C belongs to $\mathfrak{S}$ and so $\mathrm{C} \in \mathfrak{S}$, i.e.

$\bigcup(X\colon X \in \mathscr{C}) \in \mathfrak{S}$. Hence, if $A \in \mathfrak{S}$, there exists, by Theorem 1.5.2, a maximal member $M \in \mathfrak{S}$ such that $A \subseteq M$.

Here Tukey's lemma has been deduced from Zorn's lemma: it can be shown that the reverse inference is equally valid. Thus the axiom of choice, Zorn's lemma, and Tukey's lemma (as well as a number of other statements of the same general character) are all equivalent to each other.

Exercises 1.5

1. Establish the following strengthened form of Zorn's lemma. If $(X, \leqslant)$ is a partially ordered set, if $a \in X$, and if every chain in X has an upper bound, then X contains a maximal element x such that $a \leqslant x$.

2. Use Tukey's lemma to show that every linearly independent subset of a vector space is contained in some basis.

3. Prove 'Kuratowski's lemma' to the effect that every chain in a partially ordered set is contained in some maximal chain.

4. Show that every antichain in a partially ordered set is contained in some maximal antichain.

5. Let $(X, \leqslant)$ be a partially ordered set containing n elements, Show that there exists a set N of n natural numbers and a bijection $\phi\colon X \to N$ such that, for $x, y \in X$, we have $x \leqslant y$ if and only if $\phi(x)$ is a factor of $\phi(y)$.

6. Let $(X, \leqslant)$ be a partially ordered set. Show that there exists a collection $\mathfrak{F}$ of subsets of X and a bijection $\psi\colon X \to \mathfrak{F}$ such that, for any elements $x, y \in X$, we have $x \leqslant y$ if and only if $\psi(x) \subseteq \psi(y)$.

7. Let X be a set and let $<$, $<^*$ be two partial orders defined on X. We say that $<^*$ is an 'extension' of $<$ if $x < y$ implies $x <^* y$.

   Verify that the collection P of all partial orders on X extending a given partial order is partially ordered by extension. Show that every chain in P has an upper bound, and deduce that every partial order on X can be extended to a total order.

   [E. Szpilrajn (**1**)]

8. Let S be a set of n elements and denote by T the collection of the $2^n - 1$ non-empty subsets of S, partially ordered by inclusion. Show that, for any antichain A in T,

$$\sum_{x \in A} 1 \Big/ \binom{n}{|x|} \leqslant 1.$$

   Deduce that the maximum cardinal of antichains in T is equal to

$$\binom{n}{[\frac{1}{2}n]},$$

   where $[\frac{1}{2}n]$ denotes the largest integer not exceeding $\frac{1}{2}n$.

   [Lubel (**1**); cf. Sperner (**2**)]

9. Do a maximal chain and a maximal antichain in a partially ordered set necessarily have a non-empty intersection?

1.6 Tychonoff's theorem

Since combinatorial analysis is primarily concerned with objects that have little or no structure, it draws in the main upon results in the theory of sets. There is, however, one basic combinatorial theorem (Theorem 4.1.1) the proof of which can be framed in topological terms. Consequently, the present section will be devoted to recalling the pertinent definitions and results in general topology.

Let S be a non-empty set. A collection $\mathscr{G}$ of subsets of S is called a *topology* (on S) if it is closed with respect to the formation of arbitrary unions and finite intersections. Since an empty union of subsets of S is Ø and an empty intersection is S, it follows that Ø and S are members of $\mathscr{G}$. The pair (S, $\mathscr{G}$) is called a *topological space*; we say that it is finite when S is finite. (When the specification of $\mathscr{G}$ is clear from the context, we refer, rather loosely, to S as the topological space.) The members of $\mathscr{G}$ are called the *open sets* of the topological space (S, $\mathscr{G}$).

A particularly simple topological space may be constructed as follows. Let S be any (non-empty) set, and let $\mathscr{D}$ denote its power set. Then (S, $\mathscr{D}$) is plainly a topological space: we say that we have formed it by endowing S with the *discrete topology*. The space (S, $\mathscr{D}$) is called a *discrete topological space*.

Let S be a topological space. A family $\mathfrak{C} = (C_i : i \in I)$ of subsets of S is called a *cover* (of S) if $\bigcup(C_i : i \in I) = S$. This cover is said to be open if all C_i are open sets; it is said to be finite if I is finite. A subfamily $\mathfrak{C}'$ of the cover $\mathfrak{C}$ is called a *subcover* of $\mathfrak{C}$ if it is itself a cover of S. The space S is said to be *compact* if every open cover contains a finite subcover. It is clear that every finite topological space is compact.

Let S be an arbitrary (non-empty) set and $\mathscr{H}$ an arbitrary collection of subsets of S. Denote by $\mathscr{G}$ the collection of all subsets of S such that every set in $\mathscr{G}$ can be expressed as a union of finite intersections of sets in $\mathscr{H}$. It is then easily verified that $\mathscr{G}$ is a topology on S; this topology is known as the topology *generated* by $\mathscr{H}$. Plainly, it is the 'coarsest' topology in which all sets of $\mathscr{H}$ are open, i.e. if $\mathscr{G}'$ is any other topology on S with this property, then $\mathscr{G} \subseteq \mathscr{G}'$.

We shall next describe a standard procedure which enables us to combine a number of given topological spaces into a new topological space. Let $((S_i, \mathscr{G}_i) : i \in I)$ be a (non-empty) family of topological spaces, and write

$$S = \bigtimes_{i \in I} S_i$$

so that, by the axiom of choice, $S \neq Ø$. Let us, for the moment, call a subset of S 'special' if it can be written in the form $\bigtimes(G_i : i \in I)$, where $G_i \in \mathscr{G}_i$ for all i and $G_i = S_i$ for all but a finite number of i. We shall denote by $\mathscr{G}$ the topology on S generated by the 'special sets'†; then $\mathscr{G}$ is known as the *product topology* on S. The topological space (S, $\mathscr{G}$) is called the *topological product* of the spaces (S_i, $\mathscr{G}_i$), $i \in I$.

In the course of our discussion, we shall need to invoke a result which is pivotal in the entire theory of topological spaces.

THEOREM 1.6.1. (Tychonoff) *The topological product of a family of compact topological spaces is again compact.*

This implies that the topological product of a family of finite topological spaces is compact. More particularly, *the topological product of a family of finite discrete spaces is compact*. It is this last result to which we shall appeal in § 4.1.

1.7 Graphs

Possibly the most intensively cultivated area of combinatorial mathematics is the theory of graphs. Though its relation to transversal theory is intimate, our account will touch on the theory of graphs only in passing, and for our purpose it will be sufficient to introduce a small number of technical terms and to refer to two theorems.

A *graph*‡ G is a pair (N, E), where N is a non-empty set and E is a collection of subsets of N of cardinal 2. The elements of N are called *nodes*, those of E are called *edges*. The graph G is called finite or infinite according as N is finite or infinite. If $x, y \in N$, $x \neq y$ and $e = \{x, y\} \in E$, then the nodes x and y are said to be *linked* by the edge e. We also say that e and x, and equally e and y, are *incident*. Two edges which are incident with the same node are said to be *concurrent*. It is often helpful to picture a graph as a set of points (nodes) in which certain pairs of points are linked by lines (edges).

A class of graphs which is particularly important in the discussion of many problems of transversal theory is the class of bipartite graphs. A graph $G = (N, E)$ is said to be *bipartite* if there exists a partition $N = N_1 \cup N_2$ such that no two nodes in N_1, and equally no two in N_2, are linked by an edge.

Let $G = (N, E)$ be a graph. Suppose that, for a certain integer k, there exists a partition $N = N_1 \cup \ldots \cup N_k$ such that no two nodes which are linked by an edge belong to the same N_i. (To put the matter more picturesquely: we can

† It is, in fact, clear that a subset of S belongs to $\mathscr{G}$ if and only if it can be expressed as a union of 'special' sets.

‡ We shall use the term 'graph' in a somewhat restricted sense to mean an 'undirected graph without loops or multiple edges'.

paint each node with one of a set of k different colours such that no two nodes linked by an edge are painted alike.) The least integer k with this property, if one exists, is called the *chromatic number* of G.

Let $G = (\mathrm{N}, \mathrm{E})$, $G' = (\mathrm{N}', \mathrm{E}')$ be two graphs. We say that G' is a *subgraph* of G if N' is a subset of N and if E' consists of precisely those edges in E both of whose nodes belong to N'. Thus†

$$\mathrm{N}' \subseteq \mathrm{N}; \quad \mathrm{E}' = \left\{\{x, y\}_{\neq} : x, y \in \mathrm{N}', \ \{x, y\}_{\neq} \in \mathrm{E}\right\}.$$

Next, we mention two standard results in the theory of graphs.

THEOREM 1.7.1. (König) *Let G be a bipartite graph. If the minimum number of nodes such that every edge is incident with at least one of these is finite, then it is equal to the maximum number of edges such that no two of these are concurrent.*

To formulate a more general result, we need some further terminology. Let $G = (\mathrm{N}, \mathrm{E})$ be a graph, and let x, y be distinct nodes of G. By a *path*, say P, linking x and y we understand a set of edges of the type

$$\left\{\{a_1, a_2\}, \ \{a_2, a_3\}, \ldots, \ \{a_{m-1}, a_m\}\right\},$$

where $a_1 = x$, $a_m = y$, and $a_1, \ldots, a_m$ are distinct. In that case, $a_1, a_2, \ldots, a_m$ are called the nodes of P. In particular, $\{x, y\}$ is a path if the nodes x and y are linked by an edge in G. A collection of paths in G is said to be *disjoint* if no two paths have a node in common.

Let X, Y be disjoint subsets of N. A path which links a node in X and a node in Y will be said to link X and Y. A set $\mathrm{S} \subseteq \mathrm{N}$ is said to *separate* X and Y (or to be a *separating set*) if every path which links X and Y has a node in S.

We are now able to formulate Menger's celebrated graph theorem.

THEOREM 1.7.2. (K. Menger) *Let* $G = (\mathrm{N}, \mathrm{E})$ *be a graph and let* X, Y *be disjoint subsets of* N. *If the minimum number of nodes in sets separating* X *and* Y *is finite, then it is equal to the maximum number of disjoint paths linking* X *and* Y.

We note that König's theorem is simply the special case to which Menger's theorem reduces for the case of bipartite graphs.

† For some purposes, variants of this definition have to be adopted.

Notes on Chapter 1

§ 1.1 The reader who wishes to pursue the theory of sets in a systematic manner is recommended to consult the account offered by Halmos (**1**)† or that of Rotman & Kneebone (**1**).

§ 1.3. Lemma 1.3.1 is due to Knaster and Tarski; see Knaster (**1**); cf. also D. König (**3**) for a number of related results. Theorems 1.3.2 and 1.3.4—which are basic to our subsequent exposition—are taken from paper (**1**) by Perfect & Pym, but a result equivalent to Theorem 1.3.4 had appeared earlier in Ore's book (**3**, Theorem 7.4.1). Corollary 1.3.3, proved by Banach (**1**) in 1924, may be regarded as the genesis of all results in the present section. Banach's object was to exhibit the common principle underlying all proofs of the Schröder–Bernstein theorem (Theorem 1.3.6). A particularly simple proof of this latter result will be found in J. L. Kelley's treatise (**1**, 28). R. A. Brualdi (**3**) obtained generalizations of almost all results in this section.

For the theory of cardinal numbers, see the sources cited in the Notes on § 1.1.

§ 1.4. Most mathematicians are undoubtedly familiar with boolean atoms, but it is not altogether easy to give precise references. The notion is, for example, implicit in R. Rado's paper (**2**).

§ 1.5. The principle now universally referred to as 'Zorn's lemma' was formulated by M. Zorn (**1**) in 1935, though not quite in the form that has since become accepted. Kelley's book (**1**) contains a full discussion of this result and of other statements equivalent to the axiom of choice. Very brisk proofs of Zorn's lemma have been given by J. D. Weston (**1**) and, more recently, by H. Kneser (**1**).

§ 1.6. The standard account of general topology is to be found in Kelley (**1**). For a more elementary treatment, see e.g. Bushaw (**1**). Both books (and, of course, many others) contain proofs of Tychonoff's theorem. This result is now known to be equivalent to the axiom of choice.

§ 1.7. The theory of graphs is largely the creation of Dénes König, who also wrote the standard treatise on the subject (**7**). More recent accounts have been given by C. Berge (**1**), O. Ore (**3**), and F. Harary (**1**). Ore is also the author of a short elementary introduction (**4**) to this branch of mathematics.

For König's theorem 1.7.1, see (**5**), (**6**), and (**7**, 233). Theorem 1.7.2 was discovered by K. Menger (**1**). The most satisfactory proof of Menger's theorem is probably that of J. S. Pym (**2**).

† Figures in bold-face type refer to the bibliography at the end of the book.

2 Hall's Theorem and the Notion of Duality

In the present chapter we initiate the study of combinatorial problems by proving P. Hall's classical theorem on 'distinct representatives'. As the discussion proceeds, we shall recognize that the whole of transversal theory may be regarded as a natural development of Hall's investigation.

2.1 Transversals, representatives, and representing sets

Let $\mathfrak{A} = (A_i : i \in I)$ be a family of subsets of a ground set E. Suppose that it is possible to select one element x_i from each set A_i in such a way that the elements x_i, $i \in I$ so selected are distinct. Then the set $\{x_i : i \in I\}_{\neq}$ of these elements is called a 'transversal' of $\mathfrak{A}$. For example, let $\mathfrak{A} = (A_1, A_2, A_3, A_4)$, where

$$A_1 = \{2, 3\}, \qquad A_2 = \{1, 3, 4\}, \qquad A_3 = \{3, 4, 5\}, \qquad A_4 = \{1, 4\}.$$

Then $\{1, 3, 4, 5\}$ is a transversal of $\mathfrak{A}$, since

$$3 \in A_1, \qquad 4 \in A_2, \qquad 5 \in A_3, \qquad 1 \in A_4;$$

and another transversal is $\{2, 3, 4, 5\}$, since

$$2 \in A_1, \qquad 3 \in A_2, \qquad 5 \in A_3, \qquad 4 \in A_4.$$

On the other hand, if

$$A_1 = \{1, 2\}, \quad A_2 = \{3, 4\}, \quad A_3 = \{3\}, \quad A_4 = \{4\}, \quad A_5 = \{4\}, \tag{1}$$

then $(A_1, A_2, A_3, A_4, A_5)$ plainly has no transversal, for we cannot find 5 distinct elements which belong to $A_1, \ldots, A_5$ respectively as only 4 distinct elements are available in all.

To formalize the notion of a transversal, we shall say that a subset T of E is a *transversal* of $\mathfrak{A} = (A_i : i \in I)$ if there exists a bijection $\psi : T \to I$ such that $x \in A_{\psi(x)}$ for all $x \in T$. This implies, in particular, that a transversal of a family $\mathfrak{A}$ has the same cardinal as the index set of $\mathfrak{A}$. We also note that $\mathfrak{A}$ possesses a transversal precisely if there exists an injection $\theta : I \to E$ such that $\theta(i) \in A_i$ for all $i \in I$; in that case $\theta(I)$ is a transversal.

Next, we call a subset X of E a *partial transversal* (PT) of the family $\mathfrak{A} = (A_i : i \in I)$ (of subsets of E) if X is a transversal of a subfamily of $\mathfrak{A}$. This means that there exists an injection $\psi: \mathrm{X} \to \mathrm{I}$ such that $x \in \mathrm{A}_{\psi(x)}$ for all $x \in \mathrm{X}$. It is clear that a transversal is a particular instance of a partial transversal (corresponding to the case when ψ is a bijection). To emphasize the distinction, we shall sometimes refer to a transversal as a *total transversal.* By convention, we shall regard the empty set as a partial transversal of $\mathfrak{A}$. A subset of a transversal (or, indeed, of a PT) is again a PT; but, if the family is infinite, it can also happen that a proper subset of a transversal is a transversal. Again, the same set might be both a transversal and a partial transversal which is not total.

By way of illustration of the concept of a PT, let us note that the collection of PTs of the family (1) consists of the sets

$$\emptyset, \{1\}, \{2\}, \{3\}, \{4\}, \{1,3\}, \{1,4\}, \{2,3\}, \{2,4\}, \{3,4\}, \{1,3,4\}, \{2,3,4\}.$$

Let $\mathfrak{A} = (\mathrm{A}_i : i \in \mathrm{I})$ be a family of subsets of E and let d be a non-negative integer. We say that a subset X of E is a partial transversal of *defect d* of $\mathfrak{A}$ if there exists a set $\mathrm{I}_0 \subseteq \mathrm{I}$ with $|\mathrm{I} \setminus \mathrm{I}_0| = d$ such that X is a transversal of the subfamily $\mathfrak{A}(\mathrm{I}_0)$ of $\mathfrak{A}$. (To put the matter more informally: a PT X has defect d if it 'represents' all but d of the sets in $\mathfrak{A}$. Thus a PT of defect 0 is simply a transversal.) If the family is finite, then the defect of every PT X of $\mathfrak{A}$ is a uniquely determined number, namely $|\mathrm{I}| - |\mathrm{X}|$. The situation is different when I is infinite: quite simple examples show that a set X may be a PT with more than one defect (indeed, possibly with infinitely many).

Next, considering again a family $\mathfrak{A} = (\mathrm{A}_i : i \in \mathrm{I})$ of subsets of E, let us choose one element x_i in each A_i. If the elements so chosen are not necessarily distinct, their totality is best described by the use of the notion of a family. We say, then, that the family $\mathfrak{X} = (x_j : j \in \mathrm{J})$ of elements of E is a *system of representatives* of $\mathfrak{A}$ if there exists a bijection $\phi: \mathrm{J} \to \mathrm{I}$ such that $x_j \in \mathrm{A}_{\phi(j)}$ for all $j \in \mathrm{J}$. If, in addition, the x_j are distinct, then $\mathfrak{X}$ is called a *system of distinct representatives* (SDR) of $\mathfrak{A}$, and we say that the element x_j 'represents' the set $\mathrm{A}_{\phi(j)}$. It is plain that the range $\{x_j : j \in \mathrm{J}\}_{\neq}$ of the SDR $(x_j : j \in \mathrm{J})$ is a transversal of $\mathfrak{A}$. Conversely, to each transversal T of $\mathfrak{A}$, there exists a SDR $\mathfrak{X}$ of $\mathfrak{A}$ such that T is the range of $\mathfrak{X}$.

We can, of course, consider the range of a system of not necessarily distinct representatives. In this context, it is useful to introduce the notion of a representing set. Let $\mathfrak{A} = (\mathrm{A}_i : i \in \mathrm{I})$ be a family of sets. Then a set R is called a *representing set* of $\mathfrak{A}$ if $\mathrm{R} \cap \mathrm{A}_i \neq \emptyset$ for all $i \in \mathrm{I}$. If $\mathfrak{X} = (x_j : j \in \mathrm{J})$ is any system of representatives of $\mathfrak{A}$, then its range $\{x_j : j \in \mathrm{J}\}$ is clearly a representing set of $\mathfrak{A}$.

The object of transversal theory is the determination of criteria for the existence of transversals, systems of representatives, representing sets, or similar 'transversal-like' objects. We shall not be able to say a great deal about representing sets: questions concerned with them seem to be exceptionally difficult. The major part of our discussion will be devoted to (partial or total) transversals and to systems of representatives, possibly subject to certain restrictions and associated with one or more than one family. Nearly all problems we study are qualitative in the sense that they are concerned with questions of existence. Quantitative problems arise quite naturally and are of equal interest, but they are generally found to be intractable and we shall refer to them only on isolated occasions.

The study of transversals will involve the formulation of a certain property relating to families. Let $\mathfrak{A} = (A_1, \ldots, A_n)$ be a finite family. We say that $\mathfrak{A}$ satisfies *Hall's condition* (or, briefly, *condition* $\mathscr{H}$) if, for each k with $1 \leqslant k \leqslant n$, the union of any k A's contains at least k (distinct) elements. In other words, $\mathfrak{A}$ satisfies $\mathscr{H}$ if and only if

$$|A_{i_1} \cup \ldots \cup A_{i_k}| \geqslant k$$

whenever $1 \leqslant k \leqslant n$ and $1 \leqslant i_1 < \ldots < i_k \leqslant n$. Now write $J = \{i_1, \ldots, i_k\}$. Then $J \subseteq \{1, \ldots, n\}$, $|J| = k$, and

$$A_{i_1} \cup \ldots \cup A_{i_k} = \bigcup_{i \in J} A_i.$$

Hence we see that $\mathfrak{A}$ satisfies condition $\mathscr{H}$ if, for each $J \subseteq \{1, \ldots, n\}$,

$$\left|\bigcup_{i \in J} A_i\right| \geqslant |J|.$$

We shall, for brevity, write $A(J) = \bigcup(A_i : i \in J)$ and use an analogous notation for families denoted by other letters. With this notation, $\mathscr{H}$ can be expressed thus: $\mathfrak{A}$ satisfies condition $\mathscr{H}$ if, for each $J \subseteq \{1, \ldots, n\}$,

$$|A(J)| \geqslant |J|. \tag{2}$$

It is clear that (2) holds trivially for $J = \emptyset$. Now there are $2^n - 1$ non-empty subsets of $\{1, \ldots, n\}$ and so, in relation to the family $(A_1, \ldots, A_n)$, $\mathscr{H}$ has $2^n - 1$ constituent conditions. It is of some interest to observe that these conditions are independent, i.e. none of them is implied by the others. For let J_0 be a non-empty subset of $\{1, \ldots, n\}$ and define the family $\mathfrak{A} = (A_1, \ldots, A_n)$ by the equations

$$A_i = \begin{cases} \{1, 2, \ldots, |J_0| - 1\} & (i \in J_0) \\ \{1, 2, \ldots, n\} & (i \notin J_0). \end{cases}$$

Then (2) is satisfied for $J \neq J_0$ but not for $J = J_0$.

Later, it will be necessary to extend $\mathscr{H}$ to arbitrary families. We shall say that the family $\mathfrak{A} = (A_i : i \in I)$ satisfies Hall's condition $\mathscr{H}$ if the inequality (2) is valid for every *finite* subset J of I.

2.2 Proofs of the fundamental theorem for finite families

We begin our treatment with an apparently quite special result whose pivotal position in transversal theory will only emerge in the course of subsequent development.

THEOREM 2.2.1. (P. Hall) *The finite family* $\mathfrak{A} = (A_i : 1 \leqslant i \leqslant n)$ *of subsets of a set* E *possesses a transversal if and only if it satisfies Hall's condition* $\mathscr{H}$, *i.e. if and only if, for each* $I \subseteq \{1, \ldots, n\}$,

$$|A(I)| \geqslant |I|. \tag{1}$$

Suppose that $\mathfrak{A}$ possesses a transversal, i.e. there exist distinct elements $x_1, \ldots, x_n$ such that $x_1 \in A_1, \ldots, x_n \in A_n$. If $1 \leqslant k \leqslant n$, $1 \leqslant i_1 < \ldots < i_k \leqslant n$, then

$$A_{i_1} \cup \ldots \cup A_{i_k} \supseteq \{x_{i_1}, \ldots, x_{i_k}\}$$

and so

$$|A_{i_1} \cup \ldots \cup A_{i_k}| \geqslant k.$$

The necessity of Hall's condition thus holds trivially,† and we next offer three proofs of its sufficiency. Yet a further proof will be found in § 6.5.

The first proof proceeds by induction. For $n = 1$ the assertion (that Hall's condition implies the existence of a transversal) is obviously valid. Suppose, then, that $n > 1$ and that the assertion holds for every family consisting of at most $n - 1$ sets. Let $\mathfrak{A} = (A_1, \ldots, A_n)$ be a family of n sets and let it satisfy $\mathscr{H}$, i.e.

$$|A_{i_1} \cup \ldots \cup A_{i_k}| \geqslant k$$

whenever $1 \leqslant k \leqslant n$, $1 \leqslant i_1 < \ldots < i_k \leqslant n$.

Case 1. Suppose that

$$|A_{i_1} \cup \ldots \cup A_{i_k}| \geqslant k + 1$$

whenever $1 \leqslant k < n$, $1 \leqslant i_1 < \ldots < i_k \leqslant n$. Now, in view of Hall's condition, $|A_1| \geqslant 1$ and we can therefore choose an element $x_1 \in A_1$. Write

$$B_i = A_i \setminus \{x_1\} \qquad (2 \leqslant i \leqslant n).$$

† It is a common occurrence in transversel theory that, in a criterion for the existence of some object, the necessity part is trivial while the sufficiency part is comparatively difficult.

For $1 \leqslant k < n$, $2 \leqslant i_1 < \ldots < i_k \leqslant n$, we have

$$\begin{aligned} |B_{i_1} \cup \ldots \cup B_{i_k}| &= |(A_{i_1} \cup \ldots \cup A_{i_k}) \setminus \{x_1\}| \\ &\geqslant |A_{i_1} \cup \ldots \cup A_{i_k}| - |\{x_1\}| \\ &= |A_{i_1} \cup \ldots \cup A_{i_k}| - 1 \geqslant (k+1) - 1 = k. \end{aligned}$$

The family $(B_2, \ldots, B_n)$ therefore satisfies $\mathscr{H}$ and so, by the induction hypothesis, possesses a transversal; i.e. there exist distinct elements $x_2, \ldots, x_n$ such that $x_2 \in B_2, \ldots, x_n \in B_n$. Now $x_1 \notin B_2, \ldots, x_1 \notin B_n$ and so $x_1, x_2, \ldots, x_n$ are distinct. Moreover

$$x_1 \in A_1, \quad x_2 \in B_2 \subseteq A_2, \ldots, \quad x_n \in B_n \subseteq A_n,$$

and so $\{x_1, x_2, \ldots, x_n\}_{\neq}$ is a transversal of $\mathfrak{A}$.

Case 2. Suppose that, for some k with $1 \leqslant k < n$ and certain i's with $1 \leqslant i_1 < \ldots < i_k \leqslant n$, we have

$$|A_{i_1} \cup \ldots \cup A_{i_k}| = k.$$

For notational convenience, we shall take $i_1 = 1, \ldots, i_k = k$ so that, for some k with $1 \leqslant k < n$,

$$|A_1 \cup \ldots \cup A_k| = k. \tag{2}$$

Now, since $(A_1, \ldots, A_n)$ satisfies $\mathscr{H}$, so does $(A_1, \ldots, A_k)$. Hence, by the induction hypothesis, $(A_1, \ldots, A_k)$ possesses a transversal, say

$$x_1 \in A_1, \ldots, \quad x_k \in A_k,$$

where $x_1, \ldots, x_k$ are distinct. We define

$$B_i = A_i \setminus \{A_1 \cup \ldots \cup A_k\} \qquad (k+1 \leqslant i \leqslant n).$$

Let $1 \leqslant r \leqslant n-k$, $k+1 \leqslant i_1 < \ldots < i_r \leqslant n$. Then, using (2) and the fact that $A_1 \cup \ldots \cup A_k$ and any B are disjoint, we have

$$\begin{aligned} |B_{i_1} \cup \ldots \cup B_{i_r}| &= |B_{i_1} \cup \ldots \cup B_{i_r}| + |A_1 \cup \ldots \cup A_k| - k \\ &= |B_{i_1} \cup \ldots \cup B_{i_r} \cup A_1 \cup \ldots \cup A_k| - k \\ &= |A_{i_1} \cup \ldots \cup A_{i_r} \cup A_1 \cup \ldots \cup A_k| - k. \end{aligned}$$

Now the integers $i_1, \ldots, i_r, 1, \ldots, k$ are distinct and so, by $\mathscr{H}$,

$$|B_{i_1} \cup \ldots \cup B_{i_r}| \geqslant (r+k) - k = r.$$

Thus the family $(B_{k+1}, \ldots, B_n)$ satisfies $\mathscr{H}$ and so, by the induction hypothesis, possesses a transversal, say

$$x_{k+1} \in B_{k+1} \subseteq A_{k+1}, \ldots, x_n \in B_n \subseteq A_n,$$

where $x_{k+1}, \ldots, x_n$ are distinct. Further, no B contains any of the elements $x_1, \ldots, x_k$; therefore $x_1, \ldots, x_k, x_{k+1}, \ldots, x_n$ are distinct and $\{x_1, \ldots, x_n\}$ is a transversal of $\mathfrak{A}$.

The proof just given is extremely transparent and can, as we shall see, be adapted to serve in other situations (cf. e.g. § 5.2 and Ex. 6.2.3). Our next proof is equally flexible (see § 6.2) and has the advantage of even greater simplicity. To describe it, we shall need two preliminary results; and in view of subsequent applications, we formulate these for arbitrary rather than for finite families.

LEMMA 2.2.2. *Let the family* $(A_i : i \in I) + (B)$† *of sets satisfy Hall's condition. If* $|B| \geqslant 2$, *then there exists an element* $x \in B$ *such that* $(A_i : i \in I) + (B \setminus \{x\})$ *again satisfies Hall's condition.*

Let $x_1, x_2 \in B$, $x_1 \neq x_2$, and assume that neither x_1 nor x_2 can be taken as x. Then there exist finite subsets I_1, I_2 of I such that

$$|A(I_1) \cup (B \setminus \{x_1\})| \leqslant |I_1|, \qquad |A(I_2) \cup (B \setminus \{x_2\})| \leqslant |I_2|.$$

Writing

$$S_1 = A(I_1) \cup (B \setminus \{x_1\}), \qquad S_2 = A(I_2) \cup (B \setminus \{x_2\}),$$

we have

$$\begin{aligned} |I_1| + |I_2| \geqslant |S_1| + |S_2| &= |S_1 \cup S_2| + |S_1 \cap S_2| \\ &\geqslant |A(I_1) \cup A(I_2) \cup B| + |A(I_1) \cap A(I_2)| \\ &\geqslant |A(I_1 \cup I_2) \cup B| + |A(I_1 \cap I_2)| \\ &\geqslant |I_1 \cup I_2| + 1 + |I_1 \cap I_2| = |I_1| + |I_2| + 1. \end{aligned}$$

We arrive, then, at a contradiction. Hence at least one of the two elements x_1, x_2 has the required property.

LEMMA 2.2.3. *If* B *is a finite set and the family* $(A_i : i \in I) + (B)$ *satisfies Hall's condition, then so does the family* $(A_i : i \in I) + (\{x\})$ *for some* $x \in B$.

To establish this result, we simply apply the preceding lemma $|B| - 1$ times.

† To be entirely precise, we ought to add that the index set, which is naturally a singleton, of the family (B) is assumed to be disjoint from I.

We now come to our next proof of Hall's theorem. Suppose that the family $\mathfrak{A} = (A_1, ..., A_n)$ satisfies $\mathscr{H}$. Define A_i' as A_i if A_i is finite and as any subset of cardinal n of A_i if A_i is infinite. Then $(A_1', ..., A_n')$ again satisfies $\mathscr{H}$. We now apply Lemma 2.2.3 repeatedly, taking B in turn as $A_1', ..., A_n'$. We then eventually obtain a family $(\{x_1\}, ..., \{x_n\})$ of singletons which satisfies $\mathscr{H}$ and the relations $x_i \in A_i' \subseteq A_i$ $(1 \leqslant i \leqslant n)$. Hence $\{x_1, ..., x_n\}_{\neq}$ is a transversal of $\mathfrak{A}$.

It is also worth noting that Hall's theorem is an easy consequence of König's theorem 1.7.1, i.e. of Menger's theorem 1.7.2 for the case of bipartite graphs. Write the given family in the form $\mathfrak{A} = (A_i : i \in I)$, where $I = \{1, ..., n\}$ and the A_i are subsets of E; assume, as may be done without loss of generality, that $I \cap E = \emptyset$; and consider the bipartite graph $G = (N, F)$, where $N = I \cup E$ and

$$F = \left\{\{i, e\} : i \in I, \quad e \in E, \quad e \in A_i\right\}.$$

Let $I' \subseteq I$, $E' \subseteq E$ and suppose that $S = I' \cup E'$ is a set separating I and E in G. Then $A(I \setminus I') \cap (E \setminus E') = \emptyset$ and so $A(I \setminus I') \subseteq E'$. Hence, by $\mathscr{H}$,

$$|S| = |I'| + |E'| \geqslant |I'| + |A(I \setminus I')| \geqslant |I'| + |I \setminus I'| = n.$$

Thus no set separating I and E contains fewer than n nodes of G. Hence there exist n disjoint paths linking I and E, i.e. $\mathfrak{A}$ possesses a transversal.

We shall conclude the present section by deriving a result which stands in a 'dual' relation to Hall's theorem. The precise meaning of this term will be elucidated in § 2.3. The theorem in question is as follows.

THEOREM 2.2.4. *Let* $\mathfrak{A} = (A_1, ..., A_n)$ *be a family of subsets of* E. *Then* E *is a partial transversal of* $\mathfrak{A}$ *if and only if every subset* F *of* E *intersects at least* $|F|$ A's, *i.e.*

$$|\{i : 1 \leqslant i \leqslant n, \ A_i \cap F \neq \emptyset\}| \geqslant |F| \ \textit{for each} \ F \subseteq E. \tag{3}$$

The necessity of condition (3) is obvious and it only remains to establish its sufficiency. If (3) holds, then E is finite (in fact, $|E| \leqslant n$) and we shall write $E = \{x_1, ..., x_m\}_{\neq}$. We define

$$B_j = \{i : 1 \leqslant i \leqslant n, \ x_j \in A_i\} \qquad (1 \leqslant j \leqslant m),$$

so that $\mathfrak{B} = (B_1, ..., B_m)$ is a family of subsets of $\{1, ..., n\}$. Let $1 \leqslant k \leqslant m$, $1 \leqslant j_1 < ... < j_k \leqslant m$. Then

$$B_{j_1} \cup ... \cup B_{j_k} = \left\{i : 1 \leqslant i \leqslant n, \ \{x_{j_1}, ..., x_{j_k}\} \cap A_i \neq \emptyset\right\}$$

and so, by (3),

$$|B_{j_1} \cup \ldots \cup B_{j_k}| \geqslant k.$$

The family $\mathfrak{B}$ therefore satisfies condition $\mathscr{H}$ and so, by Theorem 2.2.1, possesses a transversal, say

$$i_1 \in B_1, \ldots, i_m \in B_m,$$

where $\{i_1, \ldots, i_m\}_{\neq} \subseteq \{1, \ldots, n\}$. Thus

$$x_1 \in A_{i_1}, \ldots, x_m \in A_{i_m},$$

i.e. $\{x_1, \ldots, x_m\} = E$ is a PT of $\mathfrak{A}$.

It is worth observing that Hall's theorem can, in turn, be deduced from Theorem 2.2.4. For let $\mathfrak{A} = (A_1, \ldots, A_n)$ satisfy $\mathscr{H}$. If $F = \{i_1, \ldots, i_k\}_{\neq} \subseteq \{1, \ldots, n\}$, and the B's are defined as above, then

$$\{j : 1 \leqslant j \leqslant m,\ B_j \cap F \neq \emptyset\} = \{j : 1 \leqslant j \leqslant m, x_j \in A_{i_1} \cup \ldots \cup A_{i_k}\}$$

and hence

$$|\{j : 1 \leqslant j \leqslant m, B_j \cap F \neq \emptyset\}| = |A_{i_1} \cup \ldots \cup A_{i_k}| \geqslant k = |F|.$$

Therefore, by Theorem 2.2.4 (applied to $\mathfrak{B}$), $\{1, \ldots, n\}$ is a PT of $(B_1, \ldots, B_m)$; and this means that $\mathfrak{A}$ possesses a transversal.

Finally, we mention a slight generalization of the preceding theorem.

COROLLARY 2.2.5. *Let* $\mathfrak{A} = (A_1, \ldots, A_n)$ *be a family of subsets of* E, *and let* $E^* \subseteq E$. *Then* E^* *is a partial transversal of* $\mathfrak{A}$ *if and only if every subset* F *of* E^* *intersects at least* $|F|$ A'*s*.

It is clear that E^* is a PT of $\mathfrak{A}$ if and only if it is a PT of $(A_1 \cap E^*, \ldots, A_n \cap E^*)$, and the assertion now follows at once by Theorem 2.2.4.

Exercises 2.2

1. Use Theorem 2.2.1 to show that the family $(A_1, \ldots, A_n)$ of subsets of a finite set E possesses a transversal if and only if

$$\sum_{i \in I} |A_i \cap F| \geqslant |I| + |F| - |E|$$

whenever $I \subseteq \{1, \ldots, n\}$ and $F \subseteq E$.

2. Let $\mathfrak{A} = (A_1, \ldots, A_n)$ be a family of sets which satisfies Hall's condition. Show that the additional requirement $|A_1 \cup \ldots \cup A_n| = n$ is necessary and sufficient for the transversal of $\mathfrak{A}$ to be unique.

3. Show that the family $\mathfrak{A} = (A_1, \ldots, A_n)$ of sets satisfies Hall's condition if and and only if, for each $I \subseteq \{1, \ldots, n\}$,

$$\sum_{N \cap I \neq \emptyset} |A[N]| \geq |I|,$$

where $A[N], \emptyset \subset N \subseteq \{1, \ldots, n\}$, denote the boolean atoms generated by $\mathfrak{A}$.

4. Let $A_1, \ldots, A_m, B_1, \ldots, B_n$ be the nodes of a bipartite graph (in which no two A's and equally no two B's are linked by an edge.) Let the degrees of A_i, B_j be the positive numbers a_i, b_j respectively. (The 'degree' of a node is the number of edges incident with it.) Show that, if $\min a_i \geq \max b_j$, then $m \leq n$ and among the edges of the graph there are m edges linking the A's to m of the B's.

5. Let X, Y be antichains of maximum cardinal, say n, in a finite partially ordered set. Show that $X \cup Y$ can be expressed as the union of n chains.

[J. S. Pym]

6. Give an example of a set E, a subset E′ of E, and a family $\mathfrak{A} = (A_i : i \in I)$ of finite subsets of E such that, for each non-negative integer d, E′ is a PT with defect d of $\mathfrak{A}$.

2.3 Duality

In Hall's theorem we are concerned with the relation between two collections of objects, the elements of the ground set and the sets of the family. This relation is not, however, symmetrical.† For this, there are two reasons, one intrinsic, the other notational. The first reason arises simply from the definition of a transversal of a family, an entity associated with *all* sets of the family but not necessarily with all elements of the ground set. The resulting lack of symmetry is therefore inherent in Hall's theorem, and if we wish to have a symmetric statement we must formulate a different theorem. Later in the present account such theorems will, in fact, be proved. Now it is clear that a *partial* transversal is related to sets and elements in a symmetrical way. We might therefore expect some theorems involving partial transversals to be symmetric with respect to these two collections of objects. We shall, in fact, see that Theorems 3.2.4 and 3.2.6 below are results of this kind. Another symmetric result, though of a somewhat different type, is Theorem 3.4.1.

The lack of notational symmetry, on the other hand, can be corrected by a mere change of notation (and terminology). We recall, then, that we are dealing with two classes of objects: the sets A_i, $i \in I$ of the family $\mathfrak{A} = (A_i : i \in I)$, and the elements of the ground set E. Now whereas the elements are (of course) distinct, this is not in general true of the sets. Thus, it is more natural to think primarily not of the sets in $\mathfrak{A}$ but of the elements of the index set I specifying them. The two sets of objects E and I are then linked by an 'inci-

† At this stage of the discussion, the notion of symmetry is merely intuitive: its precise meaning will emerge later.

dence relation'. Thus, if $e \in E$, $i \in I$, the statement $e \in A_i$ may or may not be valid. We have therefore two sets E and I together with certain ordered pairs (e, i), namely those for which $e \in A_i$. In other words, we have E, I, and a subset of the cartesian product $E \times I$. Much of the theory we shall discuss is best expressed in terms of such triples.

We shall next introduce the terminology that will be used frequently throughout the remainder of the work. Unless the contrary is stated explicitly, we shall not assume that any set mentioned in the discussion is necessarily finite.

A *deltoid* is a triple $\mathscr{D} = (X, \Delta, Y)$, where X, Y are sets and Δ is a subset of the cartesian product $X \times Y$. The notion of a deltoid is thus effectively the same as that of a bipartite graph†, except that in a deltoid the sets X and Y need not be disjoint. The deltoid $\mathscr{D}$ is said to be finite if both X and Y are finite. If $x \in X$, $y \in Y$, and $(x, y) \in \Delta$, we shall say that the elements x, y are *linked* and we shall write $x \leftrightarrow y$.

Let $A \subseteq X$. We shall denote by $\Delta(A)$ the set of elements of Y which are linked to some elements of A, i.e.

$$\Delta(A) = \bigcup_{a \in A} \{y \in Y : (a, y) \in \Delta\}.$$

Also, if $x \in X$, we shall write $\Delta(x)$ for $\Delta(\{x\})$. For $B \subseteq Y$, $y \in Y$, the symbols $\Delta(B)$, $\Delta(y)$ are defined analogously:‡

$$\Delta(B) = \bigcup_{b \in B} \{x \in X : (x, b) \in \Delta\}, \qquad \Delta(y) = \Delta(\{y\}).$$

Further, we shall say that $\mathscr{D}$ is *locally right-finite* if the set $\Delta(x)$ is finite for each $x \in X$; we shall say that $\mathscr{D}$ is *locally left-finite* if $\Delta(y)$ is finite for each $y \in Y$.

Let $\mathscr{D} = (X, \Delta, Y)$ be a deltoid, $A \subseteq X$, $B \subseteq Y$. The set A is said to be *admissible* if there exists an injective mapping $\theta : A \to Y$ such that $(x, \theta(x)) \in \Delta$ for all $x \in A$. In that case, the mapping θ is also called admissible. Analogously, B and an injection $\psi : B \to X$ are both said to be admissible if $(\psi(y), y) \in \Delta$ for all $y \in B$.§ An admissible set is, of course, called *maximal* if it is not properly contained in any admissible set. By convention, we shall regard the empty set as an admissible subset of both X and Y. It is, of course, clear that

† Or, strictly speaking, of a 'directed bipartite graph'.

‡ The purist may take exception to our use of the same symbol Δ for two different mappings, one of $\mathscr{P}(X)$ into $\mathscr{P}(Y)$ and the other of $\mathscr{P}(Y)$ into $\mathscr{P}(X)$. However, no confusion will arise in practice.

§ Strictly speaking, we should distinguish between 'right-admissible' and 'left-admissible' sets and mappings, for otherwise our definitions become ambiguous when $X = Y$.

every subset of an admissible set is itself admissible. Equally, every restriction of an admissible mapping is again admissible.

The subsets A, B of X, Y respectively are said to be *linked* (in symbols: $A \leftrightarrow B$) if there exists an admissible *bijection* of A into B. (In particular, linked sets are, of course, admissible.) If A, B are singletons, say $A = \{x\}$, $B = \{y\}$, then the relation $\{x\} \leftrightarrow \{y\}$ is written as $x \leftrightarrow y$. This is consistent with the usage of the symbol $\leftrightarrow$ introduced earlier. By convention, $\emptyset \leftrightarrow \emptyset$.

A subset A of X is said to be *totally admissible* (or, more briefly, a *total set*) if $A \leftrightarrow Y$. Similarly, $B(\subseteq Y)$ is a total set if $X \leftrightarrow B$.

All definitions just given relate to a fixed deltoid. However, when we operate with more than one deltoid, it may be necessary to qualify our statements by some phrase such as 'with respect to $\mathscr{D}$'.

In any statement about the deltoid $\mathscr{D} = (X, \Delta, Y)$, we can interchange the roles of X and Y. To be more precise, with each deltoid $\mathscr{D}$, we associate the *dual deltoid* $\tilde{\mathscr{D}} = (Y, \tilde{\Delta}, X)$, where

$$\tilde{\Delta} = \{(y, x) : (x, y) \in \Delta\}.$$

To each statement about $\mathscr{D}$, there corresponds then a statement about $\tilde{\mathscr{D}}$, and conversely. We shall make frequent use of this obvious fact and shall regard the two statements related in the manner described as being dual to each other. Of course, as long as we operate with abstract deltoids, dualization remains a purely notational procedure which pays no dividends. It is only when we come to deal with specific models of deltoids that it becomes significant.

Next, we introduce a one-one correspondence between families and deltoids, which is already implicit in our earlier remarks. With the family $\mathfrak{A} = (A_i : i \in I)$ of subsets of a set E, we shall associate the deltoid $\mathscr{D} = (E, \Delta, I)$, where $\Delta(\subseteq E \times I)$ is specified by the requirement that $(e, i) \in \Delta$ if and only if $e \in A_i$. Conversely, if (X, Δ, Y) is a given deltoid, we shall associate with it the family $(\Delta(y) : y \in Y)$ of subsets of X. It is useful to have a catalogue of corresponding statements about the associated objects $\mathfrak{A}$ and $\mathscr{D}$, and such a catalogue is given below. In it, E^*, I^* denote typical subsets of E, I respectively. Further, the family $\mathfrak{A}$ is said to be *restricted* if no element $e \in E$ belongs to A_i for infinitely many values of $i \in I$. Also, $\mathfrak{A}(I^*)$ is called a *maximal subfamily* of $\mathfrak{A}$ if it possesses a transversal but is not properly contained in any subfamily which has a transversal (i.e. there exists no set I^{**} with $I^* \subset I^{**} \subseteq I$ such that $\mathfrak{A}(I^{**})$ has a transversal).

$\mathfrak{A}$	$\mathscr{D}$
$\mathfrak{A}$ is a family of finite sets	$\mathscr{D}$ is locally left-finite
$\mathfrak{A}$ is restricted	$\mathscr{D}$ is locally right-finite
E^* is a PT of $\mathfrak{A}$	E^* is admissible

E^* is a maximal PT of $\mathfrak{A}$	E^* is a maximal admissible set
E^* is a transversal of $\mathfrak{A}$	E^* is total
E is a transversal of $\mathfrak{A}$	E is total (= I is total)
$\mathfrak{A}$ possesses a transversal	I is admissible
$\mathfrak{A}(I^*)$ possesses a transversal	I^* is admissible
$\mathfrak{A}(I^*)$ is a maximal subfamily of $\mathfrak{A}$	I^* is a maximal admissible set
E is a transversal of $\mathfrak{A}(I^*)$	I^* is total
E is a PT of $\mathfrak{A}$	I possesses a total subset
E^* is a transversal of $\mathfrak{A}(I^*)$	$E^* \leftrightarrow I^*$

It is clear that any statement about a family of sets possesses an equivalent deltoid form (in which the notational asymmetry of the original statement has been corrected). Again, let $\mathfrak{A} = (A_i : i \in I)$ be a family of subsets of E. Denote by $\mathscr{D}$ the associated deltoid and by $\tilde{\mathfrak{A}}$ the family associated with the dual deltoid $\tilde{\mathscr{D}}$ of $\mathscr{D}$. Thus, with each family $\mathfrak{A}$ is associated its *dual family* $\tilde{\mathfrak{A}}$ (and the dual of the dual is, of course, the original family). It should also be noted that $\tilde{\mathfrak{A}}$ can be defined without the interposition of deltoids; in fact $\tilde{\mathfrak{A}} = (\tilde{A}_e : e \in E)$, where

$$\tilde{A}_e = \{i \in I : e \in A_i\}.$$

Let S be a statement about a family $\mathfrak{A}$. The same statement applied to the dual family $\tilde{\mathfrak{A}}$ yields another statement, say $\tilde{S}$, about the original family $\mathfrak{A}$. The two statements S and $\tilde{S}$ related in this way are said to be *dual* to each other. We give here a short list of pairs of dual statements.

$\mathfrak{A}$ is restricted	$\mathfrak{A}$ is a family of finite sets
E^* is a PT of $\mathfrak{A}$	$\mathfrak{A}(I^*)$ possesses a transversal
E^* is a maximal PT of $\mathfrak{A}$	$\mathfrak{A}(I^*)$ is a maximal subfamily of $\mathfrak{A}$
E^* is a transversal of $\mathfrak{A}$	E is a transversal of $\mathfrak{A}(I^*)$
E is a PT of $\mathfrak{A}$	$\mathfrak{A}$ possesses a transversal
$\lvert A(I^*)\rvert \geqslant \lvert I^*\rvert$ for each $I^* \subset\subset I$	Each $E^* \subset\subset E$ intersects at least $\lvert E^*\rvert$ A's

To put the matter briefly, if informally, dualization of a statement about a family is the process of interchanging the roles of sets and elements. From the above table, we readily recognize that Hall's theorem 2.2.1 and Theorem 2.2.4 are dual. In §2.2, these results were deduced from each other, but the superfluity of the deductions is now clear; for the two results are not so much consequences of each other as statements identical in content though not in form. Thus the lack of (intrinsic) symmetry in a theorem such as Hall's permits us to extract additional information or, at any rate, to see old information from a new angle.

It should, at the same time, be noted that there are cases where the process of dualization yields no fresh insight. Thus statements such as 'E is a transversal of $\mathfrak{A}$' possess an intrinsic symmetry as between sets and elements, and are consequently self-dual. We note, for example, that Theorems 3.2.6 and 3.4.1 are self-dual.

Summarizing what has been said earlier, we note that each theorem about families has three equivalent formulations: the original statement, the dual statement, and the corresponding statement about the associated deltoid. The three theorems so related are substantively identical; nevertheless, a change from one form to another can sometimes be helpful. In particular, when the theorem in question is complicated, the formulation of the dual theorem may not be immediately self-evident, and it may then be useful first to state the deltoid version.

Suppose, next, that our point of departure is a theorem about deltoids. Now, with each deltoid, we can associate a family and its dual family; and the original theorem thus gives rise to two (in general, formally distinct) theorems about families of sets. The process of interpreting results concerning deltoids in terms of families will be referred to as *standard interpretation.*

There is one further interpretation of deltoid theorems that we shall occasionally encounter, and that we now mention in passing. Suppose we possess a theorem T about deltoids. Let $\mathfrak{A} = (A_i : i \in I)$ and $\mathfrak{B} = (B_j : j \in J)$ be two families of sets and associate with them the deltoid $\mathscr{D} = (I, \Delta, J)$, where $(i, j) \in \Delta$ if and only if $A_i \cap B_j \neq \emptyset$. Then, applying theorem T to $\mathscr{D}$, we obtain a result concerning the families $\mathfrak{A}$ and $\mathfrak{B}$. This process will be called *symmetric interpretation.*

In what follows, we do not propose to adopt a uniform pattern of presentation. In many cases, we shall continue to speak in terms of sets and elements; in other cases we shall state the argument in deltoid form and derive statements about families of sets as immediate corollaries; and frequently we shall leave the interpretation or dualization to the reader. It should, however, be remembered that the concept of duality pervades transversal theory and that it should never be too far from our minds.

We shall conclude the present section by stating the deltoid version of a result given earlier in more traditional language.

THEOREM 2.3.1. *Let* (X, Δ, Y) *be a deltoid and let* X', Y' *be admissible subsets of* X, Y *respectively. Then there exist linked sets* X_0, Y_0 *such that* $X' \subseteq X_0 \subseteq X$, $Y' \subseteq Y_0 \subseteq Y$.

This is the deltoid formulation of Ore's mapping theorem 1.3.4. An immediate consequence of Theorem 2.3.1 runs as follows.

COROLLARY 2.3.2. *If* (X, Δ, Y) *is a deltoid, then any maximal admissible subset of* X *and any maximal admissible subset of* Y *are linked.*

Another simple deduction from Theorem 2.3.1 is, at times, useful.

COROLLARY 2.3.3. *Let* (X, Δ, Y) *be a deltoid; let* A, B *be subsets of* X, Y *respectively; and suppose that there exist admissible injections of* A *into* B *and of* B *into* A. *Then* A *and* B *are linked.*

Denote by Δ' the set of all (x, y) in Δ subject to the conditions $x \in A$, $y \in B$; and write $\mathscr{D} = (X, \Delta, Y)$, $\mathscr{D}' = (A, \Delta', B)$. Then the $\mathscr{D}$-admissible injections specified in the Corollary are also $\mathscr{D}'$-admissible. Hence, by Theorem 2.3.1 (with $X = X' = A$, $Y = Y' = B$), we infer that A, B are linked with respect to $\mathscr{D}'$ and so also with respect to $\mathscr{D}$.

Exercises 2.3

1. Let X, X', Y, Y' be sets such that $X' \subseteq X$, $Y' \subseteq Y$, and $|X'| \leqslant |Y'|$, $|X| \geqslant |Y|$. Applying Theorem 2.3.1 to the deltoid $(X, X \times Y', Y')$, establish the existence of a set X^* with $X' \subseteq X^* \subseteq X$ and $|X^*| = |Y'|$.

2. Let (X, Δ, Y) be a deltoid, A a subset of X, B a subset of Y, and suppose that $A \leftrightarrow B$ and that X is admissible. Show that there exists a set B^* with $B \subseteq B^* \subseteq Y$ such that $X \leftrightarrow B^*$. Obtain the two standard interpretations of this result.

3. Let $\mathfrak{A}$ be a family of subsets of E. If the subsets E_1, E_2 of E are maximal PTs of $\mathfrak{A}$, show that there exists a subfamily $\mathfrak{B}$ of $\mathfrak{A}$ of which both E_1 and E_2 are transversals. Extend this result to any finite number of subsets of E.

[Mirsky & Perfect (**2**)]

4. Let (X, Δ, Y) be a deltoid and suppose that, for each $y \in Y$, there exists some $x \in X$ such that $(x, y) \in \Delta$. Let Y^* be a maximal admissible subset of Y and let it be linked to subsets X_1, X_2 of X. Show that, if $X_1 \cap X_2 = \varnothing$, then $Y^* = Y$.

5. Let (X, Δ, Y) be a deltoid and let a subset Y^* of Y be linked to some maximal admissible subset of X. Show that, for every finite collection $\mathfrak{F}$ of maximal admissible subsets of X, there exists a set $\bar{Y}$ with $Y^* \subseteq \bar{Y} \subseteq Y$ which is linked to every set in $\mathfrak{F}$.

6. Let $\mathfrak{A} = (A_i : i \in I)$ be a family of subsets of a set E. Prove that there exists a subset F of E such that $|A_i \cap F| = 1$ for all $i \in I$ if and only if the dual family $\bar{\mathfrak{A}}$ possesses a subfamily which is a partition of I.

7. Let (X, Δ, Y) be a deltoid and let Y^* be an admissible subset of Y. Show that Y^* is maximal if and only if

$$\bigcup_{y \in Y \setminus Y^*} \Delta(y) \subseteq \bigcap (X^* \subseteq X : X^* \leftrightarrow Y^*).$$

State the associated results for families of sets.

Notes on Chapter 2

§ 2.1. Transversal theory has not yet acquired a standardized terminology, and expressions such as 'system of distinct representatives', 'common transversal', 'common system of representatives' and so on are used in varying senses by different authors. Our own choice has primarily been guided by the need to avoid confusion between sets and families. It should be noted that the terminology adopted in this book is not identical with that of the survey article by Mirsky & Perfect (**1**).

The independence of the $2^n - 1$ statements which constitute condition $\mathscr{H}$ was noted by R. Rado (**3**).

§ 2.2. Theorem 2.2.1 was proved by P. Hall (**1**) in 1935. It is implicit in the earlier literature and has, for this reason, been sometimes associated with the names of Dénes König and E. Egerváry. Thus, as we have seen above, it is an almost immediate consequence of König's theorem 1.7.1 on bipartite graphs. However, it is precisely Hall's formulation that has turned out to be the master key which has unlocked many closed docrs.

Hall's original proof of Theorem 2.2.1 was comparatively difficult. Other proofs of this or of closely related results have since been given by a number of writers, among them W. Maak (**1**), Marshall Hall Jr. (**2**), Weyl (**1**), Everett & Whaples (**1**), Halmos & Vaughan (**1**), D. Gale (**2**, 143-6), and R. Rado (**11**). The first of the proofs of Hall's theorem offered above is that of Halmos & Vaughan. The very transparent argument can be adapted for coping with more general situations (cf. for example Exs. 3.3.4 and 6.2.3). The second proof, due to Rado, is equally simple and the process of 'reduction' introduced here can, when combined with Zorn's lemma, be used to establish the transfinite form of Hall's theorem (see § 4.2). It can also be used to discuss 'independent' transversals (see § 6.2). D. J. A. Welsh (**7**) exploited the method of reduction to obtain generalizations of Hall's theorem.

An efficient algorithm for identifying a transversal of a given family or else demonstrating that no transversal exists has been devised by M. Hall Jr. (**3**).

A discussion of the relation of Hall's theorem to boolean algebra will be found in Hammer & Rudeanu (**1**, 252-257).

§ 2.3. The observation that in transversal theory there is a rather obvious duality between sets and elements must have been made by almost everyone attempting research in this field. The idea is implicit in the work of Edmonds & Fulkerson (**1**) and is developed explicitly by Mirsky & Perfect (**2**). The terminology employed in the present account is taken largely from the latter paper.

For much interesting work on deltoids, see J. S. Pym (**1**).

3 The Method of 'Elementary Constructions'

The present chapter pursues the development of Hall's theorem by tracing some of its comparatively easy extensions.

3.1 'Elementary constructions'

Hall's theorem 2.2.1 is 'self-refining' in a peculiarly strong sense. For let $\mathfrak{A}$ be a given family of sets, and let $\mathfrak{A}^*$ be a family constructed in one way or another from $\mathfrak{A}$. Then, applying Hall's theorem to $\mathfrak{A}^*$, we may be able to obtain a statement about the original family $\mathfrak{A}$ which constitutes a generalization of Hall's theorem. By adopting different definitions of $\mathfrak{A}^*$, we are led to a series of such generalizations.

The new family $\mathfrak{A}^*$ is invariably chosen as an object more complex than $\mathfrak{A}$. In defining it in terms of the original family $\mathfrak{A}$, we normally employ one or other of the following set-theoretic procedures, which we call *elementary constructions*.

Adjunction. The sets in the family $\mathfrak{A}$ of subsets of E are enlarged by the adjunction of 'dummy' elements which do not belong to the ground set E.

Extension. The family $\mathfrak{A}$ is extended by the addition of further sets. This means, in fact, that $\mathfrak{A}$ is replaced by $\mathfrak{A} + \mathfrak{B}$ for a suitable family $\mathfrak{B}$. (Cf. § 1.2.)

Replication. We obtain a new family by taking a suitable number of copies of each set in $\mathfrak{A}$.

Proliferation. A certain set A in $\mathfrak{A}$ is replaced by a new set A′ defined by the formula

$$A' = \{(x, i) : x \in A, 1 \leqslant i \leqslant k_x\}.$$

In other words, each element $x \in A$ is replaced by the set of pairs $(x, 1)$, $(x, 2), \ldots, (x, k_x)$. If k_x is independent of x and is equal to, say, k, then, $A' = A \times \{1, 2, \ldots, k\}$ and in that case proliferation simply reduces to the formation of cartesian products.

Substitution. Every element of a certain subset of a set A in $\mathfrak{A}$ is replaced by an associated element of a set D which does not intersect the ground set E.

Dualization. The family $\mathfrak{A} = (A_i : i \in I)$ of subsets of E is replaced by its dual family $\tilde{\mathfrak{A}} = (\tilde{A}_e : e \in E)$, where $\tilde{A}_e = \{i \in I : e \in A_i\}$.

Our list is not intended to be either exhaustive or entirely precise. Some of the constructions we shall use could be subsumed under more than one category and others do not entirely fit into any of them. Nevertheless, the above enumeration provides an approximate idea of the processes that will be frequently employed.

We shall see that the method of 'elementary constructions' is extremely effective. Indeed, by a combination of devices such as those described here, all known results in *finite* transversal theory can be exhibited as corollaries of Hall's theorem. This fact is certainly of considerable interest. Nevertheless, we should emphasize that (in the present state of knowledge) the method of elementary constructions is not by itself adequate for transfinite theory and that even in finite cases it does not invariably provide the most illuminating approach. Later, we shall see how to supplement this method with more powerful ideas, above all those based on the notion of abstract independence. In the present chapter, however, and also at various stages of the subsequent development, we shall exploit to the full the potentialities of elementary constructions.

3.2 Transversal index

Not every family of sets possesses a transversal. In the present section we shall investigate the problem of the maximum size of *partial* transversals.

If $\mathfrak{A} = (A_1, \ldots, A_n)$ is a finite family of sets, we shall understand by its *transversal index*, denoted by t^*, the maximum cardinal of PTs of $\mathfrak{A}$. We also recall that, if X is a PT of $\mathfrak{A}$, then its defect is defined as $n - |X|$.

We begin our discussion with the 'defect form' of Hall's theorem.

THEOREM 3.2.1. (Hall–Ore) *Let* $\mathfrak{A} = (A_1, \ldots, A_n)$ *be a family of subsets of* E, *and let* $1 \leqslant r \leqslant n$. *Then* $\mathfrak{A}$ *possesses a partial transversal of cardinal r (and defect* $n - r$*) if and only if*

$$|A(I)| \geqslant |I| + r - n \quad \textit{whenever} \quad I \subseteq \{1, \ldots, n\}. \tag{1}$$

For $r = n$, this result reduces of course to Hall's theorem.

To prove the theorem, we use the process of adjunction. Let D be a set (of 'dummy' elements) such that $D \cap E = \emptyset$, $|D| = n - r$, and consider the family

$$\mathfrak{A}^* = (A_1 \cup D, \ldots, A_n \cup D).$$

By Hall's theorem, $\mathfrak{A}^*$ possesses a transversal if and only if, for each (non-empty) set $I \subseteq \{1, \ldots, n\}$,

$$\begin{aligned} |I| \leqslant \left| \bigcup_{i \in I} (A_i \cup D) \right| &= |A(I) \cup D| \\ &= |A(I)| + |D| = |A(I)| + n - r. \end{aligned}$$

Thus $\mathfrak{A}^*$ *possesses a transversal if and only if* $\mathfrak{A}$ *satisfies* (1).

Now, if $\mathfrak{A}^*$ has a transversal, then we may write

$$x_1 \in A_1 \cup D, \quad x_2 \in A_2 \cup D, \quad \ldots, \quad x_n \in A_n \cup D, \tag{2}$$

where the x's are distinct. Since $|D| = n - r$, at most $n - r$ x's are contained in D and so at least r x's are contained among the A's. After possibly renumbering the A's, we therefore have

$$x_1 \in A_1, \quad \ldots, \quad x_r \in A_r. \tag{3}$$

Thus $\mathfrak{A}$ has a PT of cardinal r, namely $\{x_1, \ldots, x_r\}_{\neq}$.

Conversely, suppose that $\mathfrak{A}$ has a PT of cardinal r; we may then assume that (3) is satisfied with distinct x's. Denote by $x_{r+1}, \ldots, x_n$ the $n - r$ elements of D (none of which can be equal to any element among $x_1, \ldots, x_r$). Then

$$x_1 \in A_1, \quad \ldots, \quad x_r \in A_r, \quad x_{r+1} \in D, \quad \ldots, \quad x_n \in D.$$

A fortiori, (2) is valid with distinct x's and so $\mathfrak{A}^*$ has a transversal.

We have thus shown that $\mathfrak{A}$ *has a* PT *of cardinal r if and only if* $\mathfrak{A}^*$ *has a transversal.* The assertion now follows by a comparison of the two italicized statements.

COROLLARY 3.2.2. *Let* $\mathfrak{A} = (A_1, \ldots, A_n)$ *be a family of subsets of* E, *and let* $E^* \subseteq E$. *Then* E^* *is a partial transversal of* $\mathfrak{A}$ *if and only if, for all* $I \subseteq \{1, \ldots, n\}$,

$$|A(I) \cap E^*| \geqslant |I| + |E^*| - n.$$

It is plain that E^* is a PT of $\mathfrak{A}$ if and only if the family $(A_1 \cap E^*, \ldots, A_n \cap E^*)$ possesses a PT of cardinal $|E^*|$. The assertion now follows by Theorem 3.2.1.

Theorem 3.2.1 enables us to give a simple formula for the transversal index of a family.

COROLLARY 3.2.3. *Let* t^* *denote the transversal index of the family* $\mathfrak{A} = (A_1, \ldots, A_n)$. *Then*

$$t^* = n + \min_{I} \{|A(I)| - |I|\},$$

where the minimum is taken with respect to all sets I *such that* $\emptyset \subseteq I \subseteq \{1, \ldots, n\}$.

By Theorem 3.2.1 we know that, for all I,

$$|A(I)| \geqslant |I| + t^* - n;$$

and therefore

$$t^* \leqslant n + \min_{\mathrm{I}} \{|\mathrm{A(I)}| - |\mathrm{I}|\}.$$

Moreover, $\mathfrak{A}$ has no PT of cardinal $t^* + 1$ and so, again by Theorem 3.2.1,

$$|\mathrm{A(I)}| < |\mathrm{I}| + t^* + 1 - n$$

for some I. Hence

$$t^* \geqslant n + |\mathrm{A(I)}| - |\mathrm{I}|$$

for some I, i.e.

$$t^* \geqslant n + \min_{\mathrm{I}} \{|\mathrm{A(I)}| - |\mathrm{I}|\}.$$

The assertion is therefore proved.

Our next aim is the derivation of certain symmetric forms of Theorem 3.2.1. We continue to be concerned with the family $\mathfrak{A} = (\mathrm{A}_1, \ldots, \mathrm{A}_n)$ of subsets of E, and we take $\mathrm{E} = \{x_1, \ldots, x_m\}_{\neq}$. The symbols

$$x_1, \ldots, x_m, \quad \mathrm{A}_1, \ldots, \mathrm{A}_n$$

(where two A's are regarded as formally distinct even when they are equal as sets) will simply be called *objects*. A collection S of objects is said to be *incidence-bound* if the relation $x_i \in \mathrm{A}_j$ implies that at least one of x_i, A_j belongs to S. Again, S is said to be *incidence-free* if $x_i \notin \mathrm{A}_j$ whenever both x_i and A_j belong to S. We shall denote by t_* the minimum number of objects in an incidence-bound collection and by $\bar{t}$ the maximum number of objects in an incidence-free collection. It is, perhaps, worth noting that a collection can be both incidence-bound and incidence-free; for example, if $\mathrm{E} = \{x_1, x_2\}_{\neq}$, $\mathrm{A}_1 = \mathrm{A}_2 = \{x_1\}$, then the collection $x_2, \mathrm{A}_1, \mathrm{A}_2$ has this property.

THEOREM 3.2.4. *The transversal index of the family $\mathfrak{A} = (\mathrm{A}_1, \ldots, \mathrm{A}_n)$ of subsets of $\{x_1, \ldots, x_m\}_{\neq}$ is equal to the minimum number of objects in an incidence-bound subcollection of $x_1, \ldots, x_m, \mathrm{A}_1, \ldots, \mathrm{A}_n$. In other words, $t^* = t_*$.*

Let $\mathrm{I} \subseteq \{1, \ldots, n\}$. For simplicity of notation, write $\mathrm{I} = \{1, \ldots, p\}$, $\mathrm{A(I)} = \{x_1, \ldots, x_q\}$. Then the collection

$$x_1, \ldots, x_q, \quad \mathrm{A}_{p+1}, \ldots, \mathrm{A}_n$$

is incidence-bound and so $t_* \leqslant q + n - p$, i.e. $|\mathrm{A(I)}| \geqslant |\mathrm{I}| + t_* - n$. Since this inequality holds for each I, it follows by Theorem 3.2.1 that $t^* \geqslant t_*$. But any incidence-bound collection must contain at least t^* objects. Hence $t^* \leqslant t^*$, and the proof is complete.

LEMMA 3.2.5. *We have* $t_* + \bar{t} = m + n$.

The proof of this result depends on the simple observation that the complement of an incidence-bound resp. incidence-free collection is incidence-free resp. incidence-bound. Let

$$x_1, \ldots, x_p, \quad A_1, \ldots, A_q \tag{4}$$

be a minimal incidence-bound collection, so that $t_* = p + q$. Then

$$x_{p+1}, \ldots, x_m, \quad A_{q+1}, \ldots, A_n \tag{5}$$

is an incidence-free collection, so that $\bar{t} \geqslant (m - p) + (n - q) = m + n - t_*$. Again, let (4) be a maximal incidence-free collection, so that $\bar{t} = p + q$. Then (5) is an incidence-bound collection and so $t_* \leqslant m + n - \bar{t}$. Hence $t_* + \bar{t} = m + n$, as required.

THEOREM 3.2.6. *The family* $\mathfrak{A} = (A_1, \ldots, A_n)$ *of subsets of* $\{x_1, \ldots, x_m\}_{\neq}$ *has a partial transversal of cardinal r if and only if no collection of* $m+n-r+1$ *objects taken from* $x_1, \ldots, x_m, A_1, \ldots, A_n$ *is incidence-free.*

This theorem asserts that $t^* \geqslant r$ if and only if $\bar{t} \leqslant m + n - r$; in other words, that $t^* + \bar{t} = m + n$. In view of Lemma 3.2.5, we now see that Theorems 3.2.4 and 3.2.6 imply each other.

We shall conclude this section by restating the last result in the language of deltoids.

COROLLARY 3.2.7. *Let* (X, Δ, Y) *be a finite deltoid, and suppose that* $X \cap Y = \varnothing$. *Then* X *and* Y *possess linked subsets of cardinal r if and only if every subset of* $X \cup Y$ *of cardinal* $|X| + |Y| - r + 1$ *contains a pair of linked elements.*

Exercises 3.2

1. Show that the family $(A_1, \ldots, A_n)$ of subsets of a finite set E possesses a PT of cardininal k if and only if, for all $F \subseteq E$,

$$|\{i\colon 1 \leqslant i \leqslant n,\ F \cap A_i \neq \varnothing\}| + |E \setminus F| \geqslant k.$$

State the deltoid version of this result.

2. Deduce Theorem 3.2.1 from Theorem 3.2.4.

3. Give a direct derivation of Theorem 3.2.6 from Theorem 3.2.1.

4. Let (X, Δ, Y) be a finite deltoid. Show that X (and so also Y) possesses an admissible subset of cardinal r if and only if $(X' \times Y') \cap \Delta \neq \varnothing$ whenever $X' \subseteq X$, $Y' \subseteq Y$, and $|X'| + |Y'| > |X| + |Y| - r$.

5. Deduce the Hall–Ore theorem 3.2.1 from König's theorem 1.7.1.

3.3 Further extensions of Hall's theorem

By continuing to invoke the device of elementary constructions, we shall obtain a series of further results each of which yields Hall's theorem as a special case.

THEOREM 3.3.1. *Let* $\mathfrak{A} = (A_1, \ldots, A_n)$ *be a family of subsets of* E, *and let* $p_1, \ldots, p_n$ *be non-negative integers. Then there exist pairwise disjoint sets* $X_1, \ldots, X_n$ *such that*

$$X_i \subseteq A_i, \qquad |X_i| = p_i \quad (1 \leqslant i \leqslant n)$$

if and only if, for each $I \subseteq \{1, \ldots, n\}$,

$$|A(I)| \geqslant \sum_{i \in I} p_i.$$

This theorem reduces to Hall's for $p_1 = \ldots = p_n = 1$. The proof depends on replication. Let $\mathfrak{A}^*$ denote the family consisting of p_i copies of A_i, $1 \leqslant i \leqslant n$. It is then clear that sets $X_1, \ldots, X_n$ with the desired properties exist if and only if $\mathfrak{A}^*$ possesses a transversal. Now a typical subfamily of $\mathfrak{A}^*$ consists of k_i copies of A_i, $i \in I$, where

$$I \subseteq \{1, \ldots, n\}, \qquad 1 \leqslant k_i \leqslant p_i \quad (i \in I). \tag{1}$$

Hence, by Hall's theorem, $\mathfrak{A}^*$ has a transversal if and only if

$$|A(I)| \geqslant \sum_{i \in I} k_i$$

whenever (1) is satisfied. It is clear that this is equivalent to the stated condition.

THEOREM 3.3.2. *Let* $\mathfrak{A} = (A_1, \ldots, A_n)$ *be a family of sets and let* r *be a positive integer. Then it is possible to partition* $\mathfrak{A}$ *into* r *subfamilies each of which possesses a transversal if and only if*

$$r|A(I)| \geqslant |I| \textit{ for all } I \subseteq \{1, \ldots, n\}. \tag{2}$$

For $r = 1$, this is simply Hall's theorem. To prove the asserted result, we argue by proliferation. Write $R = \{1, 2, \ldots, r\}$ and put

$$\mathfrak{A}^* = (A_1 \times R, \ldots, A_n \times R).$$

By Hall's theorem, $\mathfrak{A}^*$ has a transversal if and only if, for each $I \subseteq \{1, \ldots, n\}$,

$$|I| \leqslant \left| \bigcup_{i \in I} (A_i \times R) \right| = |A(I) \times R| = r|A(I)|.$$

In other words, $\mathfrak{A}^*$ has a transversal if and only if (2) is satisfied.

Now suppose that $\mathfrak{A}^*$ possesses a transversal, say

$$(x_1, k_1) \in A_1 \times R, \ldots, (x_n, k_n) \in A_n \times R,$$

where no two of the pairs (x_i, k_i) are identical. Put

$$I_j = \{i : 1 \leqslant i \leqslant n, k_i = j\} \qquad (1 \leqslant j \leqslant r),$$

so that $(I_1, \ldots, I_r)$ is a partition of $\{1, \ldots, n\}$. It is then clear that $\mathfrak{A}(I_j) = (A_i : i \in I_j)$ possesses a transversal. Thus $\mathfrak{A}$ can be partitioned into r subfamilies each of which possesses a transversal.

Conversely, suppose that $\mathfrak{A}$ can be partitioned into r subfamilies each of which possesses a transversal. In fact, let $(I_1, \ldots, I_r)$ be a partition of $\{1, \ldots, n\}$ such that $\mathfrak{A}(I_1), \ldots, \mathfrak{A}(I_r)$ all possess transversals, say $T_1, \ldots, T_r$ respectively. Then the union of all pairs

$$(x, j) \qquad (1 \leqslant j \leqslant r, \quad x \in T_j)$$

is clearly a transversal of $\mathfrak{A}^*$. The proof is thus complete.

COROLLARY 3.3.3. *Let* $\mathfrak{A} = (A_i : i \in I)$ *be a family of subsets of a finite set* E, *and let* r *be a positive integer. Then* E *can be partitioned into* r *partial transversals of* $\mathfrak{A}$ *if and only if, for each susbet* F *of* E,

$$r|\{i \in I : A_i \cap F \neq \emptyset\}| \geqslant |F|.$$

This is simply the dual of Theorem 3.3.2. A variant of Corollary 3.3.3 runs as follows.

COROLLARY 3.3.4. *Let* $\mathfrak{A} = (A_i : i \in I)$ *be a family of subsets of a finite set* E, *and let* r *be a positive integer. Then* E *can be partitioned into* r *partial transversals of* $\mathfrak{A}$ *if and only if, for each subset* F *of* E,

$$|F| \leqslant r \,.\, \rho(F), \tag{3}$$

where $\rho(F)$ *denotes the maximum cardinal of partial transversals of* $\mathfrak{A}$ *contained in* F.

Obviously, we have

$$\rho(F) \leqslant |\{i \in I : A_i \cap F \neq \emptyset\}|. \tag{4}$$

Hence, if (3) holds for all $F \subseteq E$, then, by (4) and Corollary 3.3.3, E admits of the desired partition.

Next, assume that $E = E_1 \cup \ldots \cup E_r$, where $E_1, \ldots, E_r$ are pairwise disjoint PTs of $\mathfrak{A}$. Then, for any $F \subseteq E$,

$$F = (E_1 \cap F) \cup \ldots \cup (E_r \cap F)$$

and so

$$\begin{aligned} |F| &= |E_1 \cap F| + \ldots + |E_r \cap F| \\ &= \rho(E_1 \cap F) + \ldots + \rho(E_r \cap F) \leqslant r \cdot \rho(F), \end{aligned}$$

as required.

We shall indicate, very briefly, yet another criterion for the partition of the ground set into partial transversals.

THEOREM 3.3.5. *Let $\mathfrak{A} = (A_1, \ldots, A_n)$ be a finite family of subsets of a finite set* E, *and let r be a positive integer. Then* E *can be partitioned into r partial transversals of $\mathfrak{A}$ if and only if, for all* $I \subseteq \{1, \ldots, n\}$,

$$rn - |E| \geqslant r|I| - |A(I)|.$$

Denote by $\mathfrak{A}^{(r)}$ the family consisting of r copies of each of the sets $A_1, \ldots, A_n$. It is easily verified that E can be partitioned into r PTs of $\mathfrak{A}$ if and only if E is a PT of $\mathfrak{A}^{(r)}$. By the use of Theorem 3.2.1, it now follows readily that the latter requirement is satisfied if and only if the condition stated in the theorem is valid. An alternative proof can also be based on Corollary 3.3.4.

Again, let $\mathfrak{A}$ be a family of subsets of E, and let $M \subseteq E$. We shall next seek conditions for $\mathfrak{A}$ to possess a transversal which contains M as a subset. It is customary to refer to the elements of M as 'marginal elements'.

THEOREM 3.3.6. *Let $\mathfrak{A} = (A_1, \ldots, A_n)$ be a family of subsets of a finite set* E, *and let* $M \subseteq E$. *Then $\mathfrak{A}$ possesses a transversal which contains* M *as a subset if and only if both the following conditions are satisfied.*

(i) $|A(I)| \geqslant |I|$ *for all* $I \subseteq \{1, \ldots, n\}$;

(ii) $|A(I) \cap M| \geqslant |I| + |M| - n$ *for all* $I \subseteq \{1, \ldots, n\}$.

Moreover, (i) *and* (ii) *are respectively equivalent to the following statements:* (i′) $\mathfrak{A}$ *has a transversal*; (ii′) M *is a partial transversal of* $\mathfrak{A}$.

The equivalence of (i) and (i′) is simply Hall's theorem, that of (ii) and (ii′) is Corollary 3.2.2. When $M = \emptyset$, Theorem 3.3.6 (stated in terms of (i) and (ii)) reduces to Hall's theorem.

The proof of the theorem depends on extension and replication. If $\mathfrak{A}$ has a transversal, then $|E| \geqslant n$. Equally, (i) implies this relation, which can therefore be assumed throughout the argument.

Let $\mathfrak{A}^*$ denote the family consisting of the sets $A_1, \ldots, A_n$ and $|E| - n$ copies of $E \setminus M$. Since $\mathfrak{A}^*$ comprises $|E|$ subsets of E, the transversal of $\mathfrak{A}^*$, if it exists, must be E. In that case $\mathfrak{A}$ will also, of course, have a transversal; and since no element of M belongs to $E \setminus M$, this transversal must contain M.

Suppose, on the other hand, that $\mathfrak{A}$ has a transversal which contains M. This transversal consists, then, of the set M and $n - |M|$ elements in $E \setminus M$. There remain $|E \setminus M| - (n - |M|) = |E| - n$ elements in $E \setminus M$. Hence $\mathfrak{A}^*$ possesses a transversal. We have thus shown that $\mathfrak{A}$ *has a transversal which contains* M *if and only if* $\mathfrak{A}^*$ *has a transversal.*

Let k be a non-negative integer and X a set. We shall define kX as X or Ø according as $k > 0$ or $k = 0$. By Hall's theorem, $\mathfrak{A}^*$ has a transversal if and only if

$$|A(I) \cup k(E \setminus M)| \geqslant |I| + k$$

whenever $I \subseteq \{1, \ldots, n\}$, $0 \leqslant k \leqslant |E| - n$. If $|E| = n$, then $k = 0$ and this condition reduces to (i). In that case, (ii) is a consequence of (i) since

$$\begin{aligned} |A(I) \cap M| = |A(I)| + |M| - |A(I) \cup M| &\geqslant |A(I)| + |M| - |E| \\ &\geqslant |I| + |M| - n. \end{aligned}$$

If, on the other hand, $|E| > n$, then $k = 0$ or $k > 0$. For $k = 0$, we obtain (i) as before. For $k > 0$, the requirement is that

$$|A(I) \cup (E \setminus M)| \geqslant |I| + k$$

whenever $I \subseteq \{1, \ldots, n\}$, $0 < k \leqslant |E| - n$, i.e.

$$|A(I) \cup (E \setminus M)| \geqslant |I| + |E| - r \qquad \text{(all I).} \tag{5}$$

But the left-hand side is equal to $|E| - |M| + |A(I) \cap M|$, and so (5) is equivalent to (ii). We have thus shown that $\mathfrak{A}^*$ has a transversal if and only if both (i) and (ii) are satisfied. The proof is therefore complete; and the reader may well feel that, though brief, it is not strikingly illuminating. We shall, however, gain further insight in § 6.6, where the discussion of marginal elements will be resumed in the context of the theory of abstract independence.

The conditions in Theorem 3.3.6 admit of an alternative formulation.

COROLLARY 3.3.7. *Let* $\mathfrak{A} = (A_1, \ldots, A_n)$ *be a family of subsets of a finite set* E, *and let* $M \subseteq E$. *Then* $\mathfrak{A}$ *possesses a transversal which contains* M *if and only if both the following conditions are satisfied*: (i) *each* $F \subseteq E$ *contains at most* $|F|$ A's; (ii) *each* $N \subseteq M$ *intersects at least* $|N|$ A's.

By Theorem 2.2.4, applied to the family $(A_i \cap M : 1 \leqslant i \leqslant n)$, condition (ii) is satisfied if and only if M is a PT of $\mathfrak{A}$. If we can show that (i) is equivalent to Hall's condition, then the assertion will follow by Theorem 3.3.6.

Suppose Hall's condition is satisfied. Let $F \subseteq E$, and put

$$I = \{i : 1 \leqslant i \leqslant n,\ A_i \subseteq F\}.$$

Then

$$|F| \geqslant \left|\bigcup_{i \in I} A_i\right| = |A(I)| \geqslant |I| = |\{i : 1 \leqslant i \leqslant n,\ A_i \subseteq F\}|.$$

Conversely, let condition (i) be satisfied. Then, for each $I \subseteq \{1, \ldots, n\}$, we have

$$|A(I)| \geqslant |\{i : 1 \leqslant i \leqslant n,\ A_i \subseteq A(I)\}| \geqslant |I|,$$

and the proof is complete.

Exercises 3.3

1. Show that Theorem 3.3.6 remains valid for an infinite ground set E.

2. Let $\mathfrak{A} = (A_1, \ldots, A_n)$ be a family of subsets of E; let $M \subseteq E$; and let $|M| \leqslant p \leqslant n$. Show that $\mathfrak{A}$ possesses a PT of cardinal p and containing M if and only if, for all $I \subseteq \{1, \ldots, n\}$,

$$|I| \leqslant \min\{|A(I)| + n - p,\ n - |M| + |A(I) \cap M|\}.$$

3. Let $\mathfrak{A}$ be a finite family of sets, $\mathfrak{A}'$ a subfamily of $\mathfrak{A}$, and k an integer such that $|\mathfrak{A}'| \leqslant k \leqslant |\mathfrak{A}|$. Establish the equivalence of the following statements: (i) $\mathfrak{A}$ possesses a PT of cardinal k which contains a transversal of $\mathfrak{A}'$; (ii) $\mathfrak{A}$ possesses a PT of cardinal k, and $\mathfrak{A}'$ possesses a transversal.

4. Use the technique employed in the first proof of Hall's theorem 2.2.1 to establish Theorem 3.3.6.

5. Let k, d be natural numbers and let $\mathfrak{A} = (A_i : i \in I)$ be a finite family of subsets of E. Establish the equivalence of the following statements. (i) There exists a subset $I_0 \subseteq I$ with $|I \setminus I_0| \leqslant d$ such that the subfamily $\mathfrak{A}(I_0)$ possesses a system of representatives in which no element of E occurs more than k times. (ii) For every subset J of I,

$$k|A(J)| + d \geqslant |J|.$$

6. Write out in detail the proof of Theorem 3.3.5.

3.4 A self-dual variant of Hall's theorem

We have already met a number of self-dual results closely related to Hall's theorem (Theorems 3.2.4, 3.2.6, Corollary 3.2.7). Here we shall consider another, and somewhat more recondite, variation on the same theme.

THEOREM 3.4.1. *Let* $\mathfrak{A}$ *be a finite family of subsets of a finite set* E. *Let* $\mathfrak{A}' \subseteq \mathfrak{A}$ *and* $E' \subseteq E$. *The following statements and then equivalent.*

(i) *There exists a set* E_0 *and a family* $\mathfrak{A}_0$ *with* $E' \subseteq E_0 \subseteq E$, $\mathfrak{A}' \subseteq \mathfrak{A}_0 \subseteq \mathfrak{A}$ *such that* E_0 *is a transversal of* $\mathfrak{A}_0$.

(ii) (a) $\mathfrak{A}'$ *possesses a transversal*; (b) E' *is a partial transversal of* $\mathfrak{A}$.

The implication (i) $\Rightarrow$ (ii) is trivial; and we shall therefore be concerned with the proof of (ii) $\Rightarrow$ (i).

Write $\mathfrak{A} = (A_1, \ldots, A_n)$, $\mathfrak{A}' = (A_1, \ldots, A_\nu)$, $|E| = m$. Denote by D any set such that $D \cap E = \emptyset$, $|D| = n$, and by $\mathfrak{A}^*$ the family consisting of the sets $A_1, \ldots, A_\nu, A_{\nu+1} \cup D, \ldots, A_n \cup D$ and m copies of $(E \setminus E') \cup D$.

Assume, for the moment, that $\mathfrak{A}^*$ possesses a transversal. Since $\mathfrak{A}^*$ consists of $m + n$ subsets of $E \cup D$ and since $|E \cup D| = m + n$, it follows that the transversal of $\mathfrak{A}^*$ is $E \cup D$. Hence E is a transversal of a family consisting of the sets $A_1, \ldots, A_\nu$, certain of the sets $A_{\nu+1}, \ldots, A_n$, and a number (say k) of copies of $E \setminus E'$. We now remove from E the elements representing these k copies of $E \setminus E'$. Denoting the resulting set by E_0, we see that $E' \subseteq E_0 \subseteq E$ and that E_0 is a transversal of the family $\mathfrak{A}_0$ consisting of $A_1, \ldots, A_\nu$ and certain among the sets $A_{\nu+1}, \ldots, A_n$. Thus $\mathfrak{A}' \subseteq \mathfrak{A}_0 \subseteq \mathfrak{A}$, and (i) is valid. It remains to show that (ii) implies the existence of a transversal of $\mathfrak{A}^*$.

Now, by (a) and Hall's theorem, we have

$$|A(I)| \geqslant |I| \qquad (I \subseteq \{1, \ldots, \nu\}). \tag{1}$$

Further, by (b) and Corollary 3.2.2,

$$|A(I) \cap E'| \geqslant |I| + |E'| - n \qquad (I \subseteq \{1, \ldots, n\}). \tag{2}$$

Again, by Hall's theorem, $\mathfrak{A}^*$ has a transversal precisely if

$$\left|\left(\bigcup_{i\in K} A_i\right) \cup \bigcup_{i\in L} (A_i \cup D) \cup \mu\{(E \setminus E') \cup D\}\right| \geqslant |K| + |L| + \mu \tag{3}$$

whenever $K \subseteq \{1, \ldots, \nu\}$, $L \subseteq \{\nu + 1, \ldots, n\}$, $0 \leqslant \mu \leqslant m$. We shall now verify that this set of conditions is fulfilled. If $L = \emptyset$, $\mu = 0$, then (3) reduces to (1). If $L \neq \emptyset$, $\mu = 0$, then (3) reduces to the statement that

$$|A(K) \cup A(L) \cup D| \geqslant |K| + |L|$$

for all $K \subseteq \{1, \ldots, \nu\}$, $L \subseteq \{\nu + 1, \ldots, n\}$; and this holds trivially since $|D| = n \geqslant |K| + |L|$. Finally, if $\mu > 0$, then (3) is the condition that

$$|A(K) \cup A(L) \cup (E \setminus E') \cup D| \geqslant |K| + |L| + \mu$$

whenever $K \subseteq \{1, \ldots, \nu\}$, $L \subseteq \{\nu + 1, \ldots, n\}$, $0 < \mu \leqslant m$. This, in turn, is equivalent to the requirement that the inequality

$$|A(I) \cup (E \setminus E')| \geqslant |I| + m - n$$

should hold for all $I \subseteq \{1, \ldots, n\}$. But

$$|A(I) \cup (E \setminus E')| = m - |E'| + |A(I) \cap E'|,$$

and hence the condition in question is simply (2). Thus $\mathfrak{A}^*$ possesses a transversal, and the proof is complete.

We note that the case $\mathfrak{A}' = \mathfrak{A}$ of Theorem 3.4.1 is simply Theorem 3.3.6. Again, the dual result of Theorem 3.3.6 is obtained if we take $E' = E$ in Theorem 3.4.1. The detailed statement is as follows.

COROLLARY 3.4.2. *Let $\mathfrak{A}$ be a finite family of subsets of a finite set* E, *and let $\mathfrak{A}'$ be a subfamily of $\mathfrak{A}$. Then there exists a family $\mathfrak{A}_0$ such that $\mathfrak{A}' \subseteq \mathfrak{A}_0 \subseteq \mathfrak{A}$ which has* E *as a transversal if and only if $\mathfrak{A}'$ possesses a transversal and* E *is a partial transversal of $\mathfrak{A}$.*

If we write down the deltoid form of Theorem 3.4.1, we shall recognize at once that it is identical with Theorem 2.3.1 for the case of finite deltoids. Thus the somewhat technical argument in the discussion above could have been by-passed. It is, nevertheless, of interest to push as far as possible the method depending on elementary constructions. Subsequently, when the transfinite form of Hall's theorem becomes available, we shall be able to formulate the transfinite analogue of Theorem 3.4.1 (cf. §10.1).

Exercises 3.4

1. State the deltoid version of Theorem 3.4.1.

2. By specializing the proof of Theorem 3.4.1, frame a new proof of Theorem 3.3.6.

3. Let (X, Δ, Y) be a finite deltoid; let $\overline{X} \subseteq X$, $\overline{Y} \subseteq Y$; and let r, s be positive integers. Establish the equivalence of the following statements.

(i) There exist linked sets $X_0\,(\subseteq X)$, $Y_0\,(\subseteq Y)$ such that $|X_0 \cap \overline{X}| \geqslant r$, $|Y_0 \cap \overline{Y}| \geqslant s$.

(ii) (a) $|\Delta(X^*)| + |\overline{X} \setminus X^*| \geqslant r$ for all $X^* \subseteq \overline{X}$.

(b) $|\Delta(Y^*)| + |\overline{Y} \setminus Y^*| \geqslant s$ for all $Y^* \subseteq \overline{Y}$.

Give an interpretation of this result in terms of sets and elements.

Notes on Chapter 3

§ 3.1. Nearly everyone who has worked in transversal theory must have made use of one or other of the elementary constructions. The method is described and illustrated in Mirsky's papers (**6**) and (**7**).

§ 3.2. The defect form of Hall's theorem (Theorem 3.2.1) is contained in a much more general result of O. Ore (**1**). For a generalization in a different direction, see Theorem 5.1.1 below. Theorem 3.2.4 was noted by Kuhn & Tucker (**1**, Preface).

§ 3.3. Theorem 3.3.1 is due to Halmos & Vaughan (**1**). Theorem 3.3.2 was shown to me by Professor R. Rado in 1965. Corollary 3.3.4 will be seen later to be a special case of a much more comprehensive result (Corollary 8.2.2). The problem of marginal elements was first raised by Mann & Ryser (**1**) and was settled definitively by Hoffman & Kuhn (**1**) with the aid of linear programming. The proof given here was devised by Hazel Perfect (**2**).

§ 3.4. Theorem 3.4.1 is, in essence, due to Mendelsohn & Dulmage (**1**), though these writers stated their result in terms of abstract binary relations rather than sets and elements. Their proof, if lengthy, was entirely constructive. Hazel Perfect (**1**) gave a matrix formulation of the theorem (cf. Theorem 11.1.4 below) and based her short and elegant argument on ideas drawn from linear algebra. The proof offered above is due to Mirsky (**7**).

4 Rado's Selection Principle

We shall now discuss a very general principle which enables us, in particular, to prove a transfinite analogue of Hall's theorem. However, as we shall see, its effective range is not limited to this application, nor indeed to transversal theory.

4.1 Proofs of the selection principle

Let $\mathfrak{A} = (A_i : i \in I)$ be a (generally infinite) family of subsets of E. Denote by $\mathscr{J}$ the collection of all finite subsets of I and, for $J \in \mathscr{J}$, let θ_J be a choice function of the subfamily $\mathfrak{A}(J)$ of $\mathfrak{A}$ (cf. §1.2). The functions θ_J, $J \in \mathscr{J}$, will be called the 'local choice functions' (on $\mathfrak{A}$). We should like to be able to assert the existence of a 'global' choice function θ of the entire family $\mathfrak{A}$ which should, in some sense, reflect the behaviour of the local functions. Now we cannot demand that, for each $i \in I$, θ should agree on i with *every* local choice function which is defined on i, for two local functions need not agree with each other on i. On the other hand, it would be pointless merely to seek a global choice function θ such that, for each $i \in I$, θ should agree on i with *some* local function defined on i; for such a function θ is given by $\theta(i) = \theta_{\{i\}}(i)$ $(i \in I)$. We shall, in fact, establish the existence of a global choice function which satisfies much more stringent requirements.

THEOREM 4.1.1. (Rado's selection principle) *Let $\mathfrak{A} = (A_i : i \in I)$ be a family of finite subsets of a set E. Let $\mathscr{J}$ denote the collection of all finite subsets of the index set I and, for each $J \in \mathscr{J}$, let θ_J be a choice function of the subfamily $(A_i : i \in J)$.*† *Then there exists a choice function θ of $\mathfrak{A}$ with the property that, for each $J \in \mathscr{J}$, there is a K with $J \subseteq K \in \mathscr{J}$ and $\theta | J = \theta_K | J$.*

We shall refer to θ as a 'Rado choice function' corresponding to the given system θ_J, $J \in \mathscr{J}$, of local functions.

The most succinct proof of Theorem 4.1.1 depends on properties of topological spaces. Write $X = \times(A_i : i \in I)$ and let each A_i be endowed with the discrete topology. Since the A_i are finite, they are compact topological spaces. Let X be endowed with the resulting product topology.

Let $J \in \mathscr{J}$ and denote by F_J the (obviously non-empty) set of all choice

† The existence of the choice functions implies that all A_i are non-empty.

functions θ on $\mathfrak{A}$ such that $\theta|J = \theta_K|J$ for some K with $J \subseteq K \in \mathcal{J}$. Then $X \setminus F_J$ is the set of all $\theta \in X$ such that $\theta|J \neq \theta_K|J$ whenever $J \subseteq K \in \mathcal{J}$. For $\theta \in X$, $J \in \mathcal{J}$, we write

$$S_J(\theta) = \mathop{\times}_{i \in I} \begin{cases} \{\theta(i)\} & (i \in J) \\ A_i & (i \in I \setminus J). \end{cases}$$

Then clearly $S_J(\theta) = \{\phi \in X : \phi|J = \theta|J\}$, and it is a matter of immediate verification that

$$X \setminus F_J = \bigcup_{\theta \in X \setminus F_J} S_J(\theta).$$

It follows that, for each $J \in \mathcal{J}$, $X \setminus F_J$ is open in the product topology. Now assume that

$$\bigcup_{J \in \mathcal{J}} (X \setminus F_J) = X. \tag{1}$$

By Tychonoff's theorem 1.6.1, X is a compact topological space and the open covering (1) possesses therefore a finite subcovering, say

$$(X \setminus F_{J_1}) \cup \ldots \cup (X \setminus F_{J_m}) = X,$$

where $J_1, \ldots, J_m \in \mathcal{J}$. Taking complements, we infer that

$$F_{J_1} \cap \ldots \cap F_{J_m} = \emptyset.$$

But

$$F_{J_1} \cap \ldots \cap F_{J_m} \supseteq F_{J_1 \cup \ldots \cup J_m} \neq \emptyset,$$

and the contradiction shows that (1) is false. Hence

$$\bigcup_{J \in \mathcal{J}} (X \setminus F_J) \subset X$$

and, taking complements, we see that

$$\bigcap_{J \in \mathcal{J}} F_J \neq \emptyset.$$

Any choice function in the above intersection satisfies the requirements of the theorem.

The proof just concluded is extremely short, but there may be some advantage in supplementing it with an alternative treatment which relies solely on set-theoretic ideas.

As before, let θ_J, $J \in \mathcal{J}$, be local choice functions on $\mathfrak{A}$. Denote by Ω the collection of all families $(B_i : i \in I)$ with $B_i \subseteq A_i$ $(i \in I)$ and such that, given any

$J \in \mathscr{J}$, there exists a K with $J \subseteq K \in \mathscr{J}$ and $\theta_K(i) \in B_i$ $(i \in J)$. We shall partially order Ω by declaring that

$$(B_i : i \in I) \leqslant (B_i' : i \in I)$$

if and only if $B_i \subseteq B_i'$ $(i \in I)$.

Next, we consider chains in Ω. Let Λ be an ordered set and, for each $\lambda \in \Lambda$, let $\mathfrak{B}(\lambda) = \big(B_i(\lambda) : i \in I\big) \in \Omega$. Moreover, suppose that $\mathfrak{B}(\lambda)$, $\lambda \in \Lambda$, is a chain, i.e. for each $i \in I$, $B_i(\lambda) \subseteq B_i(\lambda')$ whenever $\lambda, \lambda' \in \Lambda$, $\lambda \leqslant \lambda'$. Write

$$B_i^* = \bigcap_{\lambda \in \Lambda} B_i(\lambda) \qquad (i \in I).$$

Let $i \in I$. Assume that, given any $\lambda \in \Lambda$, there exists some $\lambda' < \lambda$ such that $B_i(\lambda') \subset B_i(\lambda)$. Then there is an infinite sequence

$$\ldots \subset B_i(\lambda'') \subset B_i(\lambda') \subset B_i(\lambda), \tag{2}$$

where $\ldots < \lambda'' < \lambda' < \lambda$; and this is impossible since (in view of the finiteness of A_i) the cardinals of all terms in (2) are positive integers. Hence there exists a λ_i such that $B_i(\lambda') = B_i(\lambda_i)$ whenever $\lambda' < \lambda_i$. It follows that

$$B_i^* = B_i(\lambda_i). \tag{3}$$

Now let $J \in \mathscr{J}$. Writing $\lambda_0 = \min(\lambda_i : i \in J)$, we infer from (3) that

$$B_i(\lambda_0) \subseteq B_i(\lambda_i) = B_i^* \subseteq B_i(\lambda_0) \qquad (i \in J).$$

Hence

$$B_i^* = B_i(\lambda_0) \qquad (i \in J).$$

Now $\big(B_i(\lambda_0) : i \in I\big) \in \Omega$ and so, corresponding to the chosen set $J \in \mathscr{J}$, we have a K with $J \subseteq K \in \mathscr{J}$ and $\theta_K(i) \in B_i(\lambda_0) = B_i^*$ $(i \in J)$. Moreover, we see by (3) that, for $i \in I$, $B_i^* = B_i(\lambda_i) \subseteq A_i$. Thus $(B_i^* : i \in I) \in \Omega$. In other words, every chain in Ω possesses a lower bound. Hence, by Zorn's lemma, Ω has a minimal element, say $\mathfrak{M} = (M_i : i \in I)$.

Let $i_0 \in I$. Clearly $M_{i_0} \neq \varnothing$. Suppose that $x_1, x_2 \in M_{i_0}$. We shall show that, in fact, $x_1 = x_2$—an inference which implies that M_{i_0} is a singleton. We define M_i' as M_i or $M_{i_0} \setminus \{x_1\}$ according as $i \in I \setminus \{i_0\}$ or $i = i_0$.

Since $\mathfrak{M} \in \Omega$, we know that, given any $J \in \mathscr{J}$, there exists some K with $J \subseteq K \in \mathscr{J}$ and $\theta_K(i) \in M_i (i \in J)$. Any such K will be called an 'associate' of J.

We now assert that *there exists a set* J_1 *with* $i_0 \in J_1 \in \mathscr{J}$ *such that* $\theta_K(i_0) = x_1$ *for every associate* K *of* J_1. For assume that this assertion is false, i.e. assume that each J with $i_0 \in J \in \mathscr{J}$ possesses some associate K with $\theta_K(i_0) \neq x_1$. For any $J \in \mathscr{J}$, write $J' = J \cup \{i_0\}$. By hypothesis, J' possesses an associate K such that $\theta_K(i_0) \neq x_1$. Hence $\theta_K(i) \in M_i'$ $(i \in J')$ and, a fortiori, $\theta_K(i) \in M_i'$ $(i \in J)$. We have therefore shown that $(M_i' : i \in I) \in \Omega$, contrary to the minimality of $\mathfrak{M}$. This establishes the italicized statement. In the same way it follows that there exists a set J_2 with $i_0 \in J_2 \in \mathscr{J}$ such that $\theta_K(i_0) = x_2$ for every associate K of J_2.

Next, let K^* be an associate of $J_1 \cup J_2$, and so of J_1 and of J_2. Then $\theta_{K^*}(i_0) = x_1$, $\theta_{K^*}(i_0) = x_2$ and therefore $x_1 = x_2$. Thus M_{i_0} is a singleton and, since i_0 is an arbitrary element of I, every M_i is a singleton. We shall write $\mathfrak{M} = (\{z_i\} : i \in I)$.

Finally, let the mapping $\theta : I \to E$ be specified by $\theta(i) = z_i$ $(i \in I)$. Since $\mathfrak{M} \in \Omega$, we know that, given any $J \in \mathscr{J}$, there exists K such that $J \subseteq K \in \mathscr{J}$ and $\theta_K(i) \in \{z_i\}$ $(i \in J)$, i.e. $\theta_K(i) = z_i = \theta(i)$ $(i \in J)$. The proof is now complete.

COROLLARY 4.1.2. *Let the assumptions and notation be as in Theorem* 4.1.1. *If all local choice functions* θ_J, $J \in \mathscr{J}$, *are injective, then so is the global choice function* θ *of* $\mathfrak{A}$.

Let $i_1, i_2 \in I$, $i_1 \neq i_2$, and take $J = \{i_1, i_2\}$. By Theorem 4.1.1, there exists a set K with $\{i_1, i_2\} \subseteq K \in \mathscr{J}$ such that

$$\theta(i_1) = \theta_K(i_1), \qquad \theta(i_2) = \theta_K(i_2).$$

But θ_K is injective, so that $\theta_K(i_1) \neq \theta_K(i_2)$. Hence $\theta(i_1) \neq \theta(i_2)$, i.e. θ is injective.

4.2 Transfinite form of Hall's theorem

We shall now investigate whether Hall's theorem can be extended to families with infinite index sets. It will be recalled that the family $\mathfrak{A} = (A_i : i \in I)$ is said to satisfy 'Hall's condition' ($\mathscr{H}$) if the inequality $|A(J)| \geqslant |J|$ is valid whenever $J \subset\subset I$ (i.e. J is a finite subset of I). A family which possesses a transversal certainly satisfies Hall's condition. However, the converse inference is false as is shown, for example, by the family consisting of the sets

$$\{1, 2, 3, \ldots\}, \ \{1\}, \ \{2\}, \ \{3\}, \ \ldots .$$

Thus the obvious generalization of Hall's theorem fails. We shall show that its validity can be restored if we confine our attention to families of finite sets.

The case of a family with a denumerable index set is particularly easy, and we consider it first. Let, then, the family $\mathfrak{A} = (A_1, A_2, A_3, \ldots)$ consist of finite sets and let it satisfy Hall's condition. For each $r \geqslant 1$, there exist (by the finite case of Hall's theorem) r distinct elements $x_{r1}, \ldots, x_{rr}$ such that

$$x_{r1} \in A_1, \quad x_{r2} \in A_2, \ \ldots, \ x_{rr} \in A_r.$$

Now the elements x_{r1} ($r = 1, 2, \ldots$) all belong to the finite set A_1. Hence there exists an infinite subsequence N_1 of natural numbers such that the elements x_{r1} ($r \in N_1$) are all equal, say to y_1. Similar reasoning establishes the existence of a subsequence N_2 of N_1 such that the elements x_{r2} ($r \in N_2$) are all equal, say to y_2. Evidently $y_1 \in A_1$, $y_2 \in A_2$, $y_1 \neq y_2$. Repetition of this process yields a sequence of distinct representatives $y_k \in A_k$ ($k = 1, 2, \ldots$).

The case of a denumerable index set is rather special and our next theorem yields more comprehensive information.

THEOREM 4.2.1. (Transfinite form of Hall's theorem) *Let* $\mathfrak{A} = (A_i : i \in I)$ *be a family of finite subsets of a set* E. *The following statements are then equivalent.*

(a) $\mathfrak{A}$ *satisfies Hall's condition.*

(b) *Every finite subfamily of* $\mathfrak{A}$ *has a transversal.*

(c) $\mathfrak{A}$ *has a transversal.*

The implication (a) $\Rightarrow$ (b) holds by the finite form of Hall's theorem (Theorem 2.2.1), and the implication (c) $\Rightarrow$ (a) is trivial. To establish (b) $\Rightarrow$ (c), let $J \subset\subset I$. The subfamily $\mathfrak{A}(J)$ possesses a transversal by (b), i.e. there exists an injective mapping $\theta_J : J \to E$ such that $\theta_J(i) \in A_i$ ($i \in J$). The functions θ_J are injective local choice functions and it follows by Theorem 4.1.1 and Corollary 4.1.2 that there exists an injective choice function θ of $\mathfrak{A}$. Thus $\theta : I \to E$ is an injective mapping such that $\theta(i) \in A_i$ ($i \in I$), and therefore $\mathfrak{A}$ possesses a transversal.

We have now shown that the step from the finite to the transfinite form of Hall's theorem can be readily taken with the aid of Rado's selection principle. At the same time it may be of some interest to sketch a direct proof of Theorem 4.2.1: this will follow in essence the set-theoretic proof of the selection principle but will naturally be somewhat simpler. We shall establish the implication (a) $\Rightarrow$ (c).

Denote by Ω the collection of all families $(B_i : i \in I)$ of subsets of E which satisfy $\mathscr{H}$ as well as the inclusion relations $B_i \subseteq A_i$ ($i \in I$). We define a partial order in Ω by declaring that

$$(B_i : i \in I) \leqslant (B_i' : i \in I)$$

if and only if $B_i \subseteq B_i'$ ($i \in I$). It is easily verified that every chain in $(\Omega, \leqslant)$ possesses a lower bound. Hence, by Zorn's lemma, Ω contains a minimal element, which we denote by $(M_i : i \in I)$. Now let $i_0 \in I$. Then, by Lemma 2.2.3, there exists an element $x \in M_{i_0}$ such that the family $(M_i : i \in I \setminus \{i_0\}) + (\{x\})$ satisfies $\mathscr{H}$. If

$$M_i' = \begin{cases} M_i & (i \in I \setminus \{i_0\}) \\ \{x\} & (i = i_0), \end{cases}$$

then $(M_i' : i \in I) \in \Omega$ and so, in view of the minimality of $(M_i : i \in I)$, we have $|M_{i_0}| = 1$. Thus M_i is a singleton for each $i \in I$. Writing $M_i = \{x_i\}$ $(i \in I)$, we see that $x_i \in A_i$ $(i \in I)$ and that all x_i, $i \in I$, are distinct. Thus $\mathfrak{A}$ has a transversal. It is worth observing that in this argument we made no use of the finite version of Hall's theorem.

As an illustration of the use of Theorem 4.2.1, we shall show that any two bases in a vector space have the same cardinal number. (The standard proof of this result operates directly with Zorn's, or Tukey's, lemma and with certain properties of cardinal numbers.)

Let $\{x_i : i \in I\}_{\neq}$, $\{y_j : j \in J\}_{\neq}$ be any two bases of a given vector space. For each $i \in I$, let A_i denote the (necessarily finite) subset of J such that $j \in A_i$ if and only if y_j occurs with a non-zero coefficient in the linear expression for x_i in terms of y's. Assume that the family $(A_i : i \in I)$ fails to satisfy condition $\mathscr{H}$. Then, for some integer k and certain distinct elements $i_1, \ldots, i_k$ of I, we have

$$|A_{i_1} \cup \ldots \cup A_{i_k}| < k.$$

Write $A_{i_1} \cup \ldots \cup A_{i_k} = \{j_1, \ldots, j_m\}_{\neq}$, so that $m < k$. We then have relations of the form

$$x_{i_1} = p_{11}\, y_{j_1} + \ldots + p_{1m}\, y_{j_m}$$
$$\ldots\ldots\ldots\ldots\ldots\ldots\ldots\ldots$$
$$x_{i_k} = p_{k1}\, y_{j_1} + \ldots + p_{km}\, y_{j_m},$$

where the p's are scalars. Hence $x_{i_1}, \ldots, x_{i_k}$ are linearly dependent, contrary to hypothesis. It follows, therefore, that $(A_i : i \in I)$ satisfies $\mathscr{H}$; hence, by Theorem 4.2.1, it possesses a transversal. In particular, there exists an injective mapping $\phi : I \to J$ (such that $\phi(i) \in A_i$ for all $i \in I$). Hence $|I| \leqslant |J|$ and, by symmetry, we also have $|J| \leqslant |I|$. It follows, by the Schröder–Bernstein theorem 1.3.6, that $|I| = |J|$; and this is the required conclusion. The proof just given will be subsequently adapted to the more general case of 'independence spaces' (cf. §7.2).

Next, we give the deltoid version of Hall's theorem. (For the relevant notation and terminology, see § 2.3).

THEOREM 4.2.2. *Let* (X, Δ, Y) *be a locally right-finite deltoid. The following statements are then equivalent.*

(a) $|\Delta(A)| \geqslant |A|$ *for every* $A \subset\subset X$.

(b) *Every finite subset of* X *is admissible.*

(c) X *is admissible.*

A proof is hardly necessary, but if a formal derivation of Theorem 4.2.2 from Theorem 4.2.1 is desired, then we consider the family $\mathfrak{F} = (\Delta(x): x \in X)$ of (finite) subsets of Y. With this definition, statements (a), (b), (c) above reduce to the corresponding statements (for the family $\mathfrak{F}$) in Theorem 4.2.1. Conversely, Theorem 4.2.1 follows if we apply Theorem 4.2.2 to the deltoid (I, Δ, E), where $(i, e) \in \Delta$ if and only if $e \in A_i$. If, on the other hand, we take the deltoid (E, Δ, I) where $(e, i) \in \Delta$ if and only if $e \in A_i$, then we obtain the dual of Hall's theorem; this reads as follows.

THEOREM 4.2.3. *Let* $\mathfrak{A} = (A_i: i \in I)$ *be a restricted family of subsets of* E. *The following statements are then equivalent.*

(a) *Every* $E^* \subset\subset E$ *intersects at least* $|E^*|$ *A's.*

(b) *Every finite subset of* E *is a partial transversal of* $\mathfrak{A}$.

(c) E *is a partial transversal of* $\mathfrak{A}$.

We note that this result is the transfinite analogue of Theorem 2.2.4.

COROLLARY 4.2.4. *Let* $\mathfrak{A} = (A_i: i \in I)$ *be a restricted family of subsets of* E. *Then the set of all partial transversals of* $\mathfrak{A}$ *has finite character.*

Let $X \subseteq E$ and write $\mathfrak{A}^* = (A_i \cap X: i \in I)$. Then $\mathfrak{A}^*$ is a restricted family of subsets of X. By Theorem 4.2.3, X is a PT of $\mathfrak{A}$, and so of $\mathfrak{A}^*$, if and only if every finite subset of X is a PT of $\mathfrak{A}^*$, and so of $\mathfrak{A}$. The assertion therefore follows.

Exercises 4.2

1. Let $\mathfrak{A} = (A_i: i \in I)$ be a family of subsets of E, let $E^* \subseteq E$, and suppose that no element of E^* belongs to infinitely many A's. Show that the following statements are equivalent. (a) Every $F^* \subset\subset E^*$ intersects at least $|F^*|$ A's. (b) Every finite subset of E^* is a PT of $\mathfrak{A}$. (c) E^* is a PT of $\mathfrak{A}$.

2. Show that Corollary 4.2.4 ceases to be valid if the qualification 'restricted' is dispensed with.

3. Deduce Theorem 4.2.3 from Theorem 4.2.1.

4. Let $\mathfrak{A} = (A_i: i \in I)$ be an infinite family of finite sets, and let d be a natural number. Show that $\mathfrak{A}$ possesses a transversal with defect d if and only if the union of any k ($\geqslant d$) A's contains at least $k - d$ elements. Also state the dual of this result.

5. Let d be a natural number, $\mathfrak{X} = (x_i: i \in I)$ an infinite family of elements of a set E, and suppose that, for each $J \subset\subset I$,

$$|\{x_i: i \in J\}| \geqslant |J| - d.$$

Show (i) by Ex. 4.2.4, (ii) by Zorn's lemma, (iii) directly (and without invoking the axiom of choice) that there exists a set $I_0 \subseteq I$ with $|I \setminus I_0| \leqslant d$ such that all elements in the family $\mathfrak{X}(I_0)$ are distinct.

6. Let $\mathfrak{A} = (A_1, A_2, \ldots)$ be a denumerable family of non-empty subsets of E. Show that $\mathfrak{A}$ possesses a system of representatives in which no element of E occurs infinitely often if and only if, for every infinite set S of natural numbers, $\bigcup(A_i: i \in S)$ is an infinite set. (This question does not depend on the use of Hall's theorem.)

[R. Rado]

7. Let $\mathfrak{A} = (A_i: i \in I)$, $\mathfrak{B} = (B_j: j \in J)$ be two families of subsets of E. Suppose that each A is finite and that it intersects only a finite number of B's. Use Rado's selection principle (Theorem 4.1.1) and the result of Ex. 4.2.1 to show that the collection of subsets I* of I for which $\mathfrak{A}(I^*)$ has a transversal which is a PT of $\mathfrak{B}$ is of finite character.

8. Deduce Hall's theorem (Theorem 4.2.1) from the following proposition. 'Let $\mathfrak{A} = (A_i: i \in I)$, $\mathfrak{B} = (B_j: j \in J)$ be families of subsets of E. Suppose that each A intersects only a finite number of B's and that, for each natural number $k \leqslant |I|$, the union of any k A's intersects at least k B's. Then there exists an injective mapping $\theta: I \to J$ such that $A_i \cap B_{\theta(i)} \neq \emptyset$ for $i \in I$. 'Also obtain this proposition from Theorem 4.2.1.

4.3 A theorem of Rado and Jung

In the transfinite form of Hall's theorem (Theorem 4.2.1), we operate with families of *finite* sets. This restriction is extremely irksome as it greatly narrows the field of possible applications of Hall's theorem, but it is not at all easy to see how it might be relaxed. The next theorem records a slight progress in this direction: here we permit just one of the sets to be infinite.

Let $\mathfrak{A} = (A_i: i \in I)$ be a family of sets. We shall call a subset J of I *critical* if it is finite and $|A(J)| = |J|$.

THEOREM 4.3.1. *The infinite family* $\mathfrak{A} = (A_i: i \in I)$ *of sets among which exactly one (say* A_{i_0}*) is infinite possesses a transversal if and only if it satisfies Hall's condition and the relation*

$$A_{i_0} \nsubseteq A(I^*), \tag{1}$$

where I* *denotes the union of all critical subsets of* I.

Let $\mathfrak{A}$ possess a transversal, say $\{x_i: i \in I\}_{\neq}$, such that $x_i \in A_i$ $(i \in I)$. Hall's condition is then, of course, satisfied. Moreover, for any critical set J,

$$|\{x_i : i \in J\}| \leqslant |A(J)| = |J| = |\{x_i : i \in J\}|$$

and therefore $A(J) = \{x_i : i \in J\}$. Hence

$$A(I^*) = \{x_i : i \in I^*\}. \tag{2}$$

Now i_0 does not belong to any critical set, and so $i_0 \notin I^*$. Thus $I^* \subseteq I \setminus \{i_0\}$ and consequently

$$x_{i_0} \in A_{i_0} \setminus \{x_i : i \in I \setminus \{i_0\}\} \subseteq A_{i_0} \setminus \{x_i : i \in I^*\}.$$

By (2) we now infer that $x_{i_0} \in A_{i_0} \setminus A(I^*)$, and this establishes (1).

Next, let Hall's condition and relation (1) be satisfied. Choose an element $x_{i_0} \in A_{i_0} \setminus A(I^*)$, and consider the family

$$\mathfrak{A}^* = (A_i : i \in I^*) + (A_i \setminus \{x_{i_0}\} : i \in I \setminus I_1),$$

where $I_1 = I^* \cup \{i_0\}$. For $K \subset\subset I^*$, $L \subset\subset I \setminus I_1$, we have

$$\left|\left(\bigcup_{i \in K} A_i\right) \cup \bigcup_{i \in L} (A_i \setminus \{x_{i_0}\})\right| \geqslant \left|\left(\bigcup_{i \in K \cup L} A_i\right) \setminus \{x_{i_0}\}\right|$$

$$\geqslant \left|\bigcup_{i \in K \cup L} A_i\right| - 1.$$

If $L \neq \emptyset$, then $K \cup L \nsubseteq I^*$ and so $K \cup L$ is not a critical set. Hence

$$\left|\bigcup_{i \in K \cup L} A_i\right| \geqslant |K \cup L| + 1$$

and therefore

$$\left|\left(\bigcup_{i \in K} A_i\right) \cup \bigcup_{i \in L} (A_i \setminus \{x_{i_0}\})\right| \geqslant |K| + |L|.$$

This inequality is clearly still valid for $L = \emptyset$. Thus $\mathfrak{A}^*$ is a family of finite sets which satisfies Hall's condition. By Theorem 4.2.1, it possesses a transversal, say $\{x_i : i \in I \setminus \{i_0\}\}$, such that

$$x_i \in A_i \quad (i \in I^*), \qquad x_i \in A_i \setminus \{x_{i_0}\} \quad (i \in I \setminus I_1).$$

Thus $x_{i_0} \neq x_i$ for all $i \in I \setminus I_1$. Moreover

$$x_{i_0} \in A_{i_0} \setminus A(I^*) \subseteq A_{i_0} \setminus \{x_i : i \in I^*\}$$

and so $x_{i_0} \neq x_i$ for all $i \in \mathrm{I}^*$. Hence $\{x_i : i \in \mathrm{I}\}_{\neq}$ is a transversal of $\mathfrak{A}$ (with $x_i \in \mathrm{A}_i$ for all $i \in \mathrm{I}$).

4.4 Dilworth's decomposition theorem

In this section, we shall establish a fundamental result on partially ordered sets. The transition from the finite to the infinite case will be effected by the use of the selection principle.

THEOREM 4.4.1. (Dilworth's decomposition theorem) *Let* S *be an arbitrary partially ordered set and let m be a natural number. If* S *contains no antichain of cardinal* $m + 1$, *then it is the union of m (pairwise disjoint) chains.*

For the case of a finite set S, we argue by induction with respect to $|\mathrm{S}|$. If $|\mathrm{S}| = 1$, the assertion is clearly true. Let $|\mathrm{S}| > 1$, and let C be a maximal chain in S. If no antichain in $\mathrm{S} \setminus \mathrm{C}$ has m elements, then, by the induction hypothesis, $\mathrm{S} \setminus \mathrm{C}$ is the union of $m - 1$ chains and therefore S the union of m chains. On the other hand, let $\mathrm{S} \setminus \mathrm{C}$ possess an antichain of m elements, say $\mathrm{A} = \{a_1, \ldots, a_m\}_{\neq}$. Put

$$\mathrm{L} = \{x \in \mathrm{S} : x \leqslant a_k \text{ for some } k\},$$
$$\mathrm{R} = \{x \in \mathrm{S} : x \geqslant a_k \text{ for some } k\}.$$

Let z be the maximal element of C. If $z \in \mathrm{L}$, then $z < a_k$ for some k, in contradiction to the maximality of C. Hence $z \notin \mathrm{L}$, $|\mathrm{L}| < |\mathrm{S}|$, and by the induction hypothesis

$$\mathrm{L} = \mathrm{L}_1 \cup \ldots \cup \mathrm{L}_m,$$

where $\mathrm{L}_1, \ldots, \mathrm{L}_m$ are chains and $a_k \in \mathrm{L}_k$ $(1 \leqslant k \leqslant m)$.

Next, let $x \in \mathrm{L}_k$ so that $x \leqslant a_j$ for some j. Hence $a_k < x$ would imply $a_k < a_j$, which is false. This means that a_k is the maximal element of L_k.

By symmetry, we infer a relation of the form

$$\mathrm{R} = \mathrm{R}_1 \cup \ldots \cup \mathrm{R}_m,$$

where R_k is a chain with minimal element a_k. Now, by hypothesis, S has no antichain of cardinal $m + 1$. Hence every element in S is comparable with some a_k, and therefore

$$\mathrm{S} = \mathrm{R} \cup \mathrm{L} = (\mathrm{R}_1 \cup \mathrm{L}_1) \cup \ldots \cup (\mathrm{R}_m \cup \mathrm{L}_m).$$

Thus S is the union of m chains, which can be taken as pairwise disjoint.†

† It will be recalled that a chain can be empty.

We have now established the assertion for finite sets. When S is infinite, we consider the family $\mathfrak{A} = (A_x : x \in S)$, where $A_x = \{1, \ldots, m\}$ for all $x \in S$. Let $T \subset\subset S$. Then, by the result already proved, there exists a partition $T = T_1 \cup \ldots \cup T_m$ of T into m chains. For each $x \in T$, there is thus a unique integer k in the range $1 \leqslant k \leqslant m$ such that $x \in T_k$. Writing $\psi_T(x) = k$, we specify a choice function ψ_T of the subfamily $(A_x : x \in T)$. If $x, x' \in T$ and $\psi_T(x) = \psi_T(x')$, then x, x' belong to the same chain and so are comparable.

Denote by ψ a Rado choice function of $\mathfrak{A}$ corresponding to the local functions ψ_T, $T \subset\subset S$. For $x, x' \in S$, there exists a set K such that

$$\{x, x'\} \subseteq K \subset\subset S$$

and

$$\psi(x) = \psi_K(x), \qquad \psi(x') = \psi_K(x').$$

If $\psi(x) = \psi(x')$, then $\psi_K(x) = \psi_K(x')$ and so x, x' are comparable. Writing, for $1 \leqslant k \leqslant m$,

$$S_k = \{x \in S : \psi(x) = k\},$$

we see that $S_1, \ldots, S_m$ are pairwise disjoint chains whose union is S.

We record the following neater formulation of the theorem just proved.

COROLLARY 4.4.2. *If the maximum number of elements in an antichain of a partially ordered set* S *is finite, then it is equal to the minimum number of pairwise disjoint chains into which* S *can be decomposed.*

An obvious deduction is as follows.

COROLLARY 4.4.3. *A partially ordered set of* $rs + 1$ *elements possesses a chain of cardinal* $r + 1$ *or an antichain of cardinal* $s + 1$.

If there is no antichain of cardinal $s + 1$, then the given set, say X, can be expressed as the union of s pairwise disjoint chains, say $X = C_1 \cup \ldots \cup C_s$. Hence

$$rs + 1 = |C_1| + \ldots + |C_s|$$

and therefore $\max |C_i| \geqslant r + 1$, as required.

We note, in particular, that *a partially ordered set of* $n \geqslant r^2 + 1$ *elements possesses a chain or an antichain of cardinal* $r + 1$. An application of this result is given in the next theorem.

THEOREM 4.4.4. *A sequence of* $n \geqslant r^2 + 1$ *real terms possesses a monotonic subsequence of* $r + 1$ *terms.*

Let the given sequence be $(a_k: 1 \leqslant k \leqslant n)$. Put

$$X = \{(k, a_k) : 1 \leqslant k \leqslant n\}$$

and let X be partially ordered by the requirement that $(k, a_k) \leqslant (j, a_j)$ if and only if $k \leqslant j$ and $a_k \leqslant a_j$. Suppose that X has an antichain of cardinal $r + 1$, say

$$(k_1, a_{k_1}), \ldots, (k_{r+1}, a_{k_{r+1}}), \tag{1}$$

where $1 \leqslant k_1 < \ldots < k_{r+1} \leqslant n$. Then $a_{k_1} > \ldots > a_{k_{r+1}}$ and we have a (strictly) decreasing subsequence of $r + 1$ terms. If, on the other hand, X has no antichain of cardinal $r + 1$, then, by Corollary 4.4.3, it has a chain of cardinal $r + 1$, say (1), again with $1 \leqslant k_1 < \ldots < k_{r+1} \leqslant n$. In that case, we have $a_{k_1} \leqslant \ldots \leqslant a_{k_{r+1}}$ and there is an increasing subsequence of $r + 1$ terms.

We shall conclude the present section by pointing out that the Hall–Ore theorem 3.2.1 (for finite sets) is an easy consequence of Dilworth's theorem. Other deductions from Dilworth's theorem will be found in Chapter 11.

Let $\mathfrak{A} = (A_1, \ldots, A_n)$ be a family of subsets of $\{x_1, \ldots, x_m\}_{\neq}$; let $1 \leqslant r \leqslant n$; and suppose that

$$|A(I)| \geqslant |I| + r - n \quad \text{for all } I \subseteq \{1, \ldots, n\}. \tag{2}$$

In the set P of 'objects'

$$x_1, \ldots, x_m, \quad A_1, \ldots, A_n,$$

we introduce a partial order by declaring that $x_i < A_j$ if and only if $x_i \in A_j$. Let s denote the maximum number of objects in an antichain, and let

$$x_1, \ldots, x_k, \quad A_1, \ldots, A_h$$

be an antichain of s objects (so that $k + h = s$). Then

$$A_1 \cup \ldots \cup A_h \subseteq \{x_{k+1}, \ldots, x_m\}$$

and therefore, by (2),

$$h + r - n \leqslant |A_1 \cup \ldots \cup A_h| \leqslant m - k.$$

Hence $m + n - s \geqslant r$.

Now, by Dilworth's theorem, P can be decomposed into s pairwise disjoint chains, say

$$\{x_1, A_1\}, \ldots, \{x_i, A_i\}, \quad \{x_{i+1}\}, \ldots, \{x_m\}, \quad \{A_{i+1}\}, \ldots, \{A_n\}$$

(with the A's and x's suitably renumbered). Then $s = m + n - i$ and we see that the transversal index t^* of $\mathfrak{A}$ satisfies the relation $t^* \geqslant i = m + n - s \geqslant r$.

Exercises 4.4

1. Let S be an arbitrary partially ordered set and let m be a natural number. Show (e.g. by induction with respect to m) that, if S has no chain of cardinal $m + 1$, then it can be expressed as the union of m antichains. Also verify that Corollary 4.4.3 is a consequence of this result.

2. (i) Show that a sequence of $rs + 1$ real terms either contains an increasing subsequence of $r + 1$ terms or a decreasing subsequence of $s + 1$ terms (or both). (ii) Give an example of a sequence of rs real terms which possesses neither an increasing subsequence of $r + 1$ terms nor a decreasing subsequence of $s + 1$ terms.

3. Let $(d_1, \ldots, d_n)$ be a sequence of positive integers. We say that r is an 'independence number' if it is possible to select r d's such that none of them divides any other. Further, we say that s is a 'decomposition number' if it is possible to partition $(d_1, \ldots, d_n)$ into s subsequences such that, of any two d's in the same subsequence, one divides the other. Show that the greatest independence number is equal to the least decomposition number.

4.5 Miscellaneous applications of the selection principle

The wide range of problems on which the selection principle can be brought to bear has already been hinted at. Here we shall consider several applications in different branches of (not necessarily combinatorial) mathematics.

Chromatic number

The notion of the chromatic number of a graph was introduced in §1.7.

THEOREM 4.5.1. *Let G be an infinite graph and k a natural number. Then the chromatic number of G does not exceed k if and only if every finite subgraph of G has this property.*

Suppose that the chromatic number of every finite subgraph of G is at most k. Denote the set of nodes of G by I and consider the family $(A_i : i \in I)$, where $A_i = \{1, 2, \ldots, k\}$ for every $i \in I$. Let $J \subset\subset I$. By hypothesis, a colour (i.e. one of the integers $1, 2, \ldots, k$) can be assigned to each node in J such that, if i, i' are any two nodes in J which are linked by an edge of G, then different colours are assigned to them. This means that the subfamily $(A_i : i \in J)$ possesses a choice function θ_J such that $\theta_J(i) \neq \theta_J(i')$ whenever $i, i' \in J$ and i, i' are linked by an edge. Let θ denote a Rado choice function of the entire family $(A_i : i \in I)$. If i, i' are any two nodes which are linked by an edge, then there exists a set K such that $\{i, i'\} \subseteq K \subset\subset I$ and $\theta(i) = \theta_K(i), \theta(i') = \theta_K(i')$. But $\theta_K(i) \neq \theta_K(i')$

and so $\theta(i) \neq \theta(i')$. Thus we can paint the nodes with a stock of k colours such that no two nodes linked by an edge are painted alike.

It is, of course, possible to interchange the roles of nodes and edges and, by an argument almost identical with that given above, prove the following proposition. *The edges of a graph can be so painted with k colours that no two concurrent edges are painted alike provided this is true of every finite subgraph.*†

Problems on partitions

Let $\mathscr{E}$ be a non-empty collection of subsets of a set E, and let k be a natural number. If a subset F of E can be expressed as a union of k pairwise disjoint sets all of which are members of $\mathscr{E}$, we shall say that F is *$(\mathscr{E}, k)$-divisible.*

THEOREM 4.5.2. *Suppose that a non-empty collection $\mathscr{E}$ of subsets of* E *has finite character. Then the collection of all $(\mathscr{E}, k)$-divisible subsets of* E *also has finite character.*

Let $E^* \subseteq E$ and suppose that every finite subset of E^* is $(\mathscr{E}, k)$-divisible. We shall verify that E^* is itself $(\mathscr{E}, k)$-divisible.

For each $x \in E^*$, let $A_x = \{1, 2, ..., k\}$. For each $F \subset\subset E^*$, there exists a partition $F = F_1 \cup ... \cup F_k$, where $F_1, ..., F_k \in \mathscr{E}$. Hence, for each $x \in F$, there exists a unique integer i such that $1 \leqslant i \leqslant k$ and $x \in F_i$. Writing $\phi_F(x) = i$, we define a choice function ϕ_F of the family $(A_x : x \in F)$ with the property that, for $1 \leqslant i \leqslant k$,

$$\{x \in F : \phi_F(x) = i\} \quad (= F_i) \in \mathscr{E}. \tag{1}$$

Denote by ϕ a choice function of the family $(A_x : x \in E^*)$ of finite sets whose existence is guaranteed by Rado's selection principle. Write

$$E_i^* = \{x \in E^* : \phi(x) = i\} \qquad (1 \leqslant i \leqslant k). \tag{2}$$

Then $E^* = E_1^* \cup ... \cup E_k^*$ is a partition. It remains to show that $E_1^*, ..., E_k^*$ are members of $\mathscr{E}$.

For each $F \subset\subset E^*$, there exists G with $F \subseteq G \subset\subset E^*$ and $\phi|F = \phi_G|F$. It follows that, for $1 \leqslant i \leqslant k$,

$$\begin{aligned}\{x \in F : \phi(x) = i\} &= \{x \in F : \phi_G(x) = i\} \\ &\subseteq \{x \in G : \phi_G(x) = i\}.\end{aligned}$$

Now, by (1), $\{x \in G : \phi_G(x) = i\} \in \mathscr{E}$. But $\mathscr{E}$ has finite character and so every

† While this result is correct as stated it is, perhaps, more convenient in the present instance to define a subgraph of $G = (N, E)$ as a graph $G' = (N', E')$ such that $E' \subseteq E$ and N' is the set of those nodes in N which are incident with edges in E'.

subset of a member of $\mathscr{E}$ is again a member of $\mathscr{E}$. Hence $\{x \in F : \phi(x) = i\} \in \mathscr{E}$. In view of (2) this implies that every finite subset of E_i^* is in $\mathscr{E}$. Hence E_i^* is in $\mathscr{E}$, and the proof is complete.

Next, we consider an application of Theorem 4.5.2 to transversal theory.

THEOREM 4.5.3. *Let* $\mathfrak{A} = (A_i : i \in I)$ *be a restricted family of subsets of an arbitrary set* E. *The following statements are then equivalent.*

(a) E *can be partitioned into k partial transversals of* $\mathfrak{A}$.

(b) *For each finite subset* F *of* E,

$$|F| \leqslant k|\{i \in I : A_i \cap F \neq \emptyset\}|.$$

(c) *For each finite subset* F *of* E, *we have* $|F| \leqslant k \,.\, \rho(F)$, *where* $\rho(F)$ *denotes the maximum cardinal of partial transversals of* $\mathfrak{A}$ *contained in* F.

Let $\mathscr{E}$ denote the set of all PTs of $\mathfrak{A}$. Then, by Corollary 4.2.4, $\mathscr{E}$ has finite character. Hence, by Theorem 4.5.2, statement (a) is valid if and only if every finite subset E^* of E can be partitioned into k PTs of $\mathfrak{A}$. In view of Corollary 3.3.3, this is the case precisely if, for each $F \subseteq E^*$, the inequality in (b) is valid. In view of Corollary 3.3.4, it is also the case precisely if, for each $F \subseteq E^*$, the inequality in (c) is valid. The desired conclusion therefore follows.

A result much more general than Theorem 4.5.3 will be discussed subsequently (cf. § 8.2).

Ordered groups

Let G be a group (and denote by the same symbol the set of its elements). We shall say that a relation '$<$' defined on G is a partial order on the group G if, in addition to satisfying the axioms of partial order (cf. §1.5), it is also compatible with the multiplicative structure of G, i.e. if $x, y, z \in G$ and $x < y$, then xz, yz and also zx, zy are comparable and $xz < yz$, $zx < zy$.

A group on which a *total* order can be defined, will be called an *O-group.* It is, of course, obvious that any subgroup of an *O*-group is itself an *O*-group.

THEOREM 4.5.4. (B. H. Neumann) *If all finitely generated subgroups of a group* G *are O-groups, then* G *is itself an O-group.*

For any set $F \subseteq G$, we shall denote by F^* the subgroup of G generated by F and by $\underset{F}{<}$ a (total) order defined on F^*. Write

$$I = \{(x, y) : x, y \in G, x \neq y\}$$

and, for any $(x, y) \in \mathrm{I}$, put $\mathrm{A}_{(x,y)} = \{0, 1\}$. We shall be concerned with the family

$$\mathfrak{A} = (\mathrm{A}_{(x,y)}: (x, y) \in \mathrm{I}).$$

For $\mathrm{J} \subset\subset \mathrm{I}$, put

$$[\mathrm{J}] = \{x \in \mathrm{G}: (x, y) \in \mathrm{J} \text{ or } (y, x) \in \mathrm{J} \text{ for some } y \in \mathrm{G}\}.$$

The order $\underset{[\mathrm{J}]}{<}$ on $[\mathrm{J}]^*$ will be written, for brevity, as $\underset{\mathrm{J}}{<}$. For $(x, y) \in \mathrm{J}$, we write

$$\theta_{\mathrm{J}}((x, y)) = \begin{cases} 0 & (x \underset{\mathrm{J}}{<} y) \\ 1 & (y \underset{\mathrm{J}}{<} x). \end{cases}$$

Then θ_{J} is a choice function of the subfamily $\mathfrak{A}(\mathrm{J})$ of $\mathfrak{A}$. We shall denote by θ a choice function of $\mathfrak{A}$ which satisfies the conclusions of Rado's selection principle. We now define a relation '$<$' on G by declaring that, for $x, y \in \mathrm{G}$, $x \neq y$, we have $x < y$ if and only if $\theta((x, y)) = 0$. It remains to show that $<$ is an order on the set G which is compatible with the multiplicative structure of the group.

Let $x, y \in \mathrm{G}$, $x \neq y$. To show that precisely one of the relations $x < y$, $y < x$ is valid, we take $\mathrm{J} = \{(x, y), (y, x)\}$. By the properties of θ, there exists K such that $\mathrm{J} \subseteq \mathrm{K} \subset\subset \mathrm{I}$ and $\theta|\mathrm{J} = \theta_{\mathrm{K}}|\mathrm{J}$. The desired conclusion now follows easily. Next, let $\{x, y, z\}_{\neq} \subseteq \mathrm{G}$ and $x < y$, $y < z$. To show that $x < z$, we take $\mathrm{J} = \{(x, y), (y, z)\}$ and again use the properties of θ. It follows that $<$ is an order on the set of elements of G.

To establish compatibility, let $x, y, z \in \mathrm{G}$ and $x < y$. We shall content ourselves with showing that $xz < yz$. Let $\mathrm{J} = \{(x, y), (xz, yz)\}$. Then for an 'associated' set K, we have x, y, xz, $yz \in [\mathrm{K}] \subseteq [\mathrm{K}]^*$ and so, since $x < y$,

$$\theta_{\mathrm{K}}((x, y)) = \theta((x, y)) = 0.$$

Hence $x \underset{\mathrm{K}}{<} y$. But $\underset{\mathrm{K}}{<}$ is an order defined on $[\mathrm{K}]^*$ which is compatible with multiplication, so that $xz \underset{\mathrm{K}}{<} yz$. Hence

$$\theta((xz, yz)) = \theta_{\mathrm{K}}((xz, yz)) = 0,$$

and therefore $xz < yz$.

COROLLARY 4.5.5. *Let G be a group. Let an order $\underset{\mathrm{F}}{<}$ be defined on every finitely generated subgroup F of G, and let $<$ denote an order on G whose existence is guaranteed by Theorem 4.5.4. Then, for any $x, y \in \mathrm{G}$ with $x \neq y$, there exists a finitely generated subgroup F_0 of G such that $x, y \in \mathrm{F}_0$, and $x \underset{\mathrm{F}_0}{<} y$ if and only if $x < y$.*

This corollary is a trivial consequence of the method used in the preceding argument, and we state it only because it is useful in the proof of the next theorem. We continue to employ our previous notation.

Take $\mathrm{J} = \{(x, y)\}$. Then there exists K with $\mathrm{J} \subseteq \mathrm{K} \subset\subset \mathrm{I}$ and $\theta|\mathrm{J} = \theta_{\mathrm{K}}|\mathrm{J}$. Put $\mathrm{F}_0 = [\mathrm{K}]^*$. Then $x, y \in \mathrm{F}_0$ and

$$\theta((x, y)) = \theta_{\mathrm{K}}((x, y)).$$

Hence $x < y$ precisely if $x \underset{\mathrm{K}}{<} y$, i.e. $x \underset{\mathrm{F}_0}{<} y$.

Next, let X be a set and let $<$, $<^*$ be partial orders defined on X. We say that $<^*$ is an *extension* of (or *extends*) $<$ if the relations $x, y \in \mathrm{X}$, $x < y$ imply $x <^* y$. The same terminology is used for partial order on a group, though it has to be remembered that any such partial order must, by definition, be compatible with the group structure. Now it is a familiar fact (cf. Exs. 1.5.7 and 4.5.6) that any partial order defined on a set can be extended to a total order. This statement is no longer true for groups. We shall say that a group G is an *O*-group* if every partial order defined on G can be extended to a total order. (In particular, then, every *O**-group is an *O*-group, for we may start with the trivial partial order for which the entire group is an antichain.) We now have the following criterion.

THEOREM 4.5.6. *If all finitely generated subgroups of a group* G *are O*-groups, then* G *itself is an O*-group.*

Let $<$ be a partial order defined on G. Then $<$ induces a partial order, say $\underset{\mathrm{F}}{<}$, on every subgroup F. By hypothesis, for each finitely generated subgroup F, $\underset{\mathrm{F}}{<}$ can be extended to a total order $\underset{\mathrm{F}}{<}^*$ on F. Let $<^*$ denote the total order on G whose existence is guaranteed by Theorem 4.5.4. It remains to show that $<^*$ is an extension of $<$.

Let $x, y \in \mathrm{G}$ and suppose that $x < y$. We know, by Corollary 4.5.5, that there exists a finitely generated subgroup F_0 of G such that $x, y \in \mathrm{F}_0$ and $x <^* y$ if and only if $x \underset{\mathrm{F}_0}{<}^* y$. But $x < y$ may be written as $x \underset{\mathrm{F}_0}{<} y$, and this relation implies that $x \underset{\mathrm{F}_0}{<}^* y$. Hence $x <^* y$, as required.

The significance of the kind of approach to questions in algebra exemplified by the preceding discussion is not that proofs become shorter (though they may well do so) but rather that certain results which involve algebraic concepts are seen to bear, in fact, little relation to algebraic structure and to have an essentially set-theoretic, or combinatorial, character (cf. Theorem 10.1.6).

A representation theorem of Stone

In this paragraph, we shall indicate very briefly the use to which the selection principle can be put in the proof of (a weak version of) Stone's celebrated representation theorem for boolean algebras.

The 'extremal' elements in any boolean algebra will be denoted by 0 and 1, and the operations by $\vee$, $\wedge$, and $'$. When we deal with an algebra of sets, these operations are interpreted as union, intersection, and (set-theoretic) complementation. The partial order in a boolean algebra will be written as $\leqslant$. (For an algebra of sets, this relation becomes that of inclusion.) A homomorphism of a boolean algebra B into a boolean algebra $\tilde{B}$ is a mapping of B into $\tilde{B}$ which preserves the three operations. A bijective homomorphism is called an isomorphism. If X is a subset of a boolean algebra B, then the boolean subalgebra generated by X is defined as the intersection of all subalgebras which contain 0, 1, and X. It can be shown easily that a finite subset of B always generates a finite subalgebra. The boolean algebra $\{0, 1\}$ will be denoted by B_0.

The proof of the following preliminary result is the crucial step in the argument leading to Stone's theorem.

LEMMA 4.5.7. *To each non-zero element x_0 of a boolean algebra* B *corresponds a homomorphism $\sigma: B \to B_0$ such that $\sigma(x_0) = 1$.*

When B is finite, we can choose a minimal element x_1 such that $0 < x_1 \leqslant x_0$. By the definition of x_1 it follows that, for every $x \in B$, $x \wedge x_1$ is x_1 or 0. Hence the equations

$$\sigma(x) = \begin{cases} 1 & (x \wedge x_1 = x_1) \\ 0 & (x \wedge x_1 = 0) \end{cases}$$

define a mapping $\sigma: B \to B_0$. Moreover, $\sigma(x_0) = 1$; and a routine verification shows that σ is a homomorphism.

In the general case, write $A_x = B_0$ for each $x \in B$ and put $\mathfrak{A} = (A_x: x \in B)$. If $C \subset\subset B$, then $C \cup \{x_0\}$ generates a finite subalgebra C^* of B. Hence, by the finite case just discussed, there exists a homomorphism $\sigma_{C^*}: C^* \to B_0$ with $\sigma_{C^*}(x_0) = 1$. We now put

$$\sigma_C = \sigma_{C^*}|C.$$

In particular, if $x_0 \in C$, then $\sigma_C(x_0) = 1$. For each $C \subset\subset B$, the mapping $\sigma_C : C \to B_0$ is a choice function of the subfamily $\mathfrak{A}(C)$ of $\mathfrak{A}$. Denote by $\sigma : B \to B_0$ a corresponding Rado choice function of $\mathfrak{A}$. For each $C \subset\subset B$, there exists a set D such that

$$C \subseteq D \subset\subset B, \qquad \sigma|C = \sigma_D|C. \tag{3}$$

If $C = \{x_0\}$, then an associated D satisfies $\sigma(x_0) = \sigma_D(x_0) = 1$; and it only remains to show that σ is a homomorphism. Let $x, y \in B$ and take

$$C = \{x,\ y,\ x \vee y,\ x \wedge y,\ x'\}.$$

Then, with an associated D subject to (3), we have

$$\sigma(x \vee y) = \sigma_D(x \vee y) = \sigma_D(x) \vee \sigma_D(y) = \sigma(x) \vee \sigma(y),$$
$$\sigma(x \wedge y) = \sigma_D(x \wedge y) = \sigma_D(x) \wedge \sigma_D(y) = \sigma(x) \wedge \sigma(y),$$
$$\sigma(x') = \sigma_D(x') = (\sigma_D(x))' = (\sigma(x))'.$$

The proof is therefore complete.

The deduction of Stone's theorem is now an easy matter.

THEOREM 4.5.8. (M. H. Stone) *Any boolean algebra is isomorphic to some boolean algebra of sets.*

Denote by H the set of all homomorphisms of the given boolean algebra B into B_0 and, for $x \in B$, write

$$H_x = \{\sigma \in H: \sigma(x) = 1\}.$$

Then $H_1 = H$ and $H_0 = \emptyset$ since, for each $\sigma \in H$, $\sigma(1) = 1$ and $\sigma(0) = 0$. Put

$$\mathfrak{H} = \{H_x: x \in B\}.$$

It can be verified at once that

$$H_{x \vee y} = H_x \cup H_y, \qquad H_{x \wedge y} = H_x \cap H_y, \qquad H_{x'} = \mathscr{C}(H_x) \tag{4}$$

(where the complement $\mathscr{C}$ is taken with respect to H). Hence $\mathfrak{H}$ is a collection of subsets of H which contains H and $\emptyset$ and which is closed under unions, intersections, and complements. Thus $\mathfrak{H}$ is a boolean algebra of sets.

From (4) it follows trivially that the (surjective) mapping $\psi: B \to \mathfrak{H}$ given by $\psi(x) = H_x$ $(x \in B)$ is a homomorphism. To show that it is an injection—and so an isomorphism—let $x, y \in B$, $x \neq y$. Then $x \wedge y' \neq 0$ or $x' \wedge y \neq 0$, say the former. Hence, by Lemma 4.5.7 and (4),

$$\emptyset \neq H_{x \wedge y'} = H_x \cap \mathscr{C}(H_y).$$

Therefore $H_x \neq H_y$, i.e. $\psi(x) \neq \psi(y)$, as required.

Exercises 4.5

1. A family $\mathfrak{A} = (A_i : i \in I)$ of subsets of E is said to be 'simple' if there exists a set $R \subseteq E$ such that $|R \cap A_i| = 1$ for all $i \in I$. Show that, if all A_i are finite and every finite subfamily of $\mathfrak{A}$ is simple, then $\mathfrak{A}$ is itself simple.

Give an example of a set E and a family $\mathfrak{A}$ of subsets of E such that (i) $\mathfrak{A}$ is not simple; (ii) every finite subfamily of $\mathfrak{A}$ is simple; (iii) exactly one set in $\mathfrak{A}$ is infinite; (iv) every element of E belongs to exactly two sets in $\mathfrak{A}$.

2. Write out in detail the proof of Theorem 4.5.4.

3. Show that the edges of a graph can be so painted with k colours that no two concurrent edges are painted alike provided this is true of every finite subgraph. (Cf. the footnote on page 65.)

[B. H. Neumenn (**1**)]

4. Establish the following propositions. (i) An infinite cyclic group is an O-group. (ii) A finitely generated torsion-free abelian group is an O-group. (iii) An abelian group is an O-group if and only if it is torsion-free.

5. A graph is called a 'comparability graph' if it is possible to define a partial order on the set of its nodes such that two nodes are comparable if and only if they are linked by an edge. Show that, if every finite subgraph of a graph G is a comparability graph, then G is itself a comparability graph.

[E. S. Wolk (**1**, **2**)]

6. Use Rado's selection principle to show that every partial order on a set can be extended to a total order. (Cf. Ex. 1.5.7.)

Notes on Chapter 4

§ 4.1. The result we have dubbed 'Rado's selection principle' was first established by R. Rado (**5**) over twenty years ago. The second proof given above was shown to me by Professor Rado in 1968. The very short demonstration based on Tychonoff's theorem is due to W. H. Gottschalk (**1**); the idea of invoking Tychonoff's theorem goes back to the treatment of the transfinite case of Hall's theorem by Halmos & Vaughan (**1**). B. L. Foster (**1**) offered another proof of the selection principle by linking the question with König's 'infinity lemma' ((**4**); see Misc. Exs. No. 21). Luxemburg (**1**) based a proof on the theory of ultrafilters. A refinement of the principle was noted recently by D. Š. Lundina (**1**).

Since Rado's principle can be derived from Tychonoff's theorem and since this latter result is known to be equivalent to the axiom of choice, it is natural to inquire about the logical status of Rado's principle. This question has been investigated by E. S. Wolk (**3**) who showed that in the presence of the axiom of choice for families of finite sets, Rado's selection principle is equivalent to Tychonoff's theorem for products of finite topological spaces. It is also of interest to note that there are results in mathematical logic (such as the 'compactness theorem') which are intimately related to Rado's selection principle; see e.g. Bell & Slomson (**1**, 44-49).

§ 4.2. I owe to Professor R. Rado the very simple proof of Hall's theorem for denumerable families indicated at the beginning of the section. For a variant, see Rado (**11**). Another proof, depending on metric topology, has been given by Everett & Whaples (**1**).

Theorem 4.2.1 (the transfinite case of Hall's theorem) was originally proved by Marshall Hall Jr. (**2**); other proofs were given subsequently by Everett & Whaples (**1**), Halmos & Vaughan (**1**), L. Henkin (**1**), O. Ore (**1**), and R. Rado (**6**, **11**). Henkin's proof depends on mathematical logic, Ore's and Rado's first proof on graph theory, Halmos & Vaughan's on general topology, and the remaining proofs on the theory of sets. Of the three set-theoretic arguments, Rado's (**11**) is particularly direct and, unlike other proofs, makes no use of the finite case of Hall's theorem. It is this proof that is reproduced here.

The demonstration given above of Löwig's theorem about the equicardinality of bases in a vector space is due to M. Hall (**4**, 66-67). For proofs independent of Hall's theorem, see Löwig (**1**) or Jacobson (**1**, vol. 2, chap. 9). The dual form of Hall's theorem (Theorem 4.2.3) is taken from paper (**2**) of Mirsky & Perfect.

§ 4.3. Theorem 4.3.1 is due to R. Rado and H. A. Jung (Rado, **11**). The existence of transversals of infinite families in which a finite number of sets can be infinite has been discussed by Brualdi & Scrimger (**1**) and by J. Folkman (**1**), but the difficulty of the work makes it impossible for us to offer an account of it here.

§ 4.4. R. P. Dilworth's original proof (**2**) of his decomposition theorem 4.4.1 was fairly complicated, and a number of further proofs have since been discovered. Thus Theorem 4.4.1 for finite sets emerges, in several ways, as an off-shoot of the theory of graphs; see e.g. Fulkerson (**1**), Gallai (**1**), Gallai & Milgram (**1**). The theorem can also be established by the method of linear programming (Dantzig & Hoffman, **1**). The extremely short and transparent proof given here is due to Tverberg (**1**); for another simple proof, see M. A. Perles (**1**). The deduction of the infinite case of Dilworth's theorem was shown to me by Professor Rado. Perles (**2**) demonstrated that Theorem 4.4.1 ceases to be valid if the given set contains arbitrarily large anti-chains. My impression is that Dilworth's theorem is a result of very considerable importance whose potentialities are as yet far from exhausted. The consequences derived from it in this book assuredly do scant justice to its power.

The result on monotonic subsequences (Theorem 4.4.4) is due to Erdös & Szekeres (**1**). Subsequently, A. Seidenberg (**1**) gave a surprisingly simple proof. I cannot track down with certainty the author of the proof presented here, but it seems to be S. K. Stein. The deduction of the Hall–Ore theorem from Dilworth's theorem, as well as other applications of Dilworth's theorem (cf. § 11.2) go back, in essence, to Dilworth's own ideas (**2**, **3**).

§ 4.5. Rado's selection principle (Rado, **5**) was originally designed for the strictly limited purpose of establishing the equicardinality of bases in an independence structure (see Theorem 7.2.9). Since then it has emerged as a tool of extraordinary power and versatility which enables us to prove numerous results which assert that, in one sense or another, some system has 'finite character'. It seems safe to predict that the use of Rado's selection principle will be increasingly found to serve as a standard argument in very diverse fields of mathematical research.

In the present section we discuss several applications of the selection principle: others will be found scattered throughout the book. Theorem 4.5.1 on the chromatic number of graphs is due to de Bruijn & Erdös (**1**). The related statement on the colouring of edges is taken from the work of B. H. Neumann (**1**), as is Theorem 4.5.4 on ordered groups which was originally discovered as a corollary of a much more general result. Further information on O-groups and O^*-groups will be found in L. Fuchs's book (**1**, chap. 3). Theorem 4.5.8 is only a weak version of M. H.

Stone's representation theorem. For a more comprehensive treatment, which includes considerations of topological structure, we refer the reader to Stone's original paper (**1**); see also Simmons (**1**, Appendix 3) and Halmos (**2**, § 18). Our proof is based on ideas of N. M. Rice (**1**), who also discussed other applications of the selection principle and its relation to Tychonoff's theorem (Rice, **2**). For background material on boolean algebras, see Jacobson (**1**, vol. 1, chap. 7) or Simmons (loc. cit.).

5 Variants, Refinements, and Applications of Hall's Theorem

The present chapter has the character of an interlude. Here we develop a series of further results intimately connected with Hall's fundamental theorem.

5.1 Disjoint partial transversals

We already possess necessary and sufficient conditions for a family of sets to have a PT with a preassigned finite defect (cf. Theorem 3.2.1 and Ex. 4.2.4). We shall now consider the more general situation of several pairwise disjoint PTs. We shall use the abbreviation $z^+ = \max(z, 0)$.

THEOREM 5.1.1. (P. J. Higgins) *Let* $\mathfrak{A} = (A_i : i \in I)$ *be a family of subsets of a set* E; *and suppose that, if* I *is infinite, then all* A_i *are finite. Further, let* $m \geqslant 1$ *and let* $d_1, \ldots, d_m$ *be non-negative integers not exceeding* $|I|$. *Then* $\mathfrak{A}$ *possesses* m *pairwise disjoint partial transversals with defects* $d_1, \ldots, d_m$ *respectively if and only if, for each natural number* $k \leqslant |I|$, *the union of any* k A's *contains at least*

$$(k - d_1)^+ + \ldots + (k - d_m)^+ \tag{1}$$

(distinct) elements.

When I is finite, the case $m = 1$ of the above result is equivalent to the Hall–Ore theorem 3.2.1; when I is infinite, it is equivalent to the 'defect form' of Theorem 4.2.1 (stated as Ex. 4.2.4).

To prove the assertion, we suppose, in the first place, that $\mathfrak{A}$ possesses pairwise disjoint PTs $T_1, \ldots, T_m$ with defects $d_1, \ldots, d_m$ respectively. Among any k A's there occur at least $(k - d_j)^+$ A's which have a non-empty intersection with T_j. Thus the union of any k A's contains at least $(k - d_j)^+$ elements from T_j, and so in all at least the number of elements specified by (1).

Suppose, next, that the stated condition is satisfied. We denote by $D_1, \ldots, D_m$ arbitrary sets such that $D_1, \ldots, D_m$, E are pairwise disjoint and $|D_j| = d_j$ $(1 \leqslant j \leqslant m)$; and we consider the family

$$\mathfrak{F} = (A_i \cup D_j : \ i \in I, \ 1 \leqslant j \leqslant m).$$

If I is infinite, then all sets of $\mathfrak{F}$ are finite. Let k be any natural number not exceeding $m|I|$ and let

$$A_{i_1} \cup D_{j_1}, \ldots, A_{i_k} \cup D_{j_k}$$

be any k sets in $\mathfrak{F}$. Denote by $r_1, \ldots, r_p$ the distinct elements of I among $i_1, \ldots, i_k$ and by $s_1, \ldots, s_q$ the distinct integers among $j_1, \ldots, j_k$. Then $pq \geqslant k$ and, in view of our hypothesis concerning $\mathfrak{A}$,

$$\begin{aligned}|(A_{i_1} \cup D_{j_1}) \cup \ldots \cup (A_{i_k} \cup D_{j_k})| &= |A_{r_1} \cup \ldots \cup A_{r_p} \cup D_{s_1} \cup \ldots \cup D_{s_q}| \\ &= |A_{r_1} \cup \ldots \cup A_{r_p}| + |D_{s_1}| + \ldots + |D_{s_q}| \\ &= |A_{r_1} \cup \ldots \cup A_{r_p}| + d_{s_1} + \ldots + d_{s_q} \\ &\geqslant (p - d_1)^+ + \ldots + (p - d_m)^+ \\ &\qquad + d_{s_1} + \ldots + d_{s_q} \\ &\geqslant (p - d_{s_1}) + \ldots + (p - d_{s_q}) \\ &\qquad + d_{s_1} + \ldots + d_{s_q} \\ &= pq \geqslant k.\end{aligned}$$

Thus the family $\mathfrak{F}$ satisfies Hall's condition and so, by Theorems 4.2.1 and 2.2.1, possesses a transversal, say

$$\{x_{ij}\colon i \in I,\ 1 \leqslant j \leqslant m\},$$

where all the x's are distinct. Hence, for every j with $1 \leqslant j \leqslant m$, there exists a set $I_j \subseteq I$ such that

$$x_{ij} \in A_i \quad (i \in I_j), \qquad x_{ij} \in D_j \quad (i \in I \setminus I_j).$$

Since the x's are distinct, $T_j = \{x_{ij}\colon i \in I_j\}$ is a PT of $\mathfrak{A}$ with defect

$$|I \setminus I_j| \leqslant |D_j| = d_j.$$

Moreover, $T_1, \ldots, T_m$ are pairwise disjoint.

COROLLARY 5.1.2. *Let $m \geqslant 1$, $n \geqslant 1$; let $r_1, \ldots, r_m$ be non-negative integers not exceeding n; and let $\mathfrak{A} = (A_1, \ldots, A_n)$ be a family of pairwise disjoint sets containing $s_1, \ldots, s_n$ elements respectively. Then $\mathfrak{A}$ possesses m pairwise disjoint partial transversals of cardinal $r_1, \ldots, r_m$ respectively if and only if*

$$\sum_{j=n-k+1}^{n} \bar{s}_j \geqslant \sum_{j=n-k+1}^{n} r_j^* \qquad (1 \leqslant k \leqslant n), \tag{2}$$

where $\bar{s}_1, \ldots, \bar{s}_n$ are the numbers $s_1, \ldots, s_n$ arranged in non-ascending order of magnitude and r_j^* denotes the number of r_i's greater than or equal to j.

We shall say that $\mathfrak{A}$ satisfies condition T if it possesses the requisite partial transversals. By Theorem 5.1.1, this is the case if and only if

$$s_{i_1} + \dots + s_{i_k} = |A_{i_1} \cup \dots \cup A_{i_k}| \geqslant \sum_{i=1}^{m} (k - n + r_i)^+$$

whenever $1 \leqslant k \leqslant n$ and $1 \leqslant i_1 < \dots < i_k \leqslant n$. Hence T is satisfied if and only if

$$\sum_{j=n-k+1}^{n} \bar{s}_j \geqslant \sum_{i=1}^{m} (k - n + r_i)^+ \qquad (1 \leqslant k \leqslant n). \tag{3}$$

Now, interpreting empty sums as 0 and recalling that all r_i are $\leqslant n$, we have

$$\sum_{i=1}^{m} (k - n + r_i)^+ = \sum_{1 \leqslant i \leqslant m} \; \sum_{n-k+1 \leqslant j \leqslant r_i} 1 = \sum_{n-k+1 \leqslant j \leqslant n} \; \sum_{\substack{1 \leqslant i \leqslant m \\ r_i \geqslant j}} 1$$

$$= \sum_{j=n-k+1}^{n} r_j^*.$$

It follows by (3) that T is satisfied if and only if (2) is valid.

We shall next need some additional notation. Let $x_1, \dots, x_n, y_1, \dots, y_n$ be real numbers. Let $\bar{x}_1, \dots, \bar{x}_n$ be the numbers $x_1, \dots, x_n$ arranged in non-ascending order of magnitude, and let $\bar{y}_1, \dots, \bar{y}_n$ be defined analogously. Suppose that

$$\bar{x}_1 + \dots + \bar{x}_k \leqslant \bar{y}_1 + \dots + \bar{y}_k$$

for $1 \leqslant k \leqslant n$ and that this relation holds with the sign of equality for $k = n$. We shall then write

$$(x_1, \dots, x_n) \prec (y_1, \dots, y_n).$$

Further, we shall call a (rectangular) matrix an *incidence matrix* if all its elements are equal to 0 or 1.

THEOREM 5.1.3. (Gale–Ryser) *Let $m \geqslant 1$, $n \geqslant 1$. Let $r_1, \dots, r_m$ be non-negative integers not exceeding n, and let $s_1, \dots, s_n$ be non-negative integers. The validity of the relation*

$$(s_1, \dots, s_n) \prec (r_1^*, \dots, r_n^*) \tag{4}$$

is necessary and sufficient for the existence of an $m \times n$ incidence matrix with row-sums r_i $(1 \leqslant i \leqslant m)$ and column-sums s_j $(1 \leqslant j \leqslant n)$.†

† By a 'row-sum' of a matrix we mean, of course, the sum of all elements belonging to a particular row; and analogously for a 'column-sum'.

We begin by noting that (4) holds if and only if both (2) and the relation

$$\sum_{j=1}^{n} \bar{s}_j = \sum_{j=1}^{n} r_j^* \tag{5}$$

are valid.

Suppose, in the first place, that there exists a matrix $C = \|c_{ij}\|$ of the type specified in the theorem. Then clearly

$$\sum_{i=1}^{m} r_i = \sum_{j=1}^{n} s_j.$$

Moreover

$$\sum_{j=1}^{n} r_j^* = \sum_{1 \leqslant j \leqslant n} \sum_{\substack{1 \leqslant i \leqslant m \\ r_i \geqslant j}} 1 = \sum_{1 \leqslant i \leqslant m} \sum_{\substack{1 \leqslant j \leqslant n \\ j \leqslant r_i}} 1 = \sum_{i=1}^{m} r_i, \tag{6}$$

and so (5) is valid. Now let $\mathfrak{A} = (A_1, \ldots, A_n)$, where

$$A_j = \{(i, j): 1 \leqslant i \leqslant m, \ c_{ij} = 1\} \qquad (1 \leqslant j \leqslant n).$$

The sets $A_1, \ldots, A_n$ are pairwise disjoint and have $s_1, \ldots, s_n$ elements respectively. Further, $\mathfrak{A}$ possesses m pairwise disjoint PTs of cardinal $r_1, \ldots, r_m$ respectively. Hence, by Corollary 5.1.2, the relation (2) is valid. Since (2) and (5) hold, so does (4).

Next, let (4) be given. Then, in particular, (2) is valid; and, by Corollary 5.1.2, any family $\mathfrak{A} = (A_1, \ldots, A_n)$ of pairwise disjoint sets with $|A_j| = s_j$ $(1 \leqslant j \leqslant n)$ possesses pairwise disjoint PTs $T_1, \ldots, T_m$ of cardinal $r_1, \ldots, r_m$ respectively. Take any such family $\mathfrak{A}$ and, for $1 \leqslant i \leqslant m$, $1 \leqslant j \leqslant n$ define

$$c_{ij} = \begin{cases} 1 & (T_i \cap A_j \neq \emptyset) \\ 0 & (T_i \cap A_j = \emptyset). \end{cases}$$

Then $C = \|c_{ij}\|$ is an $m \times n$ incidence matrix with row-sums $r_1, \ldots, r_m$. Further, its column-sums, say $\sigma_1, \ldots, \sigma_n$, satisfy the inequalities

$$\sigma_j \leqslant s_j \qquad (1 \leqslant j \leqslant n). \tag{7}$$

Now (5) holds by virtue of (4). Hence, by (6),

$$\sum_{i=1}^{m} r_i = \sum_{j=1}^{n} s_j$$

and so

$$\sum_{j=1}^{n} \sigma_j = \sum_{i=1}^{m} r_i = \sum_{j=1}^{n} s_j.$$

Therefore, by (7), $\sigma_j = s_j$ $(1 \leqslant j \leqslant n)$. The matrix C thus has the requisite properties.

The theorem just proved will be discussed in a different (and possibly more illuminating) context in §11.5.

Exercises 5.1

1. Show that, if $(x_1, \ldots, x_n) \prec (y_1, \ldots, y_n)$, then $(-x_1, \ldots, -x_n) \prec (-y_1, \ldots, -y_n)$.

2. Verify that the relation $\prec$ is a partial order on the set of all vectors with n real components.

3. Let $a_1 \geqslant \ldots \geqslant a_n$ and $b_1 \geqslant \ldots \geqslant b_n$. Show that the relation $(a_1, \ldots, a_n) \prec (b_1, \ldots, b_n)$ holds if and only if

$$a_1 x_1 + \ldots + a_n x_n \leqslant b_1 x_1 + \ldots + b_n x_n$$

whenever $x_1 \geqslant \ldots \geqslant x_n$.

4. Show that the conditions $r_1, \ldots, r_m \leqslant n$ in Theorem 5.1.3 cannot be dispensed with.

5. Let $r_1, \ldots, r_m$ be non-negative integers and $s_1, \ldots, s_n$ non-negative integers not exceeding m. Show that there exists an $m \times n$ incidence matrix with row-sums r_i $(1 \leqslant i \leqslant m)$ and column-sums s_j $(1 \leqslant j \leqslant n)$ if and only if

$$(r_1, \ldots, r_m) \prec (s_1{}^*, \ldots, s_m{}^*),$$

where $s_j{}^*$ denotes the number of s_j greater than or equal to j.

5.2 Strict systems of distinct representatives

Let $\mathfrak{A} = (A_1, \ldots, A_n)$ be a family of subsets of E. We recall that a family $\mathfrak{X} = (x_1, \ldots, x_n)$ of elements of E is called a system of representatives of $\mathfrak{A}$ if, for some permutation π of $\{1, \ldots, n\}$, we have

$$x_1 \in A_{\pi(1)}, \ldots, x_n \in A_{\pi(n)}.$$

If, in addition, the x's are distinct we call $\mathfrak{X}$ a system of distinct representatives of $\mathfrak{A}$. We now impose a further restriction. We shall call $\mathfrak{X}$ a *strict system of distinct representatives* (SSDR) of $\mathfrak{A}$ if $x_1, \ldots, x_n$ are distinct and

$$x_1 \in A_1, \ldots, x_n \in A_n.$$

It is, of course, clear that $\mathfrak{A}$ possesses a SSDR precisely if it possesses a transversal. Now, if $\mathfrak{A}$ possesses a SSDR, it will in general possess many such systems. Their precise number can be given explicitly in terms of the boolean atoms generated by $\mathfrak{A}$, but the resulting formula is unwieldy and affords no real insight into the situation. We shall derive a more interesting and more useful lower estimate for the number $N(\mathfrak{A}) = N(A_1, \ldots, A_n)$ of strict systems of distinct representatives of $\mathfrak{A}$. Two systems $(x_1, \ldots, x_n)$ and $(y_1, \ldots, y_n)$ are, of course, different if $x_i \neq y_i$ for at least one value of i.

THEOREM 5.2.1. *Let* $\mathfrak{A} = (A_1, \ldots, A_n)$ *be a family of sets which satisfies Hall's condition. If* $\min(|A_1|, \ldots, |A_n|) = t$, *then*

$$N(\mathfrak{A}) \geqslant \begin{cases} t! & (t \leqslant n) \\ t!/(t-n)! & (t \geqslant n). \end{cases} \tag{1}$$

Although this result constitutes a quantitative refinement of Hall's theorem, this last theorem is not invoked in the proof. However, the argument is based on the same case distinction that we used in the first proof of Hall's theorem.

We shall denote by $\phi(t, n)$ the expression on the right-hand side of (1). It is plain that $\phi(t, n)$ increases with t.

For $n = 1$, the assertion is clearly valid. Let $n > 1$ and assume that the assertion holds for all families consisting of fewer than n sets.

Case 1. Suppose that $|A_{i_1} \cup \ldots \cup A_{i_k}| > k$ whenever $1 \leqslant k < n$ and $1 \leqslant i_i < \ldots < i_k \leqslant n$. Since $(A_1, \ldots, A_n)$ satisfies $\mathscr{H}$, we have $A_1 \neq \emptyset$. Let $x_1 \in A_1$, and put

$$B_i = A_i \setminus \{x_1\} \qquad (2 \leqslant i \leqslant n).$$

Then, for $1 \leqslant k < n,\ 2 \leqslant i_1 < \ldots < i_k \leqslant n$,

$$\begin{aligned} |B_{i_1} \cup \ldots \cup B_{i_k}| &= |(A_{i_1} \cup \ldots \cup A_{i_k}) \setminus \{x_1\}| \\ &\geqslant |A_{i_1} \cup \ldots \cup A_{i_k}| - 1 \geqslant k \end{aligned}$$

and so $(B_2, \ldots, B_n)$ satisfies $\mathscr{H}$. Moreover, if $(x_2, \ldots, x_n)$ is any SSDR of $(B_2, \ldots, B_n)$, then $(x_1, x_2, \ldots, x_n)$ is a SSDR of $\mathfrak{A}$. Hence

$$N(\mathfrak{A}) \geqslant \sum_{x_1 \in A_1} N(B_2, \ldots, B_n). \tag{2}$$

Now we have

$$|B_i| \geqslant |A_i| - 1 \geqslant t - 1 \qquad (2 \leqslant i \leqslant n).$$

Write $u = \min(|B_2|, \ldots, |B_n|)$. Then, for $t > 1$, we have $u \geqslant t - 1$. On the other hand, if $t = 1$, then since the B's are non-empty, $u \geqslant 1$.

By (2) and the induction hypothesis, we have

$$N(\mathfrak{A}) \geqslant \sum_{x_1 \in A_1} \phi(u, n-1).$$

If $t > 1$, then

$$N(\mathfrak{A}) \geqslant \sum_{x_1 \in A_1} \phi(t-1, n-1) = |A_1|\, \phi(t-1, n-1)$$
$$\geqslant t\phi(t-1, n-1) = \phi(t, n).$$

If $t = 1$, then

$$N(\mathfrak{A}) \geqslant \sum_{x_1 \in A_1} \phi(1, n-1) = |A_1|\, \phi(1, n-1)$$
$$\geqslant \phi(1, n-1) = 1 = \phi(1, n).$$

Case 2. Suppose that, for certain values of k, $i_1, \ldots, i_k$ with $1 \leqslant k < n$, $1 \leqslant i_1 < \ldots < i_k \leqslant n$, we have

$$|A_{i_1} \cup \ldots \cup A_{i_k}| = k.$$

For notational simplicity, let

$$|A_1 \cup \ldots \cup A_k| = k \tag{3}$$

for a certain value of k with $1 \leqslant k < n$. Write

$$B_i = A_i \setminus (A_1 \cup \ldots \cup A_k) \qquad (k+1 \leqslant i \leqslant n).$$

Now let $1 \leqslant r \leqslant n-k$, $k+1 \leqslant i_1 < \ldots < i_r \leqslant n$. Then

$$|B_{i_1} \cup \ldots \cup B_{i_r}| = |B_{i_1} \cup \ldots \cup B_{i_r}| + |A_1 \cup \ldots \cup A_k| - k$$
$$= |B_{i_1} \cup \ldots \cup B_{i_r} \cup A_1 \cup \ldots \cup A_k| - k$$
$$= |A_{i_1} \cup \ldots \cup A_{i_r} \cup A_1 \cup \ldots \cup A_k| - k$$
$$\geqslant (r+k) - k = r.$$

Thus $(B_{k+1}, \ldots, B_n)$ satisfies $\mathscr{H}$ as, of course, does $(A_1, \ldots, A_k)$. If $(x_1, \ldots, x_k)$ is a SSDR of $(A_1, \ldots, A_k)$ and $(x_{k+1}, \ldots, x_n)$ is a SSDR of $(B_{k+1}, \ldots, B_n)$, then clearly $(x_1, \ldots, x_n)$ is a SSDR of $\mathfrak{A}$. Hence

$$N(\mathfrak{A}) \geqslant N(A_1, \ldots, A_k). \tag{4}$$

In view of (3), we have $|A_1| \leqslant k, \ldots, |A_k| \leqslant k$ and so $t \leqslant k$. Hence, by (4) and the induction hypothesis,

$$N(\mathfrak{A}) \geqslant \phi(t, k) = t! = \phi(t, n).$$

The induction proof is therefore complete.

Refinements of the method just described lead to stronger results. Assume, as may be done without loss of generality, that $|A_1| \leqslant \ldots \leqslant |A_n|$ and write $t = |A_1|$. It can be shown that, if $\mathfrak{A}$ satisfies Hall's condition,

$$N(\mathfrak{A}) \geqslant \prod_{k=1}^{\min(n,t)} (|A_k| - k + 1). \tag{5}$$

Theorem 5.2.1 is, of course, an obvious consequence of this inequality.

A still better estimate is possible. Assuming, as before, that $|A_1| \leqslant \ldots \leqslant |A_n|$ and that $\mathfrak{A}$ satisfies Hall's condition, we have

$$N(\mathfrak{A}) \geqslant \prod_{k=1}^{n} \max(1, |A_k| - k + 1). \tag{6}$$

Exercises 5.2

1. Show that, for a family $\mathfrak{A} = (A_1, \ldots, A_n)$ of non-empty sets,

$$N(\mathfrak{A}) \leqslant \prod_{k=1}^{n} |A_k|$$

with equality if and only if the A's are pairwise disjoint.

[R. Rado (**10**)]

2. Show that, if $\mathfrak{A} = (A_1, \ldots, A_n)$ and $|A_k| > k$ $(1 \leqslant k \leqslant n)$, then

$$N(\mathfrak{A}) \geqslant \prod_{k=1}^{n} (|A_k| - k).$$

[R. Rado (**10**)]

5.3 Latin rectangles

Let $r \leqslant n$, $s \leqslant n$. An $r \times s$ matrix whose elements in each row and in each column are distinct numbers chosen from 1, 2, ..., n is called a *Latin rectangle of type* (r, s, n). When $r = s = n$, we speak of a *Latin square of order* n.

THEOREM 5.3.1. *Let* $1 \leqslant r < n$. *Then any Latin rectangle of type* (r, n, n) *can be extended to a Latin square of order* n.

Let R be the given Latin rectangle. For $1 \leqslant i \leqslant n$, denote by A_i the set of numbers among $1, 2, \ldots, n$ which do not occur in the ith column of R. Then

$$|A_1| = \ldots = |A_n| = n - r.$$

Let $1 \leqslant k < n$, $1 \leqslant i_1 < \ldots < i_k \leqslant n$. Write $A_{i_1} \cup \ldots \cup A_{i_k} = \{x_1, \ldots, x_p\}_{\neq}$; and, for $1 \leqslant j \leqslant p$, let x_j occur in v_j of the sets $A_{i_1}, \ldots, A_{i_k}$. Then

$$v_j \leqslant n - r \qquad (1 \leqslant j \leqslant p)$$

since any $x \in \{1, \ldots, n\}$ occurs in exactly r of the columns of R and so in exactly $n - r$ of the sets $A_1, \ldots, A_n$. We therefore have

$$(n - r)k = |A_{i_1}| + \ldots + |A_{i_k}| = v_1 + \ldots + v_p \leqslant (n - r)p$$

and so $p \geqslant k$, i.e. $|A_{i_1} \cup \ldots \cup A_{i_k}| \geqslant k$. The family $(A_1, \ldots, A_n)$ therefore satisfies $\mathscr{H}$ and so possesses a transversal. Let, then, $y_1 \in A_1, \ldots, y_n \in A_n$, where the y's are distinct. If, now, $(y_1, \ldots, y_n)$ is adjoined as the $(r + 1)$-th row of R, we obtain a Latin rectangle of type $(r + 1, n, n)$. The entire process can be repeated till we arrive at a Latin square of order n.

Recalling Theorem 5.2.1, we note that an $(r + 1)$-th row can be adjoined to R in at least $(n - r)!$ ways so as to produce a Latin rectangle of type $(r + 1, n, n)$. The proof of the next theorem is based on this observation.

THEOREM 5.3.2. *Let $1 \leqslant r \leqslant n$. Then there exist at least*

$$n!(n - 1)! \ldots (n - r + 1)!$$

Latin rectangles of type (r, n, n). In particular, there exist at least

$$n!(n - 1)! \ldots 2!\,1!$$

Latin squares of order n.

There are $n!$ Latin rectangles of type $(1, n, n)$. By the remark at the end of the preceding proof, each of these may be extended in at least $(n - 1)!$ different ways to a Latin rectangle of type $(2, n, n)$. Thus there are at least $n!\,(n - 1)!$ Latin rectangles of type $(2, n, n)$. We complete the proof by repetition of the same argument.

We next consider a harder problem than that settled by Theorem 5.3.1.

THEOREM 5.3.3. *Let $1 \leqslant r, s \leqslant n$ and let R be a Latin rectangle of type (r, s, n). Then R can be extended to a Latin square of order n if and only if every integer among $1, 2, \ldots, n$ occurs at least $r + s - n$ times as an element of R.*

It is plain that Theorem 5.3.1 is contained in this result.

For $1 \leqslant i \leqslant n$, let $N(i)$ be the number of elements of R equal to i. Thus $N(i)$ is equal to the number of rows of R which contain i as an element. Denote by $M(i)$ the number of rows of R which do not contain i. It is then plain that

$$N(i) + M(i) = r \qquad (1 \leqslant i \leqslant n).$$

Suppose, in the first place, that R can be extended to a Latin square of order n. Then evidently $M(i) \leqslant n - s$ $(1 \leqslant i \leqslant n)$ and so

$$N(i) \geqslant r + s - n \qquad (1 \leqslant i \leqslant n). \tag{1}$$

Next, suppose that (1) is satisfied. For $1 \leqslant j \leqslant r$, denote by A_j the set of $n - s$ integers among $1, 2, \ldots, n$ which do not occur in the jth row of R. Thus j occurs in $M(j)$ of the A's.

Let $1 \leqslant k \leqslant r$, $1 \leqslant v_1 < \ldots < v_k \leqslant r$. For $1 \leqslant i \leqslant n$, denote by $M_0(i)$ the number of sets among $A_{v_1}, \ldots, A_{v_k}$ which contain i. Then, in view of (1),

$$M_0(i) \leqslant M(i) \leqslant n - s \qquad (1 \leqslant i \leqslant n).$$

Further, if $A_{v_1} \cup \ldots \cup A_{v_k} = \{i_1, \ldots, i_t\}_{\neq}$, then $M_0(i) = 0$ whenever $i \neq i_1, \ldots, i_t$. Hence

$$k(n - s) = |A_{v_1}| + \ldots + |A_{v_k}| = \sum_{i=1}^{n} M_0(i) \leqslant t(n - s).$$

Thus $t \geqslant k$, i.e.

$$(A_1, \ldots, A_r) \text{ satisfies } \mathscr{H}. \tag{2}$$

Next, let

$$P = \{i : 1 \leqslant i \leqslant n,\ N(i) = r + s - n\}$$

and consider a subset, say $P^* = \{i_1, \ldots, i_t\}_{\neq}$, of P. Suppose that P^* intersects just h A's, say $A_{j_1}, \ldots, A_{j_h}$. Then $M(i_1) = \ldots = M(i_t) = n - s$ and so

$$\begin{aligned} t(n - s) = M(i_1) + \ldots + M(i_t) &= |P^* \cap A_{j_1}| + \ldots + |P^* \cap A_{j_h}| \\ &\leqslant |A_{j_1}| + \ldots + |A_{j_h}| = h(n - s). \end{aligned}$$

Hence $h \geqslant t$, i.e. for each $P^* \subseteq P$

$$|\{j : 1 \leqslant j \leqslant r,\ P^* \cap A_j \neq \emptyset\}| \geqslant |P^*|. \tag{3}$$

Now, by (2) and Hall's theorem, $(A_1, \ldots, A_r)$ possesses a transversal. Further, by (3) and Corollary 2.2.5, P is a partial transversal of $(A_1, \ldots, A_r)$. Hence, by Theorem 3.3.6, $(A_1, \ldots, A_r)$ possesses a transversal which contains P. Let, then,

$$P \subseteq \{y_1, \ldots, y_r\}_{\neq} \subseteq \{1, \ldots, n\}$$

and $y_1 \in A_1, \ldots, y_r \in A_r$. Adjoining the column $(y_1, \ldots, y_r)$ to R, we obtain a Latin rectangle R' of type $(r, s + 1, n)$.

Denote by $N'(i)$ the number of elements of R' equal to i, and by $Y(i)$ the number of integers among $y_1, \ldots, y_r$ equal to i. Then

$$N'(i) = N(i) + Y(i).$$

If $i \in \mathrm{P}$, then $N(i) = r + s - n$ and $Y(i) = 1$, so that

$$N'(i) = r + (s + 1) - n.$$

If $i \notin \mathrm{P}$, then $N(i) > r + s - n$ and so

$$N'(i) \geqslant N(i) > r + s - n.$$

Hence, for all i,

$$N'(i) \geqslant r + (s + 1) - n.$$

We can now repeat the process described above until we obtain a Latin rectangle R^* of type (r, n, n). For this rectangle, $N(i) = r$ and so (1) is satisfied. Hence R^* can be extended to a Latin square of order n (a fact which also follows by virtue of Theorem 5.3.1).

5.4 Subsets with a prescribed pattern of overlaps

When we say that the family $(\mathrm{A}_i\colon 1 \leqslant i \leqslant n)$ possesses a transversal, we assert the existence of subsets of $\mathrm{A}_1, \ldots, \mathrm{A}_n$ which exhibit a particular set-theoretic structure: they are pairwise disjoint singletons. It is therefore natural to extend the discussion by seeking to determine conditions which ensure that $\mathrm{A}_1, \ldots, \mathrm{A}_n$ should possess subsets the pattern of whose overlaps is prescribed in advance. To formulate our question with greater precision, let us say that the families $(\mathrm{A}_i\colon 1 \leqslant i \leqslant n)$ and $(\mathrm{B}_i\colon 1 \leqslant i \leqslant n)$ are *combinatorially equivalent* if there exists a bijection

$$\sigma\colon \bigcup_{i=1}^{n} \mathrm{A}_i \to \bigcup_{i=1}^{n} \mathrm{B}_i \tag{1}$$

such that $\sigma(\mathrm{A}_i) = \mathrm{B}_i$ $(1 \leqslant i \leqslant n)$. We shall first prove a preliminary result which will help us to visualize the meaning of combinatorial equivalence. By a *boolean polynomial* in $\mathrm{A}_1, \ldots, \mathrm{A}_n$, we understand a finite expression involving the A's and formed by means of unions, intersections, and differences. If unions and intersections only are admitted, then we speak of a *restricted boolean polynomial*. We also recall, from §1.4, the notion of boolean atoms $\mathrm{A}[\mathrm{N}]$ generated by $(\mathrm{A}_1, \ldots, \mathrm{A}_n)$. We shall write $\mathrm{N}_0 = \{1, \ldots, n\}$.

THEOREM 5.4.1. *Let* $\mathfrak{A} = (\mathrm{A}_1, \ldots, \mathrm{A}_n)$, $\mathfrak{B} = (\mathrm{B}_1, \ldots, \mathrm{B}_n)$ *be two families of sets. Each of the following three statements then implies the other two.*

(i) *$\mathfrak{A}$ and $\mathfrak{B}$ are combinatorially equivalent.*

(ii) $|\beta(A_1, \ldots, A_n)| = |\beta(B_1, \ldots, B_n)|$ *for every boolean polynomial* β.

(iii) $|A[N]| = |B[N]|$ *whenever* $\emptyset \subset N \subseteq N_0$.

Further, if $\mathfrak{A}$, $\mathfrak{B}$ are families of finite sets, then the phrase 'boolean polynomial' in (ii) *can be replaced by 'restricted boolean polynomial'.*

Let (i) be satisfied, and let σ be an associated bijection, as in (1). Since, in particular, σ is injective, we have (cf. Exs. 1.1.3 and 1.2.1)

$$\beta(B_1, \ldots, B_n) = \beta(\sigma(A_1), \ldots, \sigma(A_n)) = \sigma(\beta(A_1, \ldots, A_n))$$

for every boolean polynomial β. Hence (i) implies (ii). Moreover, (ii) implies (iii) trivially since $A[N]$ is a boolean polynomial in $A_1, \ldots, A_n$.

Next, let (iii) be given. Then there exist bijections $\sigma_N : A[N] \to B[N]$ $(\emptyset \subset N \subseteq N_0)$. Now the atoms $A[N], \emptyset \subset N \subseteq N_0$, are pairwise disjoint; and we shall denote by σ the direct sum (cf. §1.1) of the $2^n - 1$ bijections σ_N. By Lemma 1.4.1, the union of the $A[N]$ is $A_1 \cup \ldots \cup A_n$; and an analogous statement holds for the family $\mathfrak{B}$. Hence σ is a bijection of type (1). Moreover, by Lemma 1.4.2, we have

$$\sigma(A_k) = \sigma\Bigl(\bigcup_{k \in N \subseteq N_0} A[N]\Bigr) = \bigcup_{k \in N \subseteq N_0} \sigma(A[N])$$

$$= \bigcup_{k \in N \subseteq N_0} B[N] = B_k;$$

and so (i) is valid. This completes the proof of the first part of the theorem.

Suppose, next, that $\mathfrak{A}$ and $\mathfrak{B}$ are families of finite sets, and substitute 'restricted boolean polynomial' for 'boolean polynomial' in (ii). The proofs of the implications (i) $\Rightarrow$ (ii) and (iii) $\Rightarrow$ (i) remain precisely the same as before. To establish (ii) $\Rightarrow$ (iii), we note that, for $\emptyset \subset N \subseteq N_0$,

$$A[N] = \bigcap_{i \in N} A_i \setminus \Bigl\{ \bigcup_{i \notin N} A_i \cap \bigcap_{i \in N} A_i \Bigr\}.$$

Hence

$$|A[N]| = \Bigl|\bigcap_{i \in N} A_i\Bigr| - \Bigl|\bigcup_{i \notin N} A_i \cap \bigcap_{i \in N} A_i\Bigr|$$

and so, by (ii),

$$|A[N]| = \Bigl|\bigcap_{i \in N} B_i\Bigr| - \Bigl|\bigcup_{i \notin N} B_i \cap \bigcap_{i \in N} B_i\Bigr| = |B[N]|,$$

i.e. (iii) is valid.

We are now able to answer the question raised at the beginning of the present section.

THEOREM 5.4.2. (R. Rado) *Let* $(A_i: 1 \leqslant i \leqslant n)$, $(B_i: 1 \leqslant i \leqslant n)$ *be two families of finite sets. Then there exist sets* $X_i \subseteq A_i$ $(1 \leqslant i \leqslant n)$ *such that the family* $(X_i: 1 \leqslant i \leqslant n)$ *is combinatorially equivalent to* $(B_i: 1 \leqslant i \leqslant n)$ *if and only if*

$$|\rho(A_1, \ldots, A_n)| \geqslant |\rho(B_1, \ldots, B_n)| \tag{2}$$

for every restricted boolean polynomial ρ.

The stated condition is certainly necessary. For suppose that the sets X_i with the stated properties exist. Since $X_i \subseteq A_i$, we have

$$|\rho(A_1, \ldots, A_n)| \geqslant |\rho(X_1, \ldots, X_n)|$$

for every restricted boolean polynomial ρ. Further, since $(X_1, \ldots, X_n)$ and $(B_1, \ldots, B_n)$ are combinatorially equivalent, we have, by Theorem 5.4.1, (i) and (ii),

$$|\rho(X_1, \ldots, X_n)| = |\rho(B_1, \ldots, B_n)|;$$

and so (2) holds for every ρ.

Conversely, suppose that (2) holds for every ρ. For $\emptyset \subset N \subseteq N_0$, write

$$A\{N\} = \bigcap_{k \in N} A_k, \qquad B\{N\} = \bigcap_{k \in N} B_k.$$

Let $1 \leqslant k \leqslant 2^n - 1$ and denote by $N_1, \ldots, N_k$ any k different non-empty subsets of N_0. Then, in view of our hypothesis,

$$\begin{aligned} |A\{N_1\} \cup \ldots \cup A\{N_k\}| &\geqslant |B\{N_1\} \cup \ldots \cup B\{N_k\}| \\ &\geqslant |B[N_1] \cup \ldots \cup B[N_k]| \\ &= |B[N_1]| + \ldots + |B[N_k]|. \end{aligned}$$

Hence, by Theorem 3.3.1, there exist $2^n - 1$ pairwise disjoint sets, say $X\{N\}$ $(\emptyset \subset N \subseteq N_0)$ such that

$$X\{N\} \subseteq A\{N\}, \qquad |X\{N\}| = |B[N]|.$$

Write

$$X_k = \bigcup_{k \in N \subseteq N_0} X\{N\} \qquad (1 \leqslant k \leqslant n). \tag{3}$$

Then

$$X_k \subseteq \bigcup_{k \in N \subseteq N_0} A\{N\} = \bigcup_{k \in N \subseteq N_0} \bigcap_{i \in N} A_i = A_k.$$

Now let $x \in X\{N\}$. Then $x \in X_k$ for all $k \in N$ and also, since the $X\{N\}$ are pairwise disjoint, $x \notin X_k$ for all $k \notin N$. In other words, if $X[N]$ $(\emptyset \subset N \subseteq N_0)$ are the boolean atoms generated by $(X_1, \ldots, X_n)$, we have $X\{N\} \subseteq X[N]$. Therefore, by (3) and Lemma 1.4.2,

$$X_k = \bigcup_{k \in N \subseteq N_0} X\{N\} \subseteq \bigcup_{k \in N \subseteq N_0} X[N] = X_k$$

and consequently $X\{N\} = X[N]$. Hence

$$|X[N]| = |B[N]| \qquad (\emptyset \subset N \subseteq N_0)$$

and it follows by Theorem 5.4.1, (i) and (iii), that $(X_1, \ldots, X_n)$ is combinatorially equivalent to $(B_1, \ldots, B_n)$.

It is of some interest to note that, as is clear from the above proof, the validity of (2) for *all* restricted boolean polynomials is secured by its validity for a certain *finite* subset of such polynomials. We may also mention in passing that the theorem remains true if the assumption that all A's and B's are finite is dispensed with. However, in that case, there are some additional complications and we do not enter into the details of the argument.

Finally, we indicate that Hall's theorem (for a finite family of finite sets) is a special case of the result just proved. The deduction runs as follows.† Let $(A_i: 1 \leqslant i \leqslant n)$ satisfy condition $\mathscr{H}$ and let $(B_i: 1 \leqslant i \leqslant n)$ be a family of pairwise disjoint singletons. Further, let ρ be any restricted boolean polynomial. By the repeated use of the distributive law for unions over intersections, we may express $\rho(X_1, \ldots, X_n)$ in the form

$$\rho(X_1, \ldots, X_n) = Y_1 \cap \ldots \cap Y_r,$$

where each Y_i is a finite union of X's. Suppose that $X_1 \cup \ldots \cup X_s = W$ (say) appears in each Y_i but that no expression of the form $X_1 \cup \ldots \cup X_s \cup X_{s+1}$ has this property. (Here $s = 0$ means that $X_1 \cup \ldots \cup X_s = \emptyset$). We may then write

$$\begin{aligned} \rho(X_1, \ldots, X_n) &= (W \cup Z_1) \cap \ldots \cap (W \cup Z_r) \\ &= (X_1 \cup \ldots \cup X_s) \cup (Z_1 \cap \ldots \cap Z_r), \end{aligned} \tag{4}$$

where each Z is a union of X's but no X_i appears in all the Z's. Thus, if the X's are pairwise disjoint, $Z_1 \cap \ldots \cap Z_r = \emptyset$.

Using (4) and condition $\mathscr{H}$, we infer that

$$|\rho(A_1, \ldots, A_n)| \geqslant |A_1 \cup \ldots \cup A_s| \geqslant s.$$

† The reader is reminded that Theorem 3.3.1 has been used in the proof of Theorem 5.4.2. The present argument does not, therefore, furnish a new proof of Hall's theorem.

Moreover, since the B's are pairwise disjoint singletons, we have

$$|\rho(B_1, ..., B_n)| = s.$$

Thus (2) holds for any restricted boolean polynomial ρ. By Theorem 5.4.2 there exist, therefore, sets $X_i \subseteq A_i$ $(1 \leqslant i \leqslant n)$ such that $(X_1, ..., X_n)$ is combinatorially equivalent to $(B_1, ..., B_n)$. Hence the X's, too, are pairwise disjoint singletons, say $X_i = \{x_i\}$ $(1 \leqslant i \leqslant n)$, and so $x_i \in A_i$ $(1 \leqslant i \leqslant n)$, where $x_1, ..., x_n$ are distinct. It follows that $\{x_1, ..., x_n\}$ is a transversal of $(A_1, ..., A_n)$.

Exercises 5.4

1. Show that the phrase 'boolean polynomial' in Theorem 5.4.1 cannot be replaced by 'restricted boolean polynomial' if the finiteness of the A's and B's is not postulated.

2. Exhibit Theorem 3.3.1 (for finite sets) as a special case of Theorem 5.4.2.

Notes on Chapter 5

§ 5.1. P. J. Higgins (**1**) established Theorem 5.1.1 for finite families. The extension of Higgins's result and the proof offered here are due to Mirsky (**2**). Theoren 5.1.3 was proved, simultaneously and independently, by Gale (**1**) and by Ryser (**2**); our proof is based on Higgins's ideas (op. cit.) Gale's proof of Theorem 5.1.3 depends on the 'demand–supply theorem', Ryser's on direct combinatorial reasoning. Gale also describes a constructive procedure for obtaining an incidence matrix with preassigned row-sums and column-sums. For other proofs of Theorem 5.13, see Fulkerson (**2**), Ford & Fulkerson (**2**, 76 ff.), and Vogel (**2**).

The relation $\prec$ between vectors was introduced by Hardy, Littlewood & Pólya (**1**; **2**, 45).

§ 5.2. The (lower) estimate given in Theorem 5.2.1 is due to M. Hall (**1**). A further technical refinement of Hall's argument led R. Rado (**10**) to the improved estimate (5) quoted at the end of the section. More recently, P. A. Ostrand (**1**) obtained the still sharper inequality (6). Investigations bearing on the problem of upper estimates and expressed in the language of matrix theory were carried out by H. Minc (**1**) and D. R. Fulkerson (**4**).

§ 5.3. The felicitous idea of applying P. Hall's theorem to the study of Latin rectangles is due to M. Hall (**1**, **2**), who discovered Theorems 5.3.1 and 5.3.2. The proof of the latter result depends on Theorem 5.2.1: Rado's or Ostrand's sharper inequalities (cf. the preceding Note) offer no advantage here. The estimate for the number, say $N(n, r)$, of Latin rectangles of type (r, n, n) given in Theorem 5.3.2 is far from best possible. From the work of Erdös & Kaplansky (**1**) and that of Yamamoto (**1**) it is known that, if $\alpha < \frac{1}{3}$ and $r < n^\alpha$, then $N(n, r) \sim (n!)^r \exp\{-r(r-1)/2\}$ as $n \to \infty$.

Theorem 5.3.3 is due to Ryser (**1**); the proof given here is based on the presentation of Mann & Ryser (**1**).

§ 5.4. The very striking generalization of Hall's theorem contained in Theorem 5.4.2 was discovered by Rado. The theorem, as stated here, represents the gist of Rado's conclusions in (**2**), except that in that paper cardinal numbers were replaced by a more general 'measure' on sets. Still more comprehensive results were obtained by Rado in (**4**). He has also proved that the finiteness of A's and B's need not be postulated in Theorem 5.4.2; this work has not been published.

6 Independent Transversals

In this chapter we shall introduce the notion of abstract independence and examine its repercussions on transversal theory.

6.1 Pre-independence and independence

We begin with an axiomatization of the familiar notion of linear independence in vector spaces.

Let E be a non-empty set. A non-empty collection $\mathscr{E}$ of subsets of E is called a *pre-independence structure* on E if it satisfies the following two axioms.

I(1) $\mathscr{E}$ is a 'hereditary' structure: if $A \in \mathscr{E}$ and $B \subseteq A$, then $B \in \mathscr{E}$.

I(2) If A, B are finite members of $\mathscr{E}$ and $|B| = |A| + 1$, then there exists an element $b \in B \setminus A$ such that $A \cup \{b\} \in \mathscr{E}$.

Since $\mathscr{E}$ is non-empty, it follows by I(1) that $\emptyset \in \mathscr{E}$.

Suppose that, in addition to I(1) and I(2), $\mathscr{E}$ satisfies the further axiom

I(3) $\mathscr{E}$ has finite character.

Then $\mathscr{E}$ is called an *independence structure* on E. (When the ground set E is finite, the notions of pre-independence and independence are, of course, identical.) If $\mathscr{E}$ is an independence structure resp. pre-independence structure on E, then the pair $(E, \mathscr{E})$ is called an *independence space* resp. *pre-independence space*. In either case, the members of $\mathscr{E}$ are called the *independent subsets* of E. All other subsets of E are called *dependent*.

As has been indicated, abstract independence is introduced to mimick the (set-theoretic) features of linear independence in vector spaces. Requirement I(2) is called the *replacement axiom*: it reflects Steinitz's familiar exchange lemma. We note at once an almost obvious consequence of this axiom.

LEMMA 6.1.1. *Let $\mathscr{E}$ be a pre-independence structure on* E, *and let* A, B *be finite independent sets with* $|A| \leqslant |B|$. *Then there exists a set* $B' \subseteq B \setminus A$ *such that* $A \cup B'$ *is independent and* $|A \cup B'| = |B|$.

This result follows immediately by repeated application of I(2).

There are several obvious realizations of independence. Let E be a subset of a vector space. Then the collection of all linearly independent subsets of E is an independence structure. Again, let E be an arbitrary set. Then its power set $\mathscr{P}(E)$ is an independence structure on E: this structure will be referred to as the *universal structure* on E. At the other extreme we have the *trivial structure* whose only member is the empty set.

If $(E, \mathscr{E})$ is an independence (pre-independence) space and $E' \subseteq E$, then the collection $\{X \in \mathscr{E} : X \subseteq E'\}$ is an independence (pre-independence) structure on E′. We may refer to it as the *restriction* of $\mathscr{E}$ to E′.

Let $\mathscr{E}$ be a pre-independence structure on E, let $A \subseteq E$, and let n be a natural number. If A contains an independent subset of cardinal n but no independent subset of cardinal $n + 1$, we say that A has *rank n*. If A has independent subsets of every finite cardinal, we say that its rank is infinite and we denote it by ∞, with the understanding that the symbol ∞ obeys the usual conventions. The rank of A will be denoted by $\rho(A)$; if $\rho(A)$ is finite, we call A a *rank-finite* set. The function ρ, defined on $\mathscr{P}(E)$, is called the *rank function* of the pre-independence structure $\mathscr{E}$. It is plain that, for the universal structure on E, $\rho(X) = |X|$ whenever X is a finite subset of E.

For the case of *independence* structures on infinite sets, the notion of rank will be refined in §7.2. At present, however, only the simplest results on rank are needed.

THEOREM 6.1.2. *Let $\mathscr{E}$ be an independence structure on a set* E. *Then every independent set is contained in some maximal independent set. In particular, in every independence structure there is at least one maximal independent set.*

Let $F \in \mathscr{E}$ and denote by $\mathscr{F}$ the collection of all independent subsets of E which contain F. Let $\mathscr{F}$ be partially ordered by inclusion. Since $\mathscr{E}$ has finite character, so has $\mathscr{F}$. Hence, by Tukey's lemma (Theorem 1.5.3), $\mathscr{F}$ contains a maximal set F^*. Thus F^* is a maximal independent subset of E, and $F \subseteq F^*$.

We note, in passing, a familiar special case of the result just proved: every linearly independent subset of a vector space is contained in a basis.

LEMMA 6.1.3. ('Modular inequality') *Let ρ be the rank function of a pre-independence structure on a set* E. *Then, for any finite*† *subsets* A, B *of* E, *we have*

$$\rho(A) + \rho(B) \geqslant \rho(A \cup B) + \rho(A \cap B).$$

We shall consider independent subsets of $A \cup B$. The collection of all these subsets is an independence structure whose rank function is the restriction of ρ to $\mathscr{P}(A \cup B)$ (and which we shall again denote by the same symbol ρ).

Let X be an independent subset of $A \cap B$ with $|X| = \rho(A \cap B)$. Then, by

† The assertion is still valid for arbitrary A, B; but we shall make no use of this fact.

Lemma 6.1.1, there exists an independent subset Y of $A \cup B$ with $X \subseteq Y$ and $|Y| = \rho(A \cup B)$. Thus

$$\rho(A \cup B) + \rho(A \cap B) = |X| + |Y|.$$

Write $Y = X \cup V \cup W$, where $V \subseteq A \setminus B$ and $W \subseteq B \setminus A$. Then

$$\rho(A \cup B) + \rho(A \cap B) = 2|X| + |V| + |W|.$$

Now $X \cup V$ is an independent subset of A, and $X \cup W$ is an independent subset of B. Hence

$$\rho(A) \geqslant |X| + |V|, \qquad \rho(B) \geqslant |X| + |W|,$$

and the assertion follows.

Exercises 6.1

1. Let $\mathscr{E}$ be a pre-independence structure, with rank function ρ, on a set E. Show that, for all $A, B \subset\subset E$,

$$\rho(A \cup B) \leqslant \rho(A) + \rho(B).$$

2. Let $\mathscr{E}$ be a pre-independence structure on a set E, and let k be a nutural number. Show that the collection of sets

$$\mathscr{E}' = \{X \in \mathscr{E} : |X| \leqslant k\}$$

is again a pre-independence structure. Show, further, that the rank functions ρ, ρ' of $\mathscr{E}$, $\mathscr{E}'$ respectively satisfy the relation

$$\rho'(X) = \min\{\rho(X), k\} \quad (X \subset\subset E).$$

($\mathscr{E}'$ is often referred to as the 'truncation of $\mathscr{E}$ at k'.)

3. Let $\mathfrak{M}$ be a non-empty collection of subsets of a finite set E. Suppose, further, that (i) if $X \in \mathfrak{M}$, $X' \subseteq X$, then $X' \in \mathfrak{M}$; (ii) if $A \subseteq E$ and X_1, X_2 are maximal subsets of A which are members of $\mathfrak{M}$, then $|X_1| = |X_2|$. Show that $\mathfrak{M}$ is an independence structure on E.

4. Let E_1, E_2 be disjoint, non-empty sets and let $\mathscr{E}_1, \mathscr{E}_2$ be pre-independence structures on E_1, E_2 respectively. Denoting by $\mathscr{E}$ the collection of all subsets of $E_1 \cup E_2$ expressible in the form $A_1 \cup A_2$, where $A_1 \in \mathscr{E}_1$, $A_2 \in \mathscr{E}_2$, show that $\mathscr{E}$ is a pre-independence structure on $E_1 \cup E_2$.

5. Let $\mathfrak{A}$ be a finite family of subsets of a finite set E. Is the collection of complements of all representing sets of $\mathfrak{A}$ an independence structure on E?

6. Let $\mathscr{E}$ be a non-empty collection of subsets of an arbitrary set E, and let $\mathscr{E}$ satisfy the axioms I(1) and I(3). Show that two subsets of E which are maximal members of $\mathscr{E}$ do not necessarily have the same cardinal number.

7. Show that no two of the three axioms defining independence structures imply the third axiom.

8. Show that the sign '$\geqslant$' in the modular inequality (Lemma 6.1.3) cannot be replaced by '$=$'.

9. Let $\mathscr{E}$ be an independence structure on a finite set E, let $F \in \mathscr{E}$, and write

$$\mathscr{E}_F = \{X \subseteq E \setminus F\colon X \cup F \in \mathscr{E}\}.$$

Verify that $\mathscr{E}_F$ is an independence structure on $E \setminus F$. Denoting by ρ, ρ_F the rank functions of $\mathscr{E}$, $\mathscr{E}_F$ respectively, show that

$$\rho_F(X) = \rho(X \cup F) - |F| \qquad (X \subseteq E \setminus F).$$

Use this identity and the modular inequality (Lemma 6.1.3) to prove that, for $F \subseteq G \subseteq H \subseteq E$,

$$\rho_F(H \setminus G) \geqslant \rho(H) - \rho(G).$$

10. Let $\mathscr{E}$ be an independence structure, with rank function ρ, on a finite set E; and let h be a non-negative integer. Show that the collection of sets

$$\tilde{\mathscr{E}} = \{X \subseteq E\colon \rho(X) \geqslant |X| - h\}$$

is again an independence structure, and that its rank function $\tilde{\rho}$ is given by

$$\tilde{\rho}(X) = \min\{|X|, \rho(X) + h\} \qquad (X \subseteq E).$$

6.2 Rado's theorem on independent transversals

Let $\mathscr{E}$ be a pre-independence structure on a set E, and let $\mathfrak{A}$ be a family of subsets of E. If X is a transversal resp. partial transversal of $\mathfrak{A}$ and if X is also a member of $\mathscr{E}$, then we say that X is an *independent transversal* resp. *independent partial transversal* of $\mathfrak{A}$. Hall's theorem provides a criterion for the existence of transversals: we shall now establish an analogous result for independent transversals.

THEOREM 6.2.1. (R. Rado) *Let $\mathscr{E}$ be a pre-independence structure, with rank function ρ, on a set* E. *The finite family $\mathfrak{A} = (A_1, \ldots, A_n)$ of subsets of* E *possesses an independent transversal if and only if $\mathfrak{A}$ satisfies the Hall–Rado condition: for each* $I \subseteq \{1, \ldots, n\}$,

$$\rho\big(A(I)\big) \geqslant |I|. \tag{1}$$

As we shall see, this result lies at the very heart of transversal theory. For the case when $\mathscr{E}$ is the universal structure on E, it reduces to Hall's theorem 2.2.1. The proof given below is modelled on the second proof of Hall's theorem. The necessity of (1) is obvious, and we need only discuss its sufficiency.

We begin by establishing the analogue of Lemma 2.2.2. Let the family $(A_1, \ldots, A_n, B)$ of finite subsets of E satisfy the Hall–Rado condition and suppose that $|\mathrm{B}| \geqslant 2$. Let $x, y \in \mathrm{B}$, $x \neq y$. We now assert that at least one of the two families

$$(\mathrm{A}_1, \ldots, \mathrm{A}_n, \mathrm{B} \setminus \{x\}), \qquad (\mathrm{A}_1, \ldots, \mathrm{A}_n, \mathrm{B} \setminus \{y\})$$

satisfies the Hall–Rado condition. For if this were not the case, then, for certain subsets I, J of $\{1, \ldots, n\}$, we would have

$$\rho(\mathrm{S}) \leqslant |\mathrm{I}|, \qquad \rho(\mathrm{T}) \leqslant |\mathrm{J}|,$$

where

$$\mathrm{S} = \mathrm{A}(\mathrm{I}) \cup (\mathrm{B} \setminus \{x\}), \qquad \mathrm{T} = \mathrm{A}(\mathrm{J}) \cup (\mathrm{B} \setminus \{y\}).$$

Hence, by Lemma 6.1.3,

$$|\mathrm{I}| + |\mathrm{J}| \geqslant \rho(\mathrm{S}) + \rho(\mathrm{T}) \geqslant \rho(\mathrm{S} \cup \mathrm{T}) + \rho(\mathrm{S} \cap \mathrm{T}).$$

But

$$\mathrm{S} \cup \mathrm{T} = \mathrm{A}(\mathrm{I} \cup \mathrm{J}) \cup \mathrm{B}$$
$$\mathrm{S} \cap \mathrm{T} \supseteq \mathrm{A}(\mathrm{I}) \cap \mathrm{A}(\mathrm{J}) \supseteq \mathrm{A}(\mathrm{I} \cap \mathrm{J}),$$

and therefore

$$\begin{aligned} |\mathrm{I}| + |\mathrm{J}| &\geqslant \rho\big(\mathrm{A}(\mathrm{I} \cup \mathrm{J}) \cup \mathrm{B}\big) + \rho\big(\mathrm{A}(\mathrm{I} \cap \mathrm{J})\big) \\ &\geqslant |\mathrm{I} \cup \mathrm{J}| + 1 + |\mathrm{I} \cap \mathrm{J}| = |\mathrm{I}| + |\mathrm{J}| + 1. \end{aligned}$$

We thus arrive at a contradiction. It follows that there exists an element $z \in \mathrm{B}$ such that $(\mathrm{A}_1, \ldots, \mathrm{A}_n, \mathrm{B} \setminus \{z\})$ satisfies the Hall–Rado condition.

Now suppose that the family $\mathfrak{A} = (\mathrm{A}_1, \ldots, \mathrm{A}_n)$ of *finite* sets satisfies the Hall–Rado condition. Then, by the result just proved, we can reduce in turn each A_k till we are left with a family of singletons, say $(\{x_1\}, \ldots, \{x_n\})$, which still satisfies the Hall–Rado conditions and also, of course, the relations $x_k \in \mathrm{A}_k$ $(1 \leqslant k \leqslant n)$. In particular, we have

$$\rho\big(\{x_1\} \cup \ldots \cup \{x_n\}\big) \geqslant n$$

and so $\{x_1, \ldots, x_n\}_{\neq}$ is an independent transversal of $\mathfrak{A}$.

We have now proved the theorem for the case of finite A's. In the general case, the Hall–Rado condition (1) implies that, for each $\mathrm{I} \subseteq \{1, \ldots, n\}$, the set A(I) contains an independent subset, say C(I), of cardinal $|\mathrm{I}|$. Write

$$\mathrm{C} = \bigcup_{\mathrm{I} \subseteq \{1, \ldots, n\}} \mathrm{C}(\mathrm{I}), \qquad \mathrm{A}_i^* = \mathrm{A}_i \cap \mathrm{C} \qquad (1 \leqslant i \leqslant n).$$

Then, for each $I \subseteq \{1, \ldots, n\}$, we have

$$\rho\left(\bigcup_{i \in I} A_i^*\right) = \rho\left(\bigcup_{i \in I} A_i \cap C\right) \geqslant \rho(C(I)) = |I|.$$

Now the A_i^* are finite subsets of C. Hence, by the result already established, we infer the existence of elements $x_i \in A_i^* \subseteq A_i$ $(1 \leqslant i \leqslant n)$ with

$$\{x_1, \ldots, x_n\}_{\neq} \in \mathscr{E}.$$

Rado's theorem admits of the same kind of defect version as does Hall's theorem.

THEOREM 6.2.2. *Let $\mathscr{E}$ be a pre-independence structure, with rank function ρ, on a set* E. *Let $\mathfrak{A} = (A_1, \ldots, A_n)$ be a family of subsets of* E, *and let $0 \leqslant d \leqslant n$. Then $\mathfrak{A}$ possesses an independent partial transversal of defect d if and only if, for each $I \subseteq \{1, \ldots, n\}$,*

$$\rho(A(I)) \geqslant |I| - d. \tag{2}$$

For $d = 0$, this statement reduces to Rado's theorem 6.2.1.

Let D be a set such that $|D| = d$, $D \cap E = \emptyset$. Denote by $\mathscr{E}^*$ the collection of all sets of the form $X \cup Y$, where $X \in \mathscr{E}$ and $Y \subseteq D$. It is a matter of immediate verification (cf. Ex. 6.1.4) that $\mathscr{E}^*$ is a pre-independence structure on $E \cup D$. We shall denote its rank function by ρ^*.

Let (2) be satisfied for each $I \subseteq \{1, \ldots, n\}$, and consider the family $(A_i \cup D: 1 \leqslant i \leqslant n)$ of subsets of $E \cup D$. Whenever $\emptyset \subset I \subseteq \{1, \ldots, n\}$, we have

$$\rho^*\left(\bigcup_{i \in I} (A_i \cup D)\right) = \rho^*(A(I) \cup D) = \rho(A(I)) + |D|$$
$$\geqslant (|I| - d) + d = |I|.$$

It follows by Theorem 6.2.1 that $(A_i \cup D: 1 \leqslant i \leqslant n)$ possesses a transversal which is a member of $\mathscr{E}^*$, and so $\mathfrak{A}$ possesses a PT with defect d which is a member of $\mathscr{E}$. This establishes the sufficiency of condition (2): we leave it to the reader to verify its necessity.

The next result is a simple application of Theorem 6.2.2.

COROLLARY 6.2.3. *Let $0 \leqslant m \leqslant n$ and let $A_1, \ldots, A_n$ be subsets of a vector space V. The following statements are then equivalent.*

(i) *Whenever $x_i \in A_i$ $(1 \leqslant i \leqslant n)$, the vectors $x_1, \ldots, x_n$ span a subspace of V of dimension not exceeding m.*

(ii) *There exists an integer* h *with* $0 \leqslant h \leqslant m$ *and a collection of* $h + n - m$ A'*s contained in a subspace of* V *of dimension* h.

Let $\mathscr{E}$ be the independence structure consisting of all linearly independent subsets of V, and denote its rank function by ρ. Statement (i) then means that the family $\mathfrak{A} = (A_1, \ldots, A_n)$ does not possess an independent PT of cardinal $m + 1$, i.e. of defect $n - m - 1$. In view of Theorem 6.2.2, this is the case precisely if, for some $I \subseteq \{1, \ldots, n\}$,

$$\rho(A(I)) \leqslant |I| - n + m. \tag{3}$$

Again, (ii) means that there exists an integer h with $0 \leqslant h \leqslant m$ and a set $I \subseteq \{1, \ldots, n\}$ such that $|I| = h + n - m$ and $\rho(A(I)) \leqslant h$. This clearly holds if and only if (3) is satisfied for some $I \subseteq \{1, \ldots, n\}$. The equivalence of (i) and (ii) is therefore established.

We shall now prove a transfinite extension of Theorem 6.2.1.

THEOREM 6.2.4. *Let* $\mathscr{E}$ *be an independence structure, with rank function* ρ, *on a set* E. *Let* $\mathfrak{A} = (A_i : i \in I)$ *be a family of finite subsets of* E. *The following statements are then equivalent.*

(a) $\rho(A(J)) \geqslant |J|$ *for every finite subset* J *of* I.

(b) *Every finite subfamily of* $\mathfrak{A}$ *has an independent transversal.*

(c) $\mathfrak{A}$ *has an independent transversal.*

It will be observed that while Theorem 6.2.1 is framed for pre-independence structures, in Theorem 6.2.4 we postulate that $\mathscr{E}$ is an independence structure. If we take $\mathscr{E}$ to be the universal structure on E, then Theorem 6.2.4 reduces to the transfinite form of Hall's theorem (Theorem 4.2.1).

The implication (a) ⇒ (b) holds by Theorem 6.2.1 and the implication (c) ⇒ (a) is immediate. It remains to establish (b) ⇒ (c).

Let $J \subset\subset I$. The subfamily $\mathfrak{A}(J)$ has an independent transversal by (b), i.e. there exists an injective choice function θ_J of $\mathfrak{A}(J)$ with $\theta_J(J) \in \mathscr{E}$. Denote by θ a corresponding Rado choice function of $\mathfrak{A}$ which, by Corollary 4.1.2, is injective. Write $\theta(I) = X$. Then X is a transversal of $\mathfrak{A}$ and we need only show that $X \in \mathscr{E}$. Now $\theta : I \to X$ is a bijection. Let $Y \subset\subset X$ and write $\theta^{-1}(Y) = J$. Then $J \subset\subset I$, and there exists therefore a set K with $J \subseteq K \subset\subset I$ and $\theta | J = \theta_K | J$. Thus

$$Y = \theta(J) = \theta_K(J) \subseteq \theta_K(K) \in \mathscr{E},$$

and so every finite subset of X is independent. Hence, since $\mathscr{E}$ has finite character, X is itself independent.

It is of interest to note that the theorem just proved admits of a natural sharpening in which the finiteness of the A_i is replaced by the weaker requirement of rank-finiteness. We first need a preliminary result.

LEMMA 6.2.5. *Let $\mathscr{E}$ be a pre-independence structure with rank function ρ, on a set* E; *and let* $(A_1, ..., A_n)$ *be a (finite) family of rank-finite subsets of* E. *If $B_1, ..., B_n$ are any independent sets such that $B_i \subseteq A_i$, $|B_i| = \rho(A_i)$ $(1 \leqslant i \leqslant n)$, then*

$$\rho(A_1 \cup ... \cup A_n) = \rho(B_1 \cup ... \cup B_n).$$

Write $A = A_1 \cup ... \cup A_n$, $B = B_1 \cup ... \cup B_n$. Then clearly $\rho(A) \geqslant \rho(B)$. Assume that the inequality is strict. Then there exist sets C, D with

$$C \subset\subset A, \quad C \in \mathscr{E}, \quad |C| = \rho(A),$$
$$D \subseteq B, \quad D \in \mathscr{E}, \quad |D| = \rho(B),$$

such that $|C| > |D|$. By the replacement axiom, there exists an element $x \in C \setminus D$ such that $\{x\} \cup D \in \mathscr{E}$. Now $x \notin B$ and so $x \in C \setminus B \subseteq A \setminus B$. Hence $x \in A_i \setminus B$ for some i. In particular, $x \notin B_i$ and therefore

$$\{x\} \cup B_i \notin \mathscr{E}. \tag{4}$$

Since $B_i \in \mathscr{E}$, we have $|B_i| = \rho(B_i) \leqslant \rho(B) = |D|$. Hence, by Lemma 6.1.1, there exists a set F_i such that

$$B_i \subseteq F_i \subseteq B, \quad F_i \in \mathscr{E}, \quad |F_i| = |D| = \rho(B).$$

Hence, by (4),

$$\{x\} \cup F_i \notin \mathscr{E}. \tag{5}$$

Finally, applying the replacement axiom to the independent sets F_i and $\{x\} \cup D$, we see that either $\{x\} \cup F_i \in \mathscr{E}$, which contradicts (5); or else $\{y\} \cup F_i \in \mathscr{E}$ for some $y \in D \setminus F_i \subseteq B$, which contradicts the definition of $\rho(B)$. We conclude, then, that $\rho(A) = \rho(B)$.

We are now able to formulate a refinement of Theorem 6.2.4.

COROLLARY 6.2.6. *Theorem* 6.2.4 *remains valid if the phrase 'finite subsets' is replaced by 'rank-finite subsets'.*

The implications (c) ⇒ (b) and (b) ⇒ (a) are trivial, and it suffices to establish (a) ⇒ (c). For each $i \in I$, let B_i be any (necessarily finite) subset of

A_i such that $B_i \in \mathscr{E}$, $|B_i| = \rho(A_i)$. Let $\mathfrak{B}$ be the family $(B_i: i \in I)$. Now, by Lemma 6.2.5, $\rho\big(B(J)\big) = \rho\big(A(J)\big)$ whenever $J \subset\subset I$, and so

$$\rho\big(B(J)\big) \geqslant |J| \qquad (J \subset\subset I).$$

Hence, by Theorem 6.2.4, $\mathfrak{B}$ possesses an independent transversal which is, of course, also an independent transversal of $\mathfrak{A}$.

Exercises 6.2

1. Let $1 \leqslant p \leqslant n$ and let $A_1, ..., A_n$ be subsets of a vector space V. Establish the equivalence of the following statements. (i) There exist p linearly independent vectors selected from A's with p different suffixes. (ii) No collection of k A's (where $n-p+1 \leqslant k \leqslant n$) is contained in a subspace of V having dimension $k-n+p-1$.

2. Obtain a 'defect form' of Theorem 6.2.4.

3. Prove Rado's theorem 6.2.1 by adapting the argument in the first proof of Theorem 2.2.1 and making use of Ex. 6.1.9.

[L. Mirsky (**9**)]

4. Let $\mathscr{E}$ be a pre-independence structure, with rank function ρ, on a set E; and let $(A_1, ..., A_n)$ be a family of subsets of E which satisfies the Hall–Rado condition. A non-empty subset I of $\{1, ..., n\}$ will be called *critical* if $\rho(A(I)) = |I|$. (i) Show that the union of two critical sets is again critical. (ii) Hence verify that, if there exists at least one critical set, then there exists a critical set I_0 such that I is non-critical whenever $I \nsubseteq I_0$.

[R. Rado]

5. Let $\mathscr{E}$ be a pre-independence structure, with rank function ρ, on a set E; and let $\mathfrak{A} = (A_1, ..., A_n)$ be a family of subsets of E which satisfies the Hall–Rado condition (and so possesses an independent transversal). Show that the condition $\rho(A_1 \cup ... \cup A_n) = n$ is necessary but not sufficient for the independent transversal of $\mathfrak{A}$ to be unique. (Cf. Ex. 2.2.2.)

6. Let $(E, \mathscr{E})$ be a pre-independence space; let $\mathfrak{A} = (A_1, ..., A_n)$ be a family of subsets of E; and let $p_1, ..., p_n$ be non-negative integers. Show that the following statements are equivalent. (i) There exist pairwise disjoint sets $X_1, ..., X_n$ with $X_i \subseteq A_i$, $|X_i| = p_i$ $(1 \leqslant i \leqslant n)$ and such that $X_1 \cup ... \cup X_n$ is independent. (ii) For each $I \subseteq \{1, ..., n\}$, the set A(I) contains an independent subset of cardinal $\Sigma\{p_i: i \in I\}$.

7. State the deltoid form of Theorem 6.2.2.

8. Let $\mathscr{E}$ be an independence structure, with rank function ρ, on a finite set E; let $1 \leqslant k \leqslant n$; and let $\mathfrak{A} = (A_1, ..., A_n)$ be a family of subsets of E. Show that $\mathfrak{A}$ possesses a transversal whose rank is at least k if and only if, for all $I \subseteq \{1, ..., n\}$,

$$|I| \leqslant \min\{|A(I)|, \rho(A(I)) + n - k\}.$$

Further, let $M \subseteq E$. Deduce from the above result that $\mathfrak{A}$ possesses a transversal X with $|X \cap M| \geqslant k$ if and only if, for all $I \subseteq \{1, ..., n\}$,

$$|I| \leqslant \min\{|A(I)|, |A(I) \cap M| + n - k\}.$$

6.3 A characteristic property of independence structures

Let E be an arbitrary (non-empty) set and $\mathscr{E}$ a non-empty collection of subsets of E. For an arbitrary family $\mathfrak{A} = (A_i : i \in I)$ of finite subsets of E, we shall consider the following two conditions.

(i) For each finite subset J of I, the set A(J) contains a subset of cardinal $|J|$ which is a member of $\mathscr{E}$.

(ii) $\mathfrak{A}$ possesses a transversal which is a member of $\mathscr{E}$.

We recall (Theorem 6.2.4) that if $\mathscr{E}$ is an independence structure on E, then for each family $\mathfrak{A}$ of finite sets either (i) and (ii) are both true or else both are false. We shall now establish the converse inference.

THEOREM 6.3.1. *If, for each family* $\mathfrak{A}$ *of finite subsets of* E, *either* (i) *and* (ii) *are both true or both are false, then* $\mathscr{E}$ *is an independence structure.*

This results exhibits the fundamental nature of Theorem 6.2.4 by demonstrating that the property of independence structures asserted by that theorem is, in fact, a characteristic property.

We shall now verify the three axioms.

(a) Let $Y \subseteq X \subseteq E, X \in \mathscr{E}$. It is then required to show that $Y \in \mathscr{E}$. The family $(\{x\} : x \in X)$ satisfies (ii) and so (i). A fortiori $(\{x\} : x \in Y)$ satisfies (i), and so (ii). Hence $Y \in \mathscr{E}$, i.e. $\mathscr{E}$ satisfies axiom I(1).

(b) Again, let $X = \{x_1, \ldots, x_n\}_{\neq} \in \mathscr{E}$, $Y = \{y_1, \ldots, y_{n+1}\}_{\neq} \in \mathscr{E}$. Define

$$A_k = \{x_k\} \quad (1 \leqslant k \leqslant n), \qquad A_{n+1} = Y.$$

Then the family $(A_1, \ldots, A_n)$ satisfies (ii) and therefore (i). Now consider the family $\mathfrak{A} = (A_1, \ldots, A_n, A_{n+1})$ and let $\emptyset \subset I \subseteq \{1, \ldots, n, n+1\}$. If $I \subseteq \{1, \ldots, n\}$, then A(I) contains a subset of cardinal $|I|$ which is a member of $\mathscr{E}$ since $(A_1, \ldots, A_n)$ satisfies (i). If $I \nsubseteq \{1, \ldots, n\}$, then A(I) contains Y, i.e. it contains a subset of cardinal $n + 1$ which is a member of $\mathscr{E}$. Hence, by (a), A(I) contains a subset of cardinal $|I|$ which is a member of $\mathscr{E}$. We have therefore shown that $\mathfrak{A}$ satisfies (i). Hence it also satisfies (ii), i.e. $\{x_1, \ldots, x_n, y_i\}_{\neq} \in \mathscr{E}$ for at least one value of i with $1 \leqslant i \leqslant n + 1$. Thus axiom I(2) is valid.

(c) Finally, let X be an infinite subset of E and suppose that all finite subsets of X belong to $\mathscr{E}$. Consider the family $\mathfrak{A} = (A_x : x \in X)$, where $A_x = \{x\}$. Let $Y \subset\subset X$. Then $A(Y) = Y$ and so A(Y) contains a subset of cardinal $|Y|$ which is a member of $\mathscr{E}$. Thus $\mathfrak{A}$ satisfies (i) and so (ii), i.e. $X \in \mathscr{E}$. This establishes I(3), and completes the proof of the theorem.

6.4 Finite independent partial transversals

It is easy to obtain a transfinite analogue of Theorem 6.2.2. Let $\mathscr{E}$ be an independence structure, with rank function ρ, on a set E. Let $\mathfrak{A} = (A_i : i \in I)$ be a family of finite subsets of E, and let d be a non-negative integer. It is then readily shown that $\mathfrak{A}$ possesses an independent PT with defect d if and only if $\rho(A(J)) \geqslant |J| - d$ whenever $J \subset\subset I$. We leave the proof of this result to the reader.

The problem involving PTs of finite defect admits, then, of an immediate solution. The corresponding problem concerning PTs of finite cardinal is to be discussed next. We shall call a subset J of I *cofinite* if $I \setminus J$ is finite.

THEOREM 6.4.1. *Let $\mathscr{E}$ be a pre-independence structure, with rank function ρ, on a set* E. *Let $\mathfrak{A} = (A_i : i \in I)$ be an arbitrary family of subsets of* E. *Let k be a natural number. Then $\mathfrak{A}$ possesses an independent partial transversal of cardinal k if and only if, for each cofinite subset* J *of* I,

$$\rho(A(J)) \geqslant k - |I \setminus J|. \tag{1}$$

Suppose, in the first place, that $\mathfrak{A}$ has an independent PT of cardinal k. Then there exists a set $I^* \subseteq I$ with $|I^*| = k$ such that $\mathfrak{A}(I^*)$ has an independent transversal. To establish (1), we may assume that $|I \setminus J| < k$. Then at least $k - |I \setminus J|$ of the sets in $\mathfrak{A}(J)$ are indexed by elements in I^*. Hence $\mathfrak{A}(J)$ has an independent PT of cardinal $k - |I \setminus J|$, and so (1) is valid for all cofinite $J \subseteq I$.

Suppose, next, that (1) is satisfied for all cofinite subsets J of I. Assume that $\mathfrak{A}$ has an independent PT of cardinal $r < k$ but none of cardinal $r + 1$. We shall then derive a contradiction.

Let $I^* \subseteq I$, $|I^*| = r$, and let $\mathfrak{A}(I^*)$ possess an independent PT. For simplicity, write $I^* = \{1, \ldots, r\}$. Put

$$A(I \setminus I^*) = M, \qquad \mathfrak{B} = (B_1, \ldots, B_{r+1}),$$

where $B_i = A_i$ for $1 \leqslant i \leqslant r$ and $B_{r+1} = M$. If $\mathfrak{B}$ has an independent transversal, say

$$x_1 \in B_1, \ldots, x_{r+1} \in B_{r+1},$$

where $\{x_1, \ldots, x_{r+1}\}_{\neq} \in \mathscr{E}$, then $x_{r+1} \in A_i$ for some $i \in I \setminus I^*$. Hence $\mathfrak{A}$ has an independent PT of cardinal $r + 1$, and this contradicts our initial assumption. Hence $\mathfrak{B}$ has no independent transversal and so, by Rado's theorem 6.2.1, there exists a set $I_1 \subseteq \{1, \ldots, r + 1\}$ such that

$$\rho(B(I_1)) \leqslant |I_1| - 1. \tag{2}$$

But $(B_1, \ldots, B_r) = (A_1, \ldots, A_r)$ has an independent transversal and therefore, again by Rado's theorem,

$$\rho(B(I')) \geqslant |I'| \qquad (I' \subseteq \{1, \ldots, r\}).$$

It follows that

$$r + 1 \in I_1. \tag{3}$$

Writing $I_2 = I \setminus (I^* \setminus I_1)$, we have

$$|I \setminus I_2| = r + 1 - |I_1|. \tag{4}$$

Also, in view of (3),

$$A(I_2) = B(I_1)$$

and so, by (2) and (4),

$$\rho(A(I_2)) + |I \setminus I_2| \leqslant r < k.$$

Thus (1) is violated, and the proof of the theorem is complete.

We note that, for the case of a finite family, the above argument yields an alternative deduction of the defect form of Rado's theorem (Theorem 6.2.2).

Taking $\mathscr{E}$ as the universal structure on E in Theorem 6.4.1, we obtain at once the following special case.

COROLLARY 6.4.2. *Let* $\mathfrak{A} = (A_i : i \in I)$ *be an arbitrary family of subsets of* E, *and let k be a natural number. Then* $\mathfrak{A}$ *possesses a partial transversal of cardinal k if and only if, for each cofinite subset* J *of* I,

$$|A(J)| \geqslant k - |I \setminus J|.$$

6.5 Transversal structures and independence structures

Let $\mathfrak{A}$ be a family of subsets of a set E. By the *transversal structure* of $\mathfrak{A}$ we shall understand the set of all PTs of $\mathfrak{A}$. So far, we have met only completely obvious instances of independence structures: below we shall see that transversal structures provide us with another and less trivial example.

Consider a finite family $\mathfrak{A} = (A_1, \ldots, A_n)$ of subsets of a finite set $E = \{x_1, \ldots, x_m\}_{\neq}$. We now construct an $m \times n$ matrix M by the requirement that the (i, j)-th element of M is equal to 0 if $x_i \notin A_j$ and that it is equal to an indeterminate if $x_i \in A_j$, it being understood that the indeterminates occurring in M are independent over the field of rational numbers. The matrix M so defined will be called a *formal incidence matrix* of $\mathfrak{A}$.† Again, let

$$F = \{x_{i_1}, \ldots, x_{i_k}\}_{\neq} \subseteq E.$$

† Strictly speaking, we should say 'of $\mathfrak{A}$ and E' since $\mathfrak{A}$ is a family of subsets of E* whenever $E \subseteq E^*$.

By the submatrix of M corresponding to F we mean the submatrix of M consisting of rows with suffixes $i_1, \ldots, i_k$.

LEMMA 6.5.1. *Let* $\mathfrak{A}$ *be a finite family of finite subsets of* E, *and let* M *be a formal incidence matrix of* $\mathfrak{A}$. *A subset* F *of* E *is then a partial transversal of* $\mathfrak{A}$ *if and only if the submatrix of* M *corresponding to* F *has rank* $|\mathrm{F}|$.

This implies, in particular, that the transversal index of $\mathfrak{A}$ is equal to the rank of the formal incidence matrix M.

Write $\mathrm{F} = \{x_{i_1}, \ldots, x_{i_k}\}$, where $1 \leqslant i_1 < \ldots < i_k \leqslant m$. Suppose that F is a PT of $\mathfrak{A}$. Then, among $1, 2, \ldots, n$, there exist distinct integers $j_1, \ldots, j_k$ such that

$$x_{i_1} \in \mathrm{A}_{j_1}, \ldots, x_{i_k} \in \mathrm{A}_{j_k}. \tag{1}$$

It follows that the places $(i_1, j_1), \ldots, (i_k, j_k)$ in the formal incidence matrix M of $\mathfrak{A}$ are occupied by indeterminates. Hence the submatrix specified by rows $i_1, \ldots, i_k$ and columns $j_1, \ldots, j_k$ is non-singular, i.e. it has rank $k = |\mathrm{F}|$. The submatrix of M corresponding to F therefore also has rank $|\mathrm{F}|$.

Conversely, suppose that the submatrix (say M^*) of M corresponding to F has rank $|\mathrm{F}|$. Then M^* possesses a non-singular $k \times k$ submatrix and so there are k places in M^* belonging to k different rows and k different columns and occupied by indeterminates. Let these places be $(i_1, j_1), \ldots, (i_k, j_k)$. Then (1) holds, and consequently F is a PT of $\mathfrak{A}$.

THEOREM 6.5.2. (Edmonds & Fulkerson) *Let* $\mathfrak{A}$ *be a family of subsets of an arbitrary set* E. *Then the collection of all partial transversals of* $\mathfrak{A}$ *is a pre-independence structure on* E.

Let $\mathscr{E}$ be the collection of all PTs of $\mathfrak{A}$. By convention, $\emptyset \in \mathscr{E}$ and so $\mathscr{E}$ is non-empty. Moreover, axiom I(1) holds trivially and it remains only to verify the replacement axiom I(2). It is not difficult to provide an ad hoc proof of the validity of I(2); however, we shall offer two alternative lines of reasoning. Throughout, we shall write $\mathfrak{A} = (\mathrm{A}_i : i \in \mathrm{I})$.

Assume, in the first place, that E and I are both finite; denote by M a formal incidence matrix of $\mathfrak{A}$; and let $z_1, \ldots, z_r$ be the indeterminates occurring in M. When we speak of linear independence of certain rows in M, we regard these rows as vectors over the field $\mathbf{Q}(z_1, \ldots, z_r)$ of rational functions, with rational coefficients, in the z's.

We know, by Lemma 6.5.1, that F ($\subseteq$ E) is a PT of $\mathfrak{A}$ if and only if the submatrix of M corresponding to F has rank $|\mathrm{F}|$, i.e. if and only if the rows specified by F are linearly independent. The validity of I(2) now follows at once from the validity of the replacement property in vector spaces.

Next, let E, I be arbitrary. Let X, Y be finite PTs of $\mathfrak{A}$ and suppose that $|Y| = |X| + 1$. Suppose, further, that X, Y are transversals of the subfamilies $\mathfrak{A}(I_1)$, $\mathfrak{A}(I_2)$ respectively. If

$$E' = X \cup Y, \qquad \mathfrak{A}' = (A_i \cap E' : i \in I_1 \cup I_2),$$

then X, Y are clearly PTs of $\mathfrak{A}'$. Therefore, by the result just proved, $X \cup \{y\}$ is a PT of $\mathfrak{A}'$, and therefore of $\mathfrak{A}$, for some $y \in Y \setminus X$. This completes the proof of I(2). The same idea will appear in more sophisticated form in §7.1.

An alternative demonstration of the replacement axiom can be based on Theorem 2.3.1 for the special case of finite deltoids. Let $\{a_1, \ldots, a_m\}_{\neq}$ and $\{b_1, \ldots, b_{m+1}\}_{\neq}$ be PTs of $\mathfrak{A} = (A_i : i \in I)$. Write

$$a_1 \in A_{i_1}, \ldots, a_m \in A_{i_m}; \quad b_1 \in A_{j_1}, \ldots, b_{m+1} \in A_{j_{m+1}},$$

where $i_1, \ldots, i_m$ resp. $j_1, \ldots, j_{m+1}$ are m resp. $m + 1$ distinct elements of I. Put

$$E' = \{a_1, \ldots, a_m\} \cup \{b_1, \ldots, b_{m+1}\}, \qquad I' = \{i_1, \ldots, i_m\} \cup \{j_1, \ldots, j_{m+1}\},$$

$$\mathfrak{A}' = (A_i \cap E' : i \in I').$$

We shall consider the deltoid (E', Δ, I'), where $\Delta\ (\subseteq E' \times I')$ is specified by the requirement that $(e, i) \in \Delta$ if and only if $e \in A_i$. Hence $\{a_1, \ldots, a_m\}$, $\{j_1, \ldots, j_{m+1}\}$ are admissible subsets of E', I' respectively. It follows by Theorem 2.3.1 that there exist linked sets E_0, I_0 such that

$$\{a_1, \ldots, a_m\} \subseteq E_0 \subseteq E', \qquad \{j_1, \ldots, j_{m+1}\} \subseteq I_0 \subseteq I'.$$

Hence $|E_0| = |I_0| \geqslant m + 1$ and so E' contains an admissible subset of at least $m + 1$ elements including the m a's. Hence E' contains an admissible subset of the form $\{a_1, \ldots, a_m, b_k\}_{\neq}$, where $1 \leqslant k \leqslant m + 1$. This, then, is a PT of $\mathfrak{A}'$, and so of $\mathfrak{A}$.

The expression 'pre-independence structure' in Theorem 6.5.2 cannot be replaced by 'independence structure', for the collection of PTs of a family of sets need not have finite character. Thus, if

$$\mathfrak{A} = (\{1, 2\}, \{1, 3\}, \{1, 4\}, \ldots),$$

then every finite subset – and indeed every proper subset—of $\{1, 2, 3, \ldots\}$ is a PT of $\mathfrak{A}$ but $\{1, 2, 3, \ldots\}$ is not itself a PT. However, we have the following variant of Theorem 6.5.2.

THEOREM 6.5.3. *Let* $\mathfrak{A}$ *be a restricted*† *family of subsets of* E. *Then the collection of all partial transversals of* $\mathfrak{A}$ *is an independence structure on* E.

† For the definition of this term, see page 34.

In view of Theorem 6.5.2, it suffices to show that the collection of all PTs of $\mathfrak{A}$ has finite character; and we know that this is the case by Corollary 4.2.4.

We shall conclude the present section by making yet further use of the notion of a formal incidence matrix to give an additional proof of Hall's theorem 2.2.1 (for the case of finite sets). Let $\mathfrak{A} = (A_1, \ldots, A_n)$ be a family of subsets of a finite set E and suppose that $\mathfrak{A}$ has no transversal. It suffices to show that Hall's condition does not hold.

Let M denote the formal incidence matrix of $\mathfrak{A}$, with the columns of M corresponding to the sets $A_1, \ldots, A_n$. Since $\mathfrak{A}$ has no transversal we infer, by the remark immediately following Lemma 6.5.1, that the columns of M are linearly dependent. We choose a minimal set of linearly dependent columns, say the first k columns. If $k = 1$, then $\mathfrak{A}$ has a zero column and Hall's condition is obviously violated. We may therefore suppose that $k > 1$. The submatrix N of M consisting of the first k columns of M has rank $k - 1$. Hence it possesses $k - 1$ linearly independent rows, say the first $k - 1$ rows. Write

$$M = \begin{Vmatrix} M_1 & M_3 \\ M_2 & M_4 \end{Vmatrix},$$

where M_1 is of type $(k - 1) \times k$. There exists a non-zero vector $\lambda = (\lambda_1, \ldots, \lambda_k)^T$, whose components are polynomials in the indeterminates occurring in M_1, such that $M_1\lambda = 0$. Further, every row in M_2 is a linear combination of the rows of M_1, and so $M_2\lambda = 0$. Hence $N\lambda = 0$; and if any λ_i were 0, then $k - 1$ columns in N would be connected by a non-trivial linear relation. Hence $\lambda_1, \ldots, \lambda_k$ are all non-zero. Now let $(c_1, \ldots, c_k)$ be any row in M_2. Since $M_2\lambda = 0$, we have

$$c_1\lambda_1 + \ldots + c_k\lambda_k = 0.$$

Recalling that each λ_i is a polynomial in the indeterminates occurring in M_1 and that the indeterminates occurring in M are independent, we infer that all c's vanish, so that $M_2 = O$. It follows that

$$|A_1 \cup \ldots \cup A_k| < k,$$

i.e. Hall's condition fails to hold.

Exercises 6.5

1. Establish the duals of Theorems 6.5.2 and 6.5.3.

2. Let $\mathscr{E}$ denote the collection of all PTs of a family $\mathfrak{A}$ of sets. Show that, if $\mathscr{E}$ contains arbitrarily large finite sets, then it contains an infinite set.

3. Suppose that each of the two families $(A_1, \ldots, A_n)$ and $(B_1, \ldots, B_{n+1})$ of subsets of E possesses a transversal. Show that, for some k with $1 \leqslant k \leqslant n+1$, the family $(A_1, \ldots, A_n, B_k)$ also possesses a transversal.

4. Use Theorem 6.5.2 to establish the assertion made in Ex. 3.3.2. Conversely, use this assertion to show that the collection of all PTs of a finite family of sets is an independence structure.

5. Let (X, Δ, Y) be a locally right-finite deltoid. Show that the collection of all admissible subsets of X is an independence structure, but that the omission of the phrase 'locally right-finite' invalidates the proposition.

6. Let $\mathfrak{A} = (A_1, \ldots, A_n)$ be a family of subsets of E; let $(E_1, \ldots, E_p)$ be a partition of E; let $r_1, \ldots, r_p$ be non-negative integers; and suppose that

$$|A(I) \cap E(J)| \geqslant |I| - n + \sum_{j \in J} r_j$$

whenever $I \subseteq \{1, \ldots, n\}$, $J \subseteq \{1, \ldots, p\}$. Show that, for each $J \subseteq \{1, \ldots, p\}$, the family $(A_i \cap E(J): 1 \leqslant i \leqslant n)$ possesses a PT of cardinal $\Sigma\, \{r_j : j \in J\}$. Hence, by means of the result in Ex. 6.2.6, show that $\mathfrak{A}$ possesses a PT X such that $|X \cap E_j| \geqslant r_j \; (1 \leqslant j \leqslant p)$.

6.6 Marginal elements

Let $\mathfrak{A} = (A_i : i \in I)$ be a family of subsets of E, and let $M \subseteq E$. Under what circumstances does $\mathfrak{A}$ possess a transversal which contains M? A preliminary discussion of this question had been undertaken in §3.3 (see, in particular, Theorem 3.3.6). We shall now be able to gain a more illuminating view of the situation.

Consider, in the first place, a finite family $\mathfrak{A}$ of subsets of a finite set E. The transversal structure $\mathscr{E}$ of $\mathfrak{A}$ is, as we know by Theorem 6.5.2, an independence structure; and if $\mathfrak{A}$ possesses at least one transversal, then, by Lemma 6.1.1, every PT can be extended to a transversal. Theorem 3.3.6 is a trivial consequence of this observation.

When we admit infinite sets, the approach just outlined is no longer feasible since $\mathscr{E}$ need not now be an independence structure (cf. the counter-example preceding Theorem 6.5.3). We are, nevertheless, able to reach the desired conclusion.

THEOREM 6.6.1. *If $\mathfrak{A}$ is an arbitrary family of subsets of* E *which possesses a transversal, then every partial transversal of $\mathfrak{A}$ is a subset of some transversal of $\mathfrak{A}$.*

An interesting feature of this result is that (unlike Theorem 6.5.3) it involves no restriction on cardinals.

With the given family $\mathfrak{A} = (A_i : i \in I)$, we associate the deltoid $\mathscr{D} = (E, \Delta, I)$, where $(e, i) \in \Delta$ if and only if $e \in A_i$. Let F be a PT of $\mathfrak{A}$. Then F is an admissible subset of E in $\mathscr{D}$. Moreover, since $\mathfrak{A}$ possesses a transversal, I is also admissible in $\mathscr{D}$. Hence, by Theorem 2.3.1, there exists a set F* which is linked to I and such that $F \subseteq F^* \subseteq E$. In other words, F* is a transversal of $\mathfrak{A}$, and it contains F.

The next two results give transfinite analogues of Theorem 3.3.6.

THEOREM 6.6.2. *Let* $\mathfrak{A}$ *be an arbitrary family of subsets of a set* E, *and let* $M \subseteq E$. *Then* $\mathfrak{A}$ *possesses a transversal which contains* M *if and only if both the following conditions are satisfied.* (i) $\mathfrak{A}$ *possesses a transversal.* (ii) M *is a partial transversal of* $\mathfrak{A}$.

This is, of course, an immediate consequence of Theorem 6.6.1.

THEOREM 6.6.3. (Hoffman–Kuhn–Rado) *Let* $\mathfrak{A} = (A_i : i \in I)$ *be a family of finite subsets of* E; *let* $M \subseteq E$; *and suppose that no element of* M *occurs in infinitely many* A's. *Then* $\mathfrak{A}$ *possesses a transversal which contains* M *if and only if both the following conditions are satisfied.*

(i) $|A(J)| \geqslant |J|$ *for all* $J \subset\subset I$.

(ii) $|\{i \in I : A_i \cap N \neq \emptyset\}| \geqslant |N|$ *for all* $N \subset\subset M$.

Statement (i) above is simply Hall's condition, which is equivalent to the requirement that $\mathfrak{A}$ should possess a transversal (cf. Theorem 4.2.1).

Next, M is a PT of $\mathfrak{A}$ if and only if it is a PT of the restricted family $(A_i \cap M : i \in I)$ of subsets of M. By Theorem 4.2.3, this is the case if and only if

$$|\{i \in I : (A_i \cap M) \cap N \neq \emptyset\}| \geqslant |N|$$

whenever $N \subset\subset M$. Thus (ii) is equivalent to the requirement that M should be a PT of $\mathfrak{A}$. The assertion now follows by Theorem 6.6.2.

The contrast between Theorems 6.6.2 and 6.6.3 is instructive. Both results furnish criteria for the existence of a transversal containing a prescribed subset, but the conditions in Theorem 6.6.2 are 'qualitative' and the theorem involves no restriction on cardinals whereas in Theorem 6.6.3 the criterion is 'quantitative' and certain assumptions about local finiteness have to be made.

Exercises 6.6

1. State and prove a transfinite analogue of Corollary 3.3.7.

2. Let $\mathfrak{A} = (A_1, \ldots, A_n)$ be a family of subsets of E, and let $M \subseteq E$. Show that

$\mathfrak{A}$ possesses a PT which contains M as a proper subset if and only if both the inequalities

$$|\mathrm{I}| \leqslant n - |\mathrm{M}| + |\mathrm{A}(\mathrm{I}) \cap \mathrm{M}|, \qquad |\mathrm{I}| \leqslant n - |\mathrm{M}| - 1 + |\mathrm{A}(\mathrm{I})|$$

are valid for all $\mathrm{I} \subseteq \{1, \ldots, n\}$.

6.7 Axiomatic treatment of the rank function

In this concluding section of the chapter, we shall investigate certain aspects of the theory of independence without referring to the study of transversals. For simplicity, we shall suppose that all sets considered below are subsets of a finite set E.

If ρ is the rank function of some independence structure on E, then it satisfies the following conditions (the first two of which hold trivially while the third is the 'modular inequality' of Lemma 6.1.3).

R(1) $\rho(\mathrm{A}) \leqslant |\mathrm{A}| \quad (\mathrm{A} \subseteq \mathrm{E})$.

R(2) $\rho(\mathrm{A}) \leqslant \rho(\mathrm{B}) \quad (\mathrm{A} \subseteq \mathrm{B} \subseteq \mathrm{E})$.

R(3) $\rho(\mathrm{A} \cup \mathrm{B}) + \rho(\mathrm{A} \cap \mathrm{B}) \leqslant \rho(\mathrm{A}) + \rho(\mathrm{B}) \quad (\mathrm{A}, \mathrm{B} \subseteq \mathrm{E})$.

Our object is to show that, in a rather obvious sense, the converse inference is also valid.

LEMMA 6.7.1. *Let the mapping* ρ *of* $\mathscr{P}(\mathrm{E})$ *into the set of non-negative integers satisfy condition* R(3). *If* S, T *are disjoint subsets of* E *and*

$$\rho(\mathrm{S} \cup \{t\}) \leqslant \rho(\mathrm{S})$$

for all $t \in \mathrm{T}$, *then*

$$\rho(\mathrm{S} \cup \mathrm{T}) \leqslant \rho(\mathrm{S}).$$

For $|\mathrm{T}| \leqslant 1$, the assertion holds trivially. If $|\mathrm{T}| > 1$, then taking $t \in \mathrm{T}$ and using R(3) for the sets

$$\mathrm{A} = \mathrm{S} \cup (\mathrm{T} \setminus \{t\}), \qquad \mathrm{B} = \mathrm{S} \cup \{t\},$$

we obtain

$$\begin{aligned} \rho(\mathrm{S} \cup \mathrm{T}) + \rho(\mathrm{S}) &\leqslant \rho(\mathrm{S} \cup (\mathrm{T} \setminus \{t\})) + \rho(\mathrm{S} \cup \{t\}) \\ &\leqslant \rho(\mathrm{S} \cup (\mathrm{T} \setminus \{t\})) + \rho(\mathrm{S}). \end{aligned}$$

The lemma now follows by induction with respect to $|\mathrm{T}|$.

Our next theorem furnishes an axiomatic characterization of those mappings which are rank functions of independence structures.

THEOREM 6.7.2. *Let* E *be a finite set and* ρ *a mapping of* $\mathscr{P}(E)$ *into the set of non-negative integers. Then* ρ *is the rank function of some independence structure on* E *if and only if it satisfies the conditions* R(1), R(2), R(3).

That every rank function satisfies the stated conditions has already been noted. Suppose, next, that ρ satisfies R(1), R(2), and R(3), and define

$$\mathscr{E} = \{X \subseteq E: \rho(X) = |X|\}.$$

Since $\rho(A)$ is non-negative for all $A \subseteq E$, it follows by R(1) that $\rho(\emptyset) = 0$. Hence $\emptyset \in \mathscr{E}$.

We note further that, in view of R(3),

$$\rho(A \cup B) \leqslant \rho(A) + \rho(B) \qquad (A, B \subseteq E). \tag{1}$$

Now let $X \in \mathscr{E}$, $Y \subseteq X$. Then $\rho(X) = |X|$ and so, by (1) and R(1),

$$|X| = \rho(X) \leqslant \rho(Y) + \rho(X \setminus Y) \leqslant |Y| + |X \setminus Y| = |X|.$$

Hence $\rho(Y) = |Y|$, i.e. $Y \in \mathscr{E}$.

Again, let $X, Y \in \mathscr{E}$, $|Y| = |X| + 1$. Then $\rho(X) = |X|$, $\rho(Y) = |X| + 1$. Assume that

$$\rho(X \cup \{y\}) \leqslant \rho(X)$$

for all $y \in Y \setminus X$. Then, by Lemma 6.7.1 (with $S = X$, $T = Y \setminus X$), we infer that $\rho(X \cup Y) \leqslant \rho(X)$. Hence, by R(2),

$$|X| + 1 = \rho(Y) \leqslant \rho(X \cup Y) \leqslant \rho(X) = |X|;$$

and we arrive at a contradiction. It follows that

$$\rho(X \cup \{y_0\}) > \rho(X) = |X|$$

for some $y_0 \in Y \setminus X$. But, by (1) and R(1),

$$\rho(X \cup \{y_0\}) \leqslant \rho(X) + \rho(\{y_0\}) \leqslant |X| + 1$$

and so $\rho(X \cup \{y_0\}) = |X| + 1 = |X \cup \{y_0\}|$, i.e. $X \cup \{y_0\} \in \mathscr{E}$.

We have now proved that $\mathscr{E}$ is an independence structure. Denoting its rank function by σ, we have

$$\sigma(A) = \max\{|X| : X \subseteq A,\ \rho(X) = |X|\} \qquad (A \subseteq E).$$

Let $A \subseteq E$ and write $\sigma(A) = |B|$, where B is a suitable subset of A with $\rho(B) = |B|$. Then

$$\sigma(A) = |B| = \rho(B)$$

and so, by R(2),

$$\sigma(A) \leqslant \rho(A).$$

Furthermore, no subset of A of cardinal $|B| + 1$ is a member of $\mathscr{E}$, so that

$$\rho(B \cup \{x\}) \neq |B \cup \{x\}|$$

for all $x \in A \setminus B$. Hence, by R(1),

$$\rho(B \cup \{x\}) \leqslant \rho(B) \qquad (x \in A \setminus B)$$

and therefore, by Lemma 6.7.1,

$$\rho(A) \leqslant \rho(B) = \sigma(A).$$

Thus $\rho = \sigma$, i.e. ρ is the rank function of the independence structure $\mathscr{E}$.

Corollary 6.7.3. *Let $\mathscr{E}$ be a collection of subsets of a finite set* E *such that $\varnothing \in \mathscr{E}$. Then $\mathscr{E}$ is an independence structure if and only if the mapping ρ, defined by the equation*

$$\rho(A) = \max\{|X| : X \subseteq A,\ X \in \mathscr{E}\} \qquad (A \subseteq E),$$

satisfies the 'modular inequality'.

If $\mathscr{E}$ is an independence structure, then its rank function ρ certainly satisfies R(3). Suppose, on the other hand, that ρ satisfies R(3). Trivially, it also satisfies R(1) and R(2). Moreover, $\rho(A) = |A|$ if and only if $A \in \mathscr{E}$; and so

$$\mathscr{E} = \{A \subseteq E : \rho(A) = |A|\}.$$

It follows, by the proof of Theorem 6.7.2, that $\mathscr{E}$ is an independence structure.

Exercises 6.7

1. Let E be a finite set and ρ a mapping of $\mathscr{P}(E)$ into the set of non-negative integers which satisfies the 'modular inequality'. Show that, for $A, B \subseteq E$,

$$\rho(A \cup B) \leqslant \sum_{b \in B} \rho(A \cup \{b\}) - (|B| - 1)\,\rho(A).$$

2. Show that the conditions R(1), R(2), R(3) are independent.

Notes on Chapter 6

§ 6.1. The study of independence structures is the axiomatic investigation of linear independence in vector spaces. This study was initiated some thirty-five years ago. We should mention H. Whitney's pioneering paper (**1**) in which finite independence structures (there called 'matroids') were subjected to a searching analysis. Again, in the second edition of van der Waerden's *Moderne Algebra* (**3**), the theory of linear dependence and the theory of algebraic dependence were derived from a common axiomatic source. Abstract independence has by now acquired a sizeable literature and, in particular, there exist numerous studies of the relation between different sets of axioms. For example, we may refer to the work of Birkhoff (**1**), MacLane (**2**), Lazarson (**1**), Bleicher & Preston (**1**), Dlab (**1**), Ashe (**1**), and Brualdi (**6**). Very substantial contributions to the theory of independence structures have been made by W. T. Tutte (**2, 3, 4**). Further information will be found in Rado's survey article (**9**).

In addition to structural questions, some interesting quantitative problems arise in the study of independence. Thus D. J. A. Welsh (**3**) obtained a result which implies, in particular, that there are at least 2^n 'non-isomorphic' independence structures on a set of n elements.

It should, perhaps, be pointed out that the terminology of the subject is as yet far from standardized. In addition to 'independence structure' (or 'space') and 'matroid', terms such as 'incidence geometry', 'matroid lattice', 'combinatorial geometry', and 'geometric lattice' have been or are being used in the same, or in a very similar, sense.

§ 6.2. The credit for recognizing the significance for transversal theory of the study of independence structures belongs undoubtedly to R. Rado who discovered both Theorem 6.2.1 (**3**) and Theorem 6.2.4 (**5**). A good case can be made out for regarding Theorem 6.2.1 as the central result in tranversal theory; in this connection, see Mirsky's expository paper (**6**). The proof of Theorem 6.2.1 given above is based on Dr Perfect's observation that Rado's proof of Hall's theorem (i.e. the second proof of theorem 2.2.1) can be adapted to the more general case of *independent* transversals. P. Scherk (**1**) gave an ad hoc proof of Corollary 6.2.3; the combinatorial treatment offered here was suggested to me by Professor H. Tverberg. The very interesting observation embodied in Corollary 6.2.6 is due to J. H. Mason (**1**).

Rado's theorem 6.2.1 passed almost unnoticed for many years, and its dominant position in tranversal theory has emerged only very recently. There is now no shortage either of its applications or extensions. We mention, in particular, the work of R. A. Brualdi (**2, 7, 8, 9**), Hazel Perfect (**5, 7**), and D. J. A. Welsh (**1, 4, 5**). An exceptionally interesting generalization of Rado's theorem, due to Brualdi, will be discussed in § 8.4.

§ 6.3. The result in this section is based on Rado's paper (**3**). For further investigations, see D. J. A. Welsh (**9**).

§ 6.4. Theorem 6.4.1 is due to Hazel Perfect (**5**).

§ 6.5. Matrices whose elements are indeterminates and zeros were used in combinatorial investigations by W. T. Tutte (**1**) as early as 1947. The idea of a formal incidence matrix associated with a family of sets appears in Hazel Perfect's paper (**1**). Theorem 6.5.2 was discovered, in essence, by Edmonds & Fulkerson (**1**) and, independently, by Mirsky & Perfect (**2**), to whom the proofs discussed here are due. The very elegant demonstration of Hall's theorem given at the end of the section is taken from the work of Edmonds (**3**).

§ 6.6. The history of the problem of marginal elements for the finite case is referred to in the Notes on § 3.3. The transfinite version contained in Theorem 6.6.3 was communicated to me by Professor Rado in 1965. Here I follow, more or less, the treatment of Mirsky & Perfect (**2**).

§ 6.7. Theorem 6.7.2 is implicit in the work of Ingleton (**1**).

7 Independence Structures and Linear Structures

Below we continue the study of independence structures begun in the preceding chapter. In particular, we shall analyse the relation between independence structures, transversal structures, and 'linear' structures.

7.1 A hierarchy of structures

We recall that the collection of all PTs of a family $\mathfrak{A}$ of subsets of E is called the 'transversal structure of $\mathfrak{A}$'. Let, now, $\mathscr{E}$ be a collection of subsets of E; we then call $\mathscr{E}$ a *transversal structure* if there exists a family $\mathfrak{A}$ of subsets of E such that $\mathscr{E}$ is the transversal structure of $\mathfrak{A}$. (This family need not, of course, be unique.)

Again, let $\mathscr{E}$ be a non-empty collection of subsets of E; write

$$\mathrm{E}^* = \{x \in \mathrm{E} : \{x\} \in \mathscr{E}\};$$

and let D be a division ring. We call $\mathscr{E}$ a *linear structure over D* (or say that it is *linear over D*) if there exists a vector space V over D and an injective mapping $\psi : \mathrm{E}^* \to V$ such that a subset X of E* belongs to $\mathscr{E}$ precisely if $\psi(\mathrm{X})$ is a linearly independent subset of V.†

It can be shown that, if D, D' are division rings such that $D \subseteq D'$ and if $\mathscr{E}$ is linear over D, then it is also linear over D'. This result is of particular interest for us when D, D' are fields. In that case the assertion is not difficult to prove, and we shall leave the details to the reader.

Let $\mathscr{E}$ be a collection of sets. If there exists some division ring over which $\mathscr{E}$ is linear, we say that $\mathscr{E}$ is a *linear structure* (or simply that it is *linear*). A number of alternative expressions are to be found in the literature. Thus, it is said that $\mathscr{E}$ is 'linearly representable' (over D). To put the matter more loosely, $\mathscr{E}$ is a linear structure if its members can be 'identified' with certain linearly independent subsets of a vector space. It is, of course, trivial that every linear structure is an independence structure.

Our first object is to clarify the relation between transversal structures and linear structures, and we begin by considering the case of finite transversal

† The reason for considering a mapping on E* rather than on E is plain. For suppose that $x, y \in \mathrm{E} \setminus \mathrm{E}^*$, $x \neq y$. Then each of $\psi(\{x\})$, $\psi(\{y\})$ is a linearly dependent subset of V. As they are singletons, both must be equal to the zero element of V; and this is incompatible with the injective character of ψ.

structures. We recall from the discussion in § 6.5 that every such structure is an independence structure; but on re-examining the arguments leading to Lemma 6.5.1 and Theorem 6.5.2, we realize that more has, in fact, been proved.

Denote by $\mathbf{Q}$ the field of rational numbers. Let $\mathscr{E}$ be the transversal structure of the family $\mathfrak{A} = (A_1, \ldots, A_n)$ of subsets of a finite set E. We may, of course, regard $\mathfrak{A}$ as a family of subsets of $E^* = A_1 \cup \ldots \cup A_n$. Let M be the formal incidence matrix in which the rows correspond to elements of E* and the columns to sets of $\mathfrak{A}$; and denote by $z_1, \ldots, z_r$ the indeterminates appearing in M. Let $F \subseteq E^*$; then, by Lemma 6.5.1, the rows of M corresponding to F are linearly independent over the field $\mathbf{Q}(z_1, \ldots, z_r)$ (of rational functions, with rational coefficients, in the z's) if and only if F is a PT of $\mathfrak{A}$. Hence $\mathscr{E}$ is not merely an independence structure: it is actually a linear structure over the function field $\mathbf{Q}(z_1, \ldots, z_r)$. This conclusion admits of yet a further refinement, which is contained in the next theorem.

THEOREM 7.1.1. *Every finite transversal structure is linear over the field of rational numbers.*

We continue to use the notation introduced above. The product of the determinants of all non-singular square submatrices of M is a polynomial $p(z_1, \ldots, z_r)$, with coefficients in $\mathbf{Q}$, which is different from the zero polynomial. Hence there exist distinct, non-zero numbers $t_1, \ldots, t_r$ in $\mathbf{Q}$ such that

$$p(t_1, \ldots, t_r) \neq 0.$$

Denote by $\tilde{M}$ the matrix obtained when the indeterminates $z_1, \ldots, z_r$ in M are replaced by $t_1, \ldots, t_r$ respectively while the zeros are left unchanged. Then any non-singular submatrix of M is transformed into a non-singular submatrix of $\tilde{M}$ and, trivially, any singular submatrix of M is transformed into a singular submatrix of $\tilde{M}$.

Now denote by V the vector space consisting of all n-tuples with entries in $\mathbf{Q}$. Write $E^* = \{x_1, \ldots, x_m\}_{\neq}$ and let the mapping $\psi: E^* \to V$ take x_k into the k-th row of $\tilde{M}$. Since $t_1, \ldots, t_r$ are distinct, ψ is plainly injective.

If $F \subseteq E^*$, then F is a PT of $\mathfrak{A}$ precisely if the rows of M corresponding to F are linearly independent over $\mathbf{Q}(z_1, \ldots, z_r)$, and this in turn is the case if and only if the rows of $\tilde{M}$ corresponding to F are linearly independent over $\mathbf{Q}$. The mapping ψ has therefore the required properties, and the assertion is proved.

We shall next widen the scope of our inquiry by considering transversal structures not necessarily finite. We first need a preliminary result.

THEOREM 7.1.2. *Let $\mathscr{E}$ be a transversal structure of finite character on a set* E.

Then there exists a restricted family $\mathfrak{A}$ *of subsets of* E *such that* $\mathscr{E}$ *is the collection of all partial transversals of* $\mathfrak{A}$.

Let $\mathfrak{B} = (B_i : i \in I)$ be any family of subsets of E whose transversal structure is identical with $\mathscr{E}$. Denote by E_0 the set of all elements in E which belong to infinitely many B's, and write

$$\mathfrak{A}' = (B_i \cap (E \setminus E_0) : i \in I), \qquad \mathfrak{A}'' = (\{e\} : e \in E_0).$$

We suppose, as may be done without loss of generality, that $I \cap E_0 = \emptyset$; and we put $\mathfrak{A} = \mathfrak{A}' + \mathfrak{A}''$. It is plain that $\mathfrak{A}$ is a restricted family.

Further, let X be any PT of $\mathfrak{B}$ and write $X = Y \cup Z$, where $Y \subseteq E \setminus E_0$, $Z \subseteq E_0$. Then Y is a PT of $\mathfrak{A}'$ while Z is a PT of $\mathfrak{A}''$. Hence X is a PT of $\mathfrak{A}$.

Conversely, let X be a PT of $\mathfrak{A}$. Let $X^* \subset\subset X$, and write $X^* = Y^* \cup Z^*$, where $Y^* \subseteq E \setminus E_0$, $Z^* \subseteq E_0$. Now Y^* is a PT of $\mathfrak{B}$. Also, each element in Z^* belongs to infinitely many B's, and we can therefore avoid those B's which are represented by elements of Y^*. Hence $Y^* \cup Z^* = X^*$ is a PT of $\mathfrak{B}$. Thus every finite subset of X is a PT of $\mathfrak{B}$. But the set $\mathscr{E}$ of PTs of $\mathfrak{B}$ has finite character and so X itself is a PT of $\mathfrak{B}$.

We have thus shown that $\mathfrak{A}$ and $\mathfrak{B}$ have exactly the same PTs. Consequently, $\mathscr{E}$ is the set of PTs of the restricted family $\mathfrak{A}$.

We now come to the main result of the present section.

THEOREM 7.1.3. *A transversal structure of finite character is a linear structure.*

Let $\mathscr{E}$ be the given transversal structure on a set E. In view of Theorem 7.1.2, there exists a restricted family $\mathfrak{A} = (A_i : i \in I)$ which has $\mathscr{E}$ as its set of PTs. We shall find it convenient to frame the argument in terms of this family. The method to be described constitutes a natural extension of the proof of Theorem 6.5.2, which is based on the notion of a 'formal incidence matrix'.

Write $Z = \{z_{ei} : e \in E, i \in I, e \in A_i\}$, where the z's are independent indeterminates over the field of rational numbers. Denote by K the field of rational functions, with rational coefficients, in the z's (each function involving only a finite number of indeterminates). Write $E^* = \{x \in E : \{x\} \in \mathscr{E}\}$. For each $e \in E^*$, let the mapping $\psi_e : I \to K$ be defined by the equations

$$\psi_e(i) = \begin{cases} z_{ei} & (e \in A_i) \\ 0 & (e \notin A_i). \end{cases}$$

For $\alpha_1, \alpha_2 \in K$ and $e_1, e_2 \in E^*$, let the mapping $\alpha_1 \psi_{e_1} + \alpha_2 \psi_{e_2} : I \to K$ be defined by the equation

$$(\alpha_1 \psi_{e_1} + \alpha_2 \psi_{e_2})(i) = \alpha_1 \psi_{e_1}(i) + \alpha_2 \psi_{e_2}(i) \qquad (i \in I).$$

We denote by V the set of all finite linear combinations, with coefficients in K, of the mappings $\psi_e, e \in E^*$. Then V is a vector space over K; and the mapping $\psi: E^* \to V$ defined by the equation $\psi(e) = \psi_e$ $(e \in E^*)$ is injective.

Now let $F \in \mathscr{E}$, i.e. let F be a PT of $\mathfrak{A}$. We shall show that $\psi(F)$ is a linearly independent subset of V, i.e. that for each $F^* \subset\subset F$ the set $\psi(F^*)$ is linearly independent. Write $F^* = \{e_1, \ldots, e_k\}_{\neq}$. Then there exists a set $\{i_1, \ldots, i_k\}_{\neq} \subseteq I$ such that

$$e_1 \in A_{i_1}, \ldots, e_k \in A_{i_k}. \tag{1}$$

Let M denote the $k \times k$ matrix $\|\psi_{e_r}(i_s)\|$ $(1 \leqslant r, s \leqslant k)$. Each element of M is either zero or an indeterminate in Z, and the indeterminates are independent. Moreover, in view of (1), all places on the main diagonal of M are occupied by indeterminates. Hence M is non-singular. Now assume that $\psi(F^*) = \{\psi_{e_1}, \ldots, \psi_{e_k}\}$ is a linearly dependent set, i.e.

$$\alpha_1 \psi_{e_1} + \ldots + \alpha_k \psi_{e_k} = 0$$

for certain elements $\alpha_1, \ldots, \alpha_k$, not all 0, of K. Then

$$\alpha_1 \psi_{e_1}(i_s) + \ldots + \alpha_k \psi_{e_k}(i_s) = 0 \qquad (1 \leqslant s \leqslant k),$$

and the rows of M are therefore linearly dependent (over K). We thus arrive at a contradiction and conclude that $\psi(F^*)$ is a linearly independent set. Consequently, $\psi(F)$ is also linearly independent.

Next, let $G \subseteq E^*$, $G \notin \mathscr{E}$. Since $\mathscr{E}$ has finite character, there exists a set $G^* \subset\subset G$, say $G^* = \{e_1, \ldots, e_k\}_{\neq}$, such that $G^* \notin \mathscr{E}$. Now the family $\mathfrak{A}$ is restricted and there exists, therefore, a finite subset $J = \{i_1, \ldots, i_p\}_{\neq}$ of I such that

$$e_r \notin A_i \qquad (1 \leqslant r \leqslant k; \quad i \in I \setminus J).$$

Hence

$$\psi_{e_r}(i) = 0 \qquad (1 \leqslant r \leqslant k; \quad i \in I \setminus J). \tag{2}$$

Let N be the $k \times p$ matrix $\|\psi_{e_r}(i_s)\|$ $(1 \leqslant r \leqslant k,\ 1 \leqslant s \leqslant p)$. If its rows are linearly independent (over K), then $k \leqslant p$ and N possesses a non-singular $k \times k$ submatrix, say $N' = \|\psi_{e_r}(i_s)\|$ $(1 \leqslant r, s \leqslant k)$. Hence a set of 'permutation places' in N' (say the places on the main diagonal) must be occupied by indeterminates. Thus (1) is satisfied and $G^* = \{e_1, \ldots, e_k\}$ is a PT of $\mathfrak{A}$, i.e. $G^* \in \mathscr{E}$, contrary to hypothesis. It follows therefore that the rows of N are linearly dependent, i.e. there exists elements $\alpha_1, \ldots, \alpha_k$ of K, not all 0, such that

$$\alpha_1 \psi_{e_1}(i) + \ldots + \alpha_k \psi_{e_k}(i) = 0 \tag{3}$$

for all $i \in J$. Moreover, in view of (2), the relation (3) is also valid for all $i \in I \setminus J$. Thus it holds for all $i \in I$, and so

$$\alpha_1 \psi_{e_1} + \ldots + \alpha_k \psi_{e_k} = 0.$$

In other words, $\psi(G^*) = \{\psi_{e_1}, \ldots, \psi_{e_k}\}$ is a linearly dependent set; and, a fortiori, $\psi(G)$ is also linearly dependent. The proof is therefore complete.

COROLLARY 7.1.4. *The collection of partial transversals of every restricted family of sets is a linear structure.*

The collection in question has finite character by Corollary 4.2.4. The assertion therefore follows by Theorem 7.1.3.

Let us now go back to Theorem 7.1.3 and take stock of the situation. This theorem states that every transversal structure of finite character is a linear structure. Moreover, every linear structure is obviously an independence structure. (These two conclusions imply, in particular, that every transversal structure of finite character is an independence structure—a result which is also implicit in Theorem 6.5.2.) However, the two statements do not give a complete account of the situation: what they do not tell us is that the inclusions are strict. This is shown by the next theorem.

THEOREM 7.1.5. (i) *There exists a linear structure which is not a transversal structure.* (ii) *There exists an independence structure which is not a linear structure.*

To prove (i), let $E = \{1, 2, \ldots, 6\}$ and let $\mathscr{E}$ be the collection of all subsets of E of cardinal at most 2, with the exception of the sets $\{1, 2\}, \{3, 4\}, \{5, 6\}$.

Denote by V the vector space of ordered pairs of rational numbers. Then the mapping $\phi: E \to V$, defined by

$$\phi(1) = (1, 0), \qquad \phi(3) = (0, 1), \qquad \phi(5) = (1, 1),$$
$$\phi(2) = (2, 0), \qquad \phi(4) = (0, 2), \qquad \phi(6) = (2, 2),$$

is clearly injective. Moreover, if $x \in E$, then $\{\phi(x)\}$ is a linearly independent subset of V, and if $\{x, y\}_{\neq} \subseteq E$, then $\{\phi(x), \phi(y)\}$ is a linearly independent subset of V except when $\{x, y\}$ is one of the pairs $\{1, 2\}, \{3, 4\}, \{5, 6\}$. Thus $\mathscr{E}$ is a linear structure.

Next, assume that $\mathscr{E}$ is a transversal structure. Then there exists a family $\mathfrak{A} = (A_1, A_2, \ldots)$ such that $\mathscr{E}$ is the transversal structure of $\mathfrak{A}$. Write $1 \in A_1$. Since $\{1, 2\} \notin \mathscr{E}$, this implies that $2 \in A_1$. Next, $\{1, 3\} \in \mathscr{E}$ and so there exists some A_k with $k \geqslant 2$ which contains 3, say $3 \in A_2$. Since $\{3, 4\} \notin \mathscr{E}$, it follows that $4 \in A_2$. Further, $\{1, 5\} \in \mathscr{E}$ and so there exists some A_k with $k \geqslant 2$ which contains 5. If $5 \in A_k$, where $k \geqslant 3$, then $\{1, 3, 5\} \in \mathscr{E}$, which is false. Hence $5 \in A_2$; and since $\{5, 6\} \notin \mathscr{E}$, we also have $6 \in A_2$. Now $\{3, 5\} \in \mathscr{E}$ and so either 3 or 5 must belong to some A in addition to A_2; and if so, then either $\{3, 4\} \in \mathscr{E}$ or $\{5, 6\} \in \mathscr{E}$. We thus arrive at a contradiction and conclude that $\mathscr{E}$ is not a transversal structure.

To prove (ii), we take $E = \{a_1, a_2, b_1, b_2, c_1, c_2, d_1, d_2\}_{\neq}$ and let $\mathscr{E}$ consist of all subsets of E of cardinal at most 4, with the exception of the sets

$$\left.\begin{array}{lll}\{a_1, a_2, b_1, b_2\}, & \{a_1, a_2, c_1, c_2\}, & \{a_1, a_2, d_1, d_2\}, \\ \{b_1, b_2, c_1, c_2\}, & \{b_1, b_2, d_1, d_2\}. & \end{array}\right\} \quad (4)$$

To show that $\mathscr{E}$ is an independence structure, we need only verify the replacement axiom. Indeed, it suffices to show that, if $\{x_1, x_2, x_3\}_{\neq} \in \mathscr{E}$ and $\{y_1, y_2, y_3, y_4\}_{\neq} \in \mathscr{E}$, then $\{x_1, x_2, x_3, y_k\}_{\neq} \in \mathscr{E}$ for some k.

If $\{x_1, x_2, x_3\}$ is not a subset of any set in (4), then the desired conclusion obviously holds. Suppose, next, that $\{x_1, x_2, x_3\}$ is a subset of one of the sets in (4). This implies that among x_1, x_2, x_3 there occur precisely 2 of the symbols a, b, c, d.

Case 1. Among y_1, y_2, y_3, y_4 there occur at least 3 of the symbols a, b, c, d. The conclusion then follows at once.

Case 2. Among y_1, y_2, y_3, y_4 there occur only 2 of the symbols a, b, c, d. Then, since $\{y_1, y_2, y_3, y_4\} \in \mathscr{E}$, we have $\{y_1, y_2, y_3, y_4\} = \{c_1, c_2, d_1, d_2\}$; and the conclusion follows once again.

$\mathscr{E}$ is, then, an independence structure. Assume that it is linear over a division ring D. We shall denote the vector associated with an element of E by the same symbol as the original element. Elements of D will be denoted by Greek letters. Since all subsets of E of cardinal 3 belong to $\mathscr{E}$ while the five sets in (4) do not, we have the following linear relations in which all coefficients are non-zero.

$$\lambda_1 a_1 + \lambda_2 a_2 + \mu_1 b_1 + \mu_2 b_2 = 0, \quad (5)$$

$$\lambda_1' a_1 + \lambda_2' a_2 + \nu_1 c_1 + \nu_2 c_2 = 0, \quad (6)$$

$$\lambda_1'' a_1 + \lambda_2'' a_2 + \rho_1 d_1 + \rho_2 d_2 = 0, \quad (7)$$

$$\mu_1' b_1 + \mu_2' b_2 + \nu_1' c_1 + \nu_2' c_2 = 0, \quad (8)$$

$$\mu_1'' b_1 + \mu_2'' b_2 + \rho_1' d_1 + \rho_2' d_2 = 0. \quad (9)$$

From (5) and (6), we infer a relation of the form

$$(\lambda_1' - \lambda_2' \lambda_2^{-1} \lambda_1) a_1 + \sigma_1 b_1 + \sigma_2 b_2 + \nu_1 c_1 + \nu_2 c_2 = 0.$$

If $\lambda_1' - \lambda_2' \lambda_2^{-1} \lambda_1 \neq 0$, then a_1 can be expressed linearly in terms of

b_1, b_2, c_1, c_2 and, in view of (8), there is therefore a non-trivial linear relation between a_1, b_1, b_2, c_1 — a conclusion which contradicts our hypothesis that $\{a_1, b_1, b_2, c_1\} \in \mathscr{E}$. It follows that

$$\lambda_1' \lambda_1^{-1} = \lambda_2' \lambda_2^{-1}.$$

Similarly, by (5), (7), and (9), we obtain

$$\lambda_1'' \lambda_1^{-1} = \lambda_2'' \lambda_2^{-1}.$$

Hence

$$\lambda_1'' \lambda_1'^{-1} = \lambda_2'' \lambda_2'^{-1}$$

or, say, $\lambda_1'' = \kappa \lambda_1'$, $\lambda_2'' = \kappa \lambda_2'$. Hence, by (6) and (7), there exists a non-trivial linear relation between c_1, c_2, d_1, d_2; and this is contrary to our hypothesis that $\{c_1, c_2, d_1, d_2\} \in \mathscr{E}$. It therefore follows that $\mathscr{E}$ is not a linear structure.

Exercises 7.1

1. Let F, F' be fields such that $F \subseteq F'$; let E be a set; and let $\mathscr{E}$ be a collection of subsets of E which is linear over F. Show that $\mathscr{E}$ is linear over F'.

2. Let $\mathfrak{A}$ be a finite family of subsets of a finite set E, and let $\mathscr{E}$ be the set of all PTs of $\mathfrak{A}$. Further, let $\mathrm{M} \in \mathscr{E}$ and denote by $\mathscr{E}'$ the collection of all subsets Y of $\mathrm{E} \setminus \mathrm{M}$ such that $\mathrm{Y} \cup \mathrm{M} \in \mathscr{E}$. Show that $\mathscr{E}'$ (which is known to be an independence structure by Ex. 6.1.9) is not necessarily a transversal structure.

3. Let $\mathrm{E} = \{1, 2, \ldots, 7\}$ and let $\mathscr{E}$ be the collection of all subsets X of E such that $|\mathrm{X}| \leqslant 3$, with the exception of the 7 sets $\{1, 2, 4\}$, $\{1, 3, 5\}$, $\{1, 6, 7\}$, $\{2, 3, 6\}$, $\{2, 5, 7\}$, $\{3, 4, 7\}$, $\{4, 5, 6\}$. Show that $\mathscr{E}$ is linear over the field of 2 elements but is not linear over the field of rational numbers.

[H. Whitney (**1**)]

4. Let $\mathrm{E} = \{1, 2, \ldots, 10\}$ and let $\mathscr{E}$ be the collection of all subsets X of E such that $|\mathrm{X}| \leqslant 3$, with the exception of the 9 sets $\{1, 4, 10\}$, $\{2, 5, 10\}$, $\{3, 6, 10\}$, $\{2, 3, 7\}$, $\{5, 6, 7\}$, $\{1, 3, 8\}$, $\{4, 6, 8\}$, $\{1, 2, 9\}$, $\{4, 5, 9\}$. Show that $\mathscr{E}$ is an independence structure but not a linear structure.

[A. W. Ingleton (**1**)]

5. Let $\mathscr{E}$ be an independence structure on a finite set E, and suppose that the cardinal of every independent set is at most 2. Show that $\mathscr{E}$ is linear over the field of rational numbers.

6. Let E_1, E_2 be disjoint sets and $\mathscr{E}_1$, $\mathscr{E}_2$ independence structures on E_1, E_2 respectively. Suppose that $\mathscr{E}_1$ is linear over all division rings of characteristic 2 and no others, while $\mathscr{E}_2$ is linear over all division rings of characteristic 3 and no others. Show that the collection of sets

$$\{\mathrm{X}_1 \cup \mathrm{X}_2 : \mathrm{X}_1 \in \mathscr{E}_1, \mathrm{X}_2 \in \mathscr{E}_2\},$$

which is obviously an independence structure, is not a linear structure.

7.2 Bases of independence spaces

From the fact (cf. Corollary 7.1.4) that the set of PTs of a restricted family is a linear structure, several properties of this set follow at once since they reflect the corresponding properties of vector spaces. However, Corollary 7.1.4 is a comparatively difficult result, and there is an alternative approach to the question. This will be explored in the present section. Here we do not use the fact that the PTs of a restricted family constitute a linear structure but only the much easier result (Theorem 6.5.3) that they form an independence structure. We shall below investigate a number of general properties of independence structures and, in particular, apply our conclusions to transversal structures of restricted families.

We shall, throughout, denote by $(E, \mathscr{E})$ a fixed independence space.

If $x \in E$ and $A \subseteq E$, we say that x *depends on* A (in symbols: $x|A$) if either $x \in A$ or $x \notin A$ and $\{x\} \cup B \notin \mathscr{E}$ for some independent subset B of A. If $A \in \mathscr{E}$, we can reframe this definition by saying that $x|A$ means that either $x \in A$ or $x \notin A$ and $\{x\} \cup A \notin \mathscr{E}$.

If $A, B \subseteq E$ and if every $a \in A$ satisfies the relation $a|B$, then we say that A *depends on* B (in symbols: $A|B$). The negation of $x|A$ is written as $x \nmid A$, the negation of $A|B$ as $A \nmid B$.

A *basis* of an independence space $(E, \mathscr{E})$ (or, more briefly, of $\mathscr{E}$ or of E) is a maximal independent subset of E.

LEMMA 7.2.1. *If* A *is a finite subset of* E *and* B, B′ *are maximal independent subsets of* A, *then* $|B| = |B'|$.

Assume that $|B| < |B'|$ and let B^* be a subset of B' such that $|B^*| = |B| + 1$. By the replacement axiom, there exists an element $x \in B^* \setminus B \subseteq A \setminus B$ such that $B \cup \{x\}$ is independent, and this is contrary to the hypothesis that B is a maximal independent subset of A. Hence $|B'| \leqslant |B|$ and, by symmetry, $|B| \leqslant |B'|$.

One of the objects of the discussion below is to demonstrate that Lemma 7.2.1 remains valid in the case when the cardinal of A is unrestricted.

LEMMA 7.2.2. *If* $x|B$ *and* $B \subseteq A$, *then* $x|A$.

If $x \in A$, then $x|A$ holds trivially. If $x \notin A$, then $x \notin B$ and hence there exists an independent set $X \subseteq B \subseteq A$ such that $\{x\} \cup X \notin \mathscr{E}$. Thus, once again, $x|A$.

LEMMA 7.2.3. *The relation* $x|A$ *is valid if and only if* $x|B$ *for some finite subset* B *of* A.

If B is any subset of A and $x|B$, then $x|A$ by Lemma 7.2.2. Next, suppose that $x|A$. If $x \in A$, we may take $B = \{x\}$. If $x \notin A$, then there exists an independent subset C of A such that $\{x\} \cup C \notin \mathscr{E}$. Since $\mathscr{E}$ has finite character, this implies that there exists a finite subset B of C such that $\{x\} \cup B \notin \mathscr{E}$. Hence $x|B$.

LEMMA 7.2.4. *Let* B *be a finite subset of* E, *and let* $x \in E$. *If* $x|B$ *and if* D *is any maximal independent subset of* B, *then* $x|D$.

We shall assume that $x \nmid D$ and derive a contradiction.

Suppose, in the first place, that $x \in B$. Since $x \nmid D$, we have $x \notin D$ and $\{x\} \cup D \in \mathscr{E}$, and this contradicts the maximality of D in B.

Suppose, next, that $x \notin B$. Since $x|B$, there exists a set $C \subseteq B$ such that $C \in \mathscr{E}$ and $\{x\} \cup C \notin \mathscr{E}$. Let D' be a maximal independent subset of B such that $C \subseteq D'$. Then $\{x\} \cup D' \notin \mathscr{E}$. Since D, D' are maximal independent subsets of the finite set B, we have, by Lemma 7.2.1, $|D| = |D'|$. We now consider the subsets $\{x\} \cup D$ and D' of $\{x\} \cup B$. Since $x \nmid D$, we have $x \notin D$ and $\{x\} \cup D \in \mathscr{E}$. Hence, by the replacement axiom, either $\{x\} \cup D' \in \mathscr{E}$ or else $\{d\} \cup D' \in \mathscr{E}$ for some $d \in D \setminus D'$; and both alternatives lead to contradictions.

LEMMA 7.2.5. *Let* $x \in E$ *and* $A, B \subseteq E$. *If* $x|A$ *and* $A|B$, *then* $x|B$.

We shall begin by considering the case of finite A and B. Let B' be any maximal independent subset of B. Then, by Lemma 7.2.4,

$$a|B' \qquad (a \in A). \tag{1}$$

Hence B' is a maximal independent subset of $A \cup B$; for otherwise there exists an element $a_0 \in A$ such that $a_0 \notin B'$ and $\{a_0\} \cup B' \in \mathscr{E}$, and this contradicts (1). Therefore, since $x|A \cup B$ by Lemma 7.2.2, we have $x|B'$ by Lemma 7.2.4. Consequently, $x|B$.

Next, let A, B be arbitrary sets. By Lemma 7.2.3, $x|C$ for some finite subset C of A. Also, since $A|B$, we have $C|B$. Let $c \in C$. Then $c|B$ and so, again by Lemma 7.2.3, $c|B_c$ where B_c is a suitable finite subset of B. Write

$$B^* = \bigcup_{c \in C} B_c \quad (\subseteq B).$$

B^* is, of course, finite and $c|B^*$ $(c \in C)$ by Lemma 7.2.2. Hence $C|B^*$. We thus have $x|C$, $C|B^*$ and so, by the finite case already dealt with, $x|B^*$. Hence $x|B$.

LEMMA 7.2.6. *Suppose that* X, Y *are finite independent sets such that* $X|Y$. *Then* $|X| \leqslant |Y|$.

Assume that $|X| > |Y|$. Then X possesses a subset X′ such that $|X'| = |Y|+1$ and also, of course, $X'|Y$. Therefore, by the replacement axiom, there exists some element $x_0 \in X' \subseteq X$ such that $x_0 \notin Y$ and $\{x_0\} \cup Y \in \mathscr{E}$. Hence $x_0 \nmid Y$, and this contradicts the hypothesis $X|Y$.

THEOREM 7.2.7. *Let* $(E, \mathscr{E})$ *be an independence space, and let* A *be an independent subset of* E. *Then there exists a basis* B *of* E *such that* $A \subseteq B$.

This result is simply a restatement of Theorem 6.1.2.

COROLLARY 7.2.8. *Let* $(E, \mathscr{E})$ *be an independence space, and let* $X \subseteq E$. *Then* X *contains a maximal independent subset.*

We consider the independence structure $\mathscr{E}'$ consisting of all independent subsets of X. By Theorem 7.2.7, the independence space $(X, \mathscr{E}')$ possesses a basis.

THEOREM 7.2.9. *Any two bases in an independence space have the same cardinal number.*

Let B, B′ be two bases in an independence space $(E, \mathscr{E})$. In view of the maximality of B′, it is obvious that $x|B'$ for each $x \in E$. Hence, by Lemma 7.2.3, for each $b \in B$ there exists a finite subset S_b of B′ such that $b|S_b$. Write $\mathfrak{S} = (S_b : b \in B)$. Let $B^* \subset\subset B$ and put

$$S = \bigcup_{b \in B^*} S_b.$$

In view of Lemma 7.2.2, we have $B^*|S$. Hence, by Lemma 7.2.6, $|B^*| \leqslant |S|$. Thus the family $\mathfrak{S}$ (of finite subsets of B′) satisfies Hall's condition and so possesses a transversal. In other words, there exists an injection $\theta: B \to B'$ (such that $\theta(b) \in S_b$ for all $b \in B$). Therefore $|B| \leqslant |B'|$. By symmetry, we have $|B'| \leqslant |B|$; and the assertion now follows by the Schröder–Bernstein theorem 1.3.6.

COROLLARY 7.2.10. *Let* $(E, \mathscr{E})$ *be an independence space and* $A \subseteq E$. *Then any two maximal independent subsets of* A *have the same cardinal number.*

The collection of independent subsets of A is an independence structure, and the Corollary therefore follows by Theorem 7.2.9.

Corollary 7.2.10 enables us to frame a new definition of rank which, for independence structures, supersedes the cruder definition given in § 6.1 for the wider class of pre-independence structures. If $(E, \mathscr{E})$ is an independence space and $A \subseteq E$, then the rank of A can now be defined as the common cardinal of all maximal independent subsets of A. (For a finite set A, the new definition of rank coincides, of course, with that introduced earlier.) In particular, if $\mathscr{E}$ is the universal structure on E, then, for any subset (finite or infinite) A of E, the rank of A is simply $|A|$.

COROLLARY 7.2.11. *Let* $(E, \mathscr{E})$ *be an independence space. If* X *is an independent subset and* Y *a basis of* E, *then* $|X| \leqslant |Y|$.

By Theorem 7.2.7, there exists a basis Z with $X \subseteq Z$ so that $|X| \leqslant |Z|$. But, by Theorem 7.2.9, $|Z| = |Y|$; and therefore $|X| \leqslant |Y|$.

Next, we establish a result stronger than Theorem 7.2.7.

THEOREM 7.2.12. *Let* $(E, \mathscr{E})$ *be an independence space; and let* A *be an independent subset and* B *a basis of* E. *Then there exists a basis* B^* *such that* $A \subseteq B^* \subseteq A \cup B$.

For the particular case when $(E, \mathscr{E})$ is a vector space of arbitrary dimension, this result reduces to the familiar Steinitz replacement theorem.

Let $\mathscr{E}'$ denote the collection of all independent subsets of $A \cup B$. Then, by Theorem 7.2.7, the independence space $(A \cup B, \mathscr{E}')$ possesses a basis B^* such that $A \subseteq B^* \subseteq A \cup B$. Now, since B is a basis of E, $x|(A \cup B)$ for all $x \in E$. Further, since B^* is a basis of $A \cup B$, we have $(A \cup B)|B^*$. Hence, by Lemma 7.2.5, $x|B^*$ for all $x \in E$; and consequently B^* is a basis of $(E, \mathscr{E})$.

We recall (from Theorem 6.5.3) that the set of all PTs of a restricted family is an independence structure. The bases in this structure are the maximal PTs, i.e. PTs which are not proper subsets of other PTs. It is, perhaps, worth emphasizing that a transversal of the family (if one exists) is not necessarily a maximal PT. Thus, in the restricted family

$$(\{1, 2\}, \{2, 4\}, \{3, 6\}, \{4, 8\}, \ldots),$$

the transversal $\{2, 4, 6, \ldots\}$ is a proper subset of the transversal $\{1, 2, 3, \ldots\}$.

From Theorems 6.5.3, 7.2.9, and 7.2.12, we now obtain at once the following result.

THEOREM 7.2.13. *Let* $\mathfrak{A}$ *be a restricted family of subsets of* E. *Then* (i) *every partial transversal of* $\mathfrak{A}$ *is contained in some maximal partial transversal*; (ii) *two maximal partial transversals have the same cardinal*; (iii) *if* A *is a partial transversal and* B *a maximal partial transversal of* $\mathfrak{A}$, *there exists a maximal partial transversal* B^* *such that* $A \subseteq B^* \subseteq A \cup B$.

The hypothesis that $\mathfrak{A}$ should be restricted cannot be dispensed with in Theorem 7.2.13. For consider the family $\mathfrak{A} = (A_1, A_2, A_3, \ldots)$, where each A_k is the set of all real numbers. Then a set of real numbers is a PT of $\mathfrak{A}$ if and only if it is finite or denumerable. Hence $\mathfrak{A}$ possesses no maximal PT.

However, there remains this further question: if a family $\mathfrak{A}$ of sets possesses at least one maximal PT, does it follow that every PT is contained in some maximal PT? The answer is No, as is shown by the following example. Let N denote the set of all non-zero integers and let $\mathfrak{A} = (A_k : k \geqslant 1)$, where $n \in N$ satisfies $n \in A_k$ if and only if $1 \leqslant n \leqslant k$ or $n = -k$. Then $\{-1, -2, -3, \ldots\}$ is a maximal PT. Again $\{1, 2, 3, \ldots\}$ is a PT. Further, any subset X of N which contains $\{1, 2, 3, \ldots\}$ may be written in the form

$$X = \{1, 2, 3, \ldots\} \cup \{-n_1, -n_2, -n_3, \ldots\},$$

where (n_i) is a (finite or infinite) increasing sequence of positive integers. It is clear that X is a PT of $\mathfrak{A}$ if and only if there exist infinitely many positive integers different from all n_i. Hence $\{1, 2, 3, \ldots\}$ is not contained in any maximal PT.

We shall conclude this section by stating the dual results of (i) and (ii) in Theorem 7.2.13.

COROLLARY 7.2.14. *Let* $\mathfrak{A}$ *be a family of finite sets. Then* (i) *each subfamily of* $\mathfrak{A}$ *which possesses a transversal is contained in some maximal subfamily*†; (ii) *any two maximal subfamilies in* $\mathfrak{A}$ *have the same cardinal number.*

Exercises 7.2

(By a *spanning set* in an independence space is meant a set which contains a basis.)

1. Let (E, $\mathscr{E}$) be an independence space. Further, let $B \in \mathscr{E}$ and E|B. Verify that B is a basis of $\mathscr{E}$.

2. Let (E, $\mathscr{E}$) be an independence space and S a spanning set. Show that any maximal independent subset of S is a basis of E.

3. Show that a basis of an independence space is an independent spanning set, and conversely.

4. Show that a basis of an independence space is a minimal spanning set, and conversely.

5. Let (E, $\mathscr{E}$) be an independence space and $S \subseteq E$. Show that S is a spanning set if and only if E|S.

6. Let (E, $\mathscr{E}$) be an independence space, A an independent set, S a spanning set, and $A \subseteq S$. Establish the existence of a basis B such that $A \subseteq B \subseteq S$.

† This term was defined on page 34.

7. Let $(E, \mathscr{E})$ be an independence space. Let $A, B \in \mathscr{E}$ and $|A| < |B|$. Prove that there exists an independent set C such that $A \subseteq C \subseteq A \cup B$ and $|A| < |C|$.

8. Let $(E, \mathscr{E})$ be an independence space. Let $A, B \in \mathscr{E}$ and $|A| \leqslant |B|$. Use the result of Ex. 2.3.1 to show that there exists an independent set C such that $A \subseteq C \subseteq A \cup B$ and $|C| = |B|$.

9. Let $\mathscr{E}$ be an independence structure, with rank function ρ, defined on a set E. Show that

$$\rho(X) + \rho(Y) = \rho(X \cup Y) + \rho(X \cap Y)$$

for all $X, Y \subset\subset E$ if and only if there exists a set $B \subseteq E$ such that $B \in \mathscr{E}$ and $\rho(E \setminus B) = 0$.

10. Let $\mathscr{E}$ be an independence structure, with rank function ρ, on a set E; let $A_1, \ldots, A_n$ be rank-finite subsets of E and, for $1 \leqslant k \leqslant n$, let B_k be a maximal independent subset of A_k. Show that

$$\rho(A_1 \cup \ldots \cup A_n) = \rho(B_1 \cup \ldots \cup B_n).$$

7.3 Totally admissible sets

In the preceding section we were concerned with bases of independence structures and, more particularly, with maximal partial transversals of a family of sets. We shall now deal with total transversals. For greater clarity, we shall begin by formulating our results in the language of deltoids.

Let (X, Δ, Y) be a deltoid, and let $A \subseteq X$, $B \subseteq Y$. We recall (cf. §2.3) that the set A is called 'totally admissible' (or just 'total') if $A \leftrightarrow Y$.

THEOREM 7.3.1. *Let* (X, Δ, Y) *be a deltoid. If* X *possesses a total subset, then every admissible subset of* X *is contained in some total subset.*

Let A be an admissible subset of X. If X possesses a total subset, then Y is admissible. Hence, by Theorem 2.3.1 (with $X' = A$, $Y' = Y$), we infer the existence of a set A_0 such that $A \subseteq A_0 \subseteq X$ and $A_0 \leftrightarrow Y$. Thus A_0 is a total set which contains the admissible set A.

It is possible to strengthen the result just established by formulating a theorem which bears a close formal resemblance to Theorem 7.2.12.

THEOREM 7.3.2. *Let* (X, Δ, Y) *be a deltoid; and let* A, B *denote an admissible subset and a total subset, respectively, of* X. *Then there exists a total set* B^* *such that* $A \subseteq B^* \subseteq A \cup B$.

Write $\mathscr{D} = (X, \Delta, Y)$, $\mathscr{D}' = (A \cup B, \Delta', Y)$, where

$$\Delta' = \{(x, y) : (x, y) \in \Delta,\ x \in A \cup B\}.$$

Since A is an admissible subset of X in $\mathscr{D}$, it is also an admissible subset of $A \cup B$ in $\mathscr{D}'$. Again, since B is a total subset of X in $\mathscr{D}$, it is a total subset of $A \cup B$ in $\mathscr{D}'$. Hence, applying Theorem 7.3.1 to the deltoid $\mathscr{D}'$, we infer the existence of a total subset B^* of $A \cup B$ in $\mathscr{D}'$ such that $A \subseteq B^*$. Since, moreover, B^* is also total in $\mathscr{D}$, our assertion is proved.

Now let $\mathfrak{A} = (A_i : i \in I)$ be a family of subsets of E. When we apply the standard interpretation to deltoids, a 'total set' becomes a 'transversal'. The interpretation of Theorem 7.3.1 yields an already familiar result, namely Theorem 6.6.1. We state its dual.

COROLLARY 7.3.3. *Let* $\mathfrak{A}$ *be a family of subsets of* E. *If* E *is a partial transversal of* $\mathfrak{A}$ *and if the subfamily* $\mathfrak{A}'$ *possesses a transversal, then* $\mathfrak{A}'$ *is contained in some subfamily of* $\mathfrak{A}$ *which has* E *as a transversal.*

For the interpretation of Theorem 7.3.2, we have the following result.

THEOREM 7.3.4. *Let* $\mathfrak{A}$ *be a family of sets. If* A *is a partial transversal and* B *a transversal of* $\mathfrak{A}$, *then there exists a transversal* B^* *such that* $A \subseteq B^* \subseteq A \cup B$.

In spite of the obvious formal similarity between this statement and Theorem 7.2.13 (iii), the two results are quite distinct.

Exercises 7.3

1. Let (X, Δ, Y) be a deltoid. Let A be an admissible subset and B a total subset of X. Show that $|A| \leqslant |B|$.

2. Let (X, Δ, Y) be a deltoid. Use Theorem 2.3.1 and the Schröder-Bernstein theorem 1.3.6 to show that any two maximal admissible subsets of X have the same cardinal number.

3. Let $\mathfrak{A}$ be a family of subsets of E and suppose that some maximal subfamily of $\mathfrak{A}$ possesses only a finite number of tranversals. Show that there exists a subset of E which is a transversal of every maximal subfamily of $\mathfrak{A}$.

[Mirsky & Perfect (**2**)]

7.4 Set-theoretic models of independence structures

Theorem 6.5.3 describes a non-trivial model for independence structures framed in purely set-theoretic terms (as contrasted with the model derived from vector spaces, which involves algebraic notions). It is, of course, of interest to discover other models of this kind, and in the present section we shall briefly consider this problem.

THEOREM 7.4.1. (H. Perfect) *Let* $\mathscr{D} = (X, \Delta, Y)$ *be a locally right-finite deltoid, and let* $\mathfrak{Y}$ *be an independence structure on* Y. *Then the collection* $\mathfrak{X}$ *of admissible subsets of* X *which are linked to members of* $\mathfrak{Y}$ *is an independence structure on* X.

For the case when $\mathfrak{Y}$ is the universal structure on Y, this result states that, if (X, Δ, Y) is a locally right-finite deltoid, then the collection of admissible subsets of X is an independence structure. This statement is, of course, simply the deltoid form of Theorem 6.5.3.

To prove the assertion, we note that $\mathfrak{X}$ is non-empty since $\emptyset \leftrightarrow \emptyset$ and $\emptyset \in \mathfrak{Y}$. Again, axiom I(1) holds trivially. To verify the replacement axiom I(2), let A, B be finite members of $\mathfrak{X}$ such that $|B| = |A| + 1$. Then there exist admissible injections $\theta: A \to Y$ and $\psi: B \to Y$ such that $\theta(A), \psi(B) \in \mathfrak{Y}$. Let ψ be chosen arbitrarily, and let θ be then chosen in such a way that the number

$$N(\theta) = |\{z \in A \cap B : \theta(z) = \psi(z)\}|$$

is maximal. By the replacement axiom (applied to $\mathfrak{Y}$), there exists an element $b^* \in B$ such that $\psi(b^*) \notin \theta(A)$ and

$$\theta(A) \cup \{\psi(b^*)\} \in \mathfrak{Y}.$$

If $b^* \in A$, then the equations

$$\sigma(a) = \begin{cases} \theta(a) & (a \in A \setminus \{b^*\}) \\ \psi(a) & (a = b^*) \end{cases}$$

define an admissible injection $\sigma: A \to Y$ such that $\sigma(A) \in \mathfrak{Y}$. Moreover, $N(\sigma) > N(\theta)$ since $\sigma(b^*) = \psi(b^*)$ whereas $\theta(b^*) \neq \psi(b^*)$. This contradicts the maximality of $N(\theta)$; and we conclude that, in fact, $b^* \notin A$. Now $(a, \theta(a)) \in \Delta$ $(a \in A)$, $(b^*, \psi(b^*)) \in \Delta$, and $\psi(b^*) \notin \theta(A)$; and it follows that $A \cup \{b^*\} \in \mathfrak{X}$. Hence I(2) is valid. It will have been noted that so far neither the right-finiteness of $\mathscr{D}$ nor the finite character of $\mathfrak{Y}$ has been used in the argument.

It remains to show that $\mathfrak{X}$ has finite character. Let $A \subseteq X$ and suppose that every finite subset of A belongs to $\mathfrak{X}$. We shall verify that A itself belongs to $\mathfrak{X}$. Let $B \subset\subset A$, so that $B \in \mathfrak{X}$. Then there exists an admissible injection $\theta_B: B \to Y$ such that $\theta_B(B) \in \mathfrak{Y}$. In other words, the subfamily $\mathfrak{A}(B) = (\Delta(x): x \in B)$ of $\mathfrak{A} = (\Delta(x): x \in A)$ possesses an injective choice function θ_B such that $\theta_B(B) \in \mathfrak{Y}$. Denote by θ a corresponding Rado choice function of $\mathfrak{A}$ (whose existence is guaranteed by Theorem 4.1.1). In view of Corollary 4.1.2, θ is again injective. Also $\theta(x) \in \Delta(x)$ $(x \in A)$, i.e. $(x, \theta(x)) \in \Delta$ $(x \in A)$, and this means that $\theta: A \to Y$ is an admissible injection. Now any finite subset of $\theta(A)$

can be expressed in the form $\theta(B)$, where $B \subset\subset A$. There exists a set C such that $B \subseteq C \subset\subset A$ and $\theta|B = \theta_C|B$. But $\theta_C(C) \in \mathfrak{Y}$ and, a fortiori, $\theta(B) = \theta_C(B) \in \mathfrak{Y}$. Thus every finite subset of $\theta(A)$ is a member of $\mathfrak{Y}$, and so $\theta(A)$ is also a member of $\mathfrak{Y}$. It follows that $A \in \mathfrak{X}$. The proof is now complete.

We know, then, that the collection $\mathfrak{X}$ in Theorem 7.4.1 is an independence structure. It is easy to determine its rank function. For the sake of simplicity, we shall now work with finite sets.

COROLLARY 7.4.2. *Let* (X, Δ, Y) *be a finite deltoid and let* $\mathfrak{Y}$ *be an independence structure, with rank function* σ, *defined on* Y. *Denote by* ρ *the rank function of the independence structure* $\mathfrak{X}$ *of admissible subsets of* X *which are linked to members of* $\mathfrak{Y}$. *Then, for every* $A \subseteq X$,

$$\rho(A) = \min_{B \subseteq A} \{\sigma(\Delta(B)) + |A \setminus B|\}.$$

Let k be a natural number. The statement $\rho(A) \geqslant k$ means that A contains an admissible subset of cardinal k which is linked to a member of $\mathfrak{Y}$. Now, by the deltoid form of Theorem 6.2.2, this is the case precisely if

$$\sigma(\Delta(B)) + |A \setminus B| \geqslant k$$

for all $B \subseteq A$. The assertion therefore follows.

We conclude with an interpretation of Theorem 7.4.1.

THEOREM 7.4.3. *Let* $\mathfrak{A} = (A_i : i \in I)$, $\mathfrak{B} = (B_j : j \in J)$ *be families of subsets of* E, *and suppose that all sets in* $\mathfrak{A}$ *are finite while* $\mathfrak{B}$ *is restricted. Then the collection of all subsets* I^* *of* I *such that* $\mathfrak{A}(I^*)$ *possesses a transversal which at the same time is a partial transversal of* $\mathfrak{B}$† *is an independence structure on* I.

We consider the deltoid (I, Δ, E), where $(i, e) \in \Delta$ precisely if $e \in A_i$. Since all A_i are finite, this deltoid is locally right-finite. The collection $\mathscr{E}$ of all PTs of $\mathfrak{B}$ is, by Theorem 6.5.3, an independence structure on E. Now a subset I^* of I is admissible if and only if $\mathfrak{A}(I^*)$ has a transversal, say E^*. Further, $E^* \in \mathscr{E}$ precisely if E^* is a PT of $\mathfrak{B}$. Thus I^* is linked to a member of $\mathscr{E}$ precisely if $\mathfrak{A}(I^*)$ has a transversal which is also a PT of $\mathfrak{B}$. The collection of all such I^* is, by Theorem 7.4.1, an independence structure.

† In the brisker terminology of § 9.1, this phrase would read as 'the collection of all subsets I^* of I such that $\mathfrak{A}(I^*)$ has a common transversal with a subfamily of $\mathfrak{B}$'.

Exercises 7.4

1. Let (E, $\mathscr{E}$) be a finite independence space and $\mathfrak{A}$ a finite family of subsets of E. Show that the collection of independent PTs of $\mathfrak{A}$ is not, in general, an independence structure.

2. Let $\mathscr{E}$ be an independence structure on a set E, and let $\mathfrak{A} = (A_i : i \in I)$ be a family of finite subsets of E. Show that the collection of all subsets I* of I such that $\mathfrak{A}(I^*)$ possesses an independent transversal is an independence structure on I.

3. $\mathfrak{A} = (A_i : i \in I)$ is a family of finite subsets of E, $\mathfrak{B} = (B_j : j \in J)$ a restricted family of subsets of E, and $\mathscr{J}$ the collection of all those subsets I* of I for which $\mathfrak{A}(I^*)$ has a transversal which is, at the same time, a PT of $\mathfrak{B}$. Use Theorem 4.2.3 and Rado's selection principle (Theorem 4.1.1) to show that $\mathscr{J}$ has finite character.

Notes on Chapter 7

§ 7.1. Theorem 7.1.1 is due to P. Vámos, who has in addition shown that every transversal structure of finite character is linear over any field of sufficiently large cardinal—a result which was also discovered independently by Piff & Welsh (**1**). I owe to Dr Hazel Perfect the proof of Theorem 7.1.2. The principal result of § 7.1, namely Theorem 7.1.3, is taken from paper (**2**) of Mirsky & Perfect. The second example in the proof of Theorem 7.1.5 is due to Vámos, but earlier models of independence structures which are not linear were exhibited by MacLane (**1**), Lazarson (**2**), and Ingleton (**1**). Vámos's construction is based on a set of 8 elements, and the number 8 is best possible since J.-C. Fournier (**1**) has shown that every independence structure on a set of 7 elements is linear. Earlier, R. Rado had proved that every independence structure on a set of 6 elements is linear over **Q**. Professor R. A. Brualdi noted that the first example in the proof of Theorem 7.1.5 is minimal in the sense that every independence structure on a set of 5 elements is transversal.

The problem of linearity of independence structures, formulated (at any rate implicitly) by H. Whitney (**1**) in 1935, may be stated thus: what conditions are necessary and sufficient for an independence structure to be linear (or linear over a specified division ring)? This problem has attracted a good deal of attention and we may refer, in particular, to the work of Fournier (**1**), Ingleton (**2**), Kertész (**1**), Lazarson (**1**, **2**), Rado (**7**, **9**), and Tutte (**5**). For linearity over *fields*, a complete solution (algorithmic in the case of finite structures) was given recently by P. Vámos (**1**). However, the general problem involving division rings has so far defied all comers.

The question of the necessary and sufficient conditions for an independence structure to be a transversal structure is, unlike Whitney's problem, purely set-theoretic and consequently more tractable. One criterion will be found in § 8.3.

§ 7.2. The core of this section is Theorem 7.2.9, which was originally proved by Rado (**5**). The approach here is somewhat different in detail and is based on an adaptation of M. Hall's proof of the theorem for the special case of vector spaces (cf. § 4.2). It is possible to develop the topic in the context of purely algebraic ideas (see e.g. P. M. Cohn (**1**), chap. 7, § 2); but, in a book devoted to combinatorial analysis, the treatment given here seems preferable. A very recent result due to R. A. Brualdi (**6**), of which Theorem 7.2.9 is an immediate consequence, may be noted: for any two bases B_1, B_2 of an independence structure, there exists an injection $\sigma: B_1 \to B_2$ such that $(B_2 \setminus \{\sigma(x)\}) \cup \{x\}$ is a basis for each $x \in B_1$.

Theorem 7.2.9 falls into a general category of results which assert that any two 'distinguished' subsets of a given system have the same cardinal numt er. Another result of this kind is a theorem due to Robertson & Weston (**1**). Dr J. S. Pym has informed me that this theorem can be derived from M. Hall's theoren 4.2.1, although this approach does not seem to shorten the argument. There are other results of this type, such as those due to Kertész (**1**) and P. J. Chase (**1**). I do not know how much light Hall's theorem may ultimately throw on questions such as these, but the topic seems certainly to merit further scrutiny.

§ 7.3. The results discussed here are taken from Mirsky & Perfect's paper (**2**).

§ 7.4. The powerful Theorem 7.4.1 is due to Hazel Perfect (**3**); it was also found independently by R. A. Brualdi (**1**). Hazel Perfect's very interesting work contains a wealth of realizations of independence structures, and will repay close study.

8 The Rank Formula of Nash-Williams

In the present chapter we describe a procedure for combining several independence structures into a new independence structure. We determine the rank function of the new structure, and we exhibit a number of consequences of our rank formula. For the sake of simplicity we shall, unless the contrary is stated, operate throughout with finite sets.

8.1 Sums of independence structures

Let $\mathscr{E}_1, \ldots, \mathscr{E}_k$ be k independence structures defined on a finite set E. We shall write

$$\mathscr{E}_1 + \ldots + \mathscr{E}_k = \{X_1 \cup \ldots \cup X_k : X_1 \in \mathscr{E}_1, \ldots, X_k \in \mathscr{E}_k\}$$

and shall refer to $\mathscr{E}_1 + \ldots + \mathscr{E}_k$ as the *sum* of $\mathscr{E}_1, \ldots, \mathscr{E}_k$.

THEOREM 8.1.1. *The sum of a (finite) number of independence structures is again an independence structure.*

It is sufficient to establish the assertion for the sum of two independence structures; it will then follow for k structures by trivial induction. Let, then, $\mathscr{E}_1, \mathscr{E}_2$ be independence structures on a finite set E. To show that $\mathscr{E}_1 + \mathscr{E}_2$ is also an independence structure, we need only verify the replacement property.

Let $N, N' \in \mathscr{E}_1 + \mathscr{E}_2$, $|N'| = |N| + 1$. Write

$$N = X_1 \cup X_2, \qquad N' = Y_1 \cup Y_2,$$

where $X_1, Y_1 \in \mathscr{E}_1$; $X_2, Y_2 \in \mathscr{E}_2$; $X_1 \cap X_2 = \emptyset$; $Y_1 \cap Y_2 = \emptyset$. Moreover, we choose X_1, X_2, Y_1, Y_2 in such a way that

$$|X_1 \cap Y_2| + |X_2 \cap Y_1| \tag{1}$$

is a minimum. Now

$$|X_1| + |X_2| = |N| < |N'| = |Y_1| + |Y_2|$$

and so $|X_1| < |Y_1|$ or $|X_2| < |Y_2|$; say, for definiteness, that $|X_1| < |Y_1|$. Since $X_1, Y_1 \in \mathscr{E}_1$, there exists therefore an element $y \in Y_1 \setminus X_1$ such that $X_1 \cup \{y\} \in \mathscr{E}_1$.

Assume that $y \in N$, so that $y \in X_2$; and consider the representation

$$N = X_1^* \cup X_2^*,$$

where $X_1^* = X_1 \cup \{y\} \in \mathscr{E}_1$, $X_2^* = X_2 \setminus \{y\} \in \mathscr{E}_2$. Then

$$|X_1^* \cap Y_2| = |X_1 \cap Y_2|, \qquad |X_2^* \cap Y_1| = |X_2 \cap Y_1| - 1$$

and so

$$|X_1^* \cap Y_2| + |X_2^* \cap Y_1| = |X_1 \cap Y_2| + |X_2 \cap Y_1| - 1,$$

contrary to the minimality of (1). It follows, therefore, that $y \notin N$, i.e. $y \in N' \setminus N$. Furthermore

$$N \cup \{y\} = (X_1 \cup \{y\}) \cup X_2 \in \mathscr{E}_1 + \mathscr{E}_2.$$

The replacement property is therefore established and the theorem is proved.

The next question suggests itself immediately: it is to determine the rank function of the sum of several independence structures in terms of the rank functions of the summands.

THEOREM 8.1.2. (Nash-Williams) *Let $\mathscr{E}_1, \ldots, \mathscr{E}_k$ be independence structures defined on a finite set* E. *If $\rho_1, \ldots, \rho_k, \rho$ denote the rank functions of $\mathscr{E}_1, \ldots, \mathscr{E}_k$, $\mathscr{E}_1 + \ldots + \mathscr{E}_k$ respectively, then, for each* $F \subseteq E$, *we have*

$$\rho(F) = \min_{X \subseteq F} \{\rho_1(X) + \ldots + \rho_k(X) + |F \setminus X|\}.$$

One half of this rank formula is almost trivial. For, when $X \subseteq F \subseteq E$, we have

$$\rho(F) \leqslant \rho(X) + \rho(F \setminus X) \leqslant \rho(X) + |F \setminus X|.$$

Let $\rho(X) = |Y|$, where $Y \subseteq X$, $Y \in \mathscr{E}_1 + \ldots + \mathscr{E}_k$. Hence $Y = Y_1 \cup \ldots \cup Y_k$, where $Y_1, \ldots, Y_k$ are pairwise disjoint sets which are members of $\mathscr{E}_1, \ldots, \mathscr{E}_k$ respectively. It follows that

$$\rho(X) = |Y_1| + \ldots + |Y_k| \leqslant \rho_1(X) + \ldots + \rho_k(X)$$

and so

$$\rho(F) \leqslant \rho_1(X) + \ldots + \rho_k(X) + |F \setminus X| \qquad (X \subseteq F).$$

To complete the proof, we shall need to establish the harder inequality

$$\rho(F) \geqslant \min_{X \subseteq F} \{\rho_1(X) + \ldots + \rho_k(X) + |F \setminus X|\} \qquad (F \subseteq E).$$

We shall prove this for the case $k = 2$: the general case is dealt with in precisely the same way but involves a more complicated notation. For brevity, we shall write

$$\sigma(\mathrm{F}) = \min_{\mathrm{X} \subseteq \mathrm{F}} \{\rho_1(\mathrm{X}) + \rho_2(\mathrm{X}) + |\mathrm{F} \setminus \mathrm{X}|\} \qquad (\mathrm{F} \subseteq \mathrm{E}).$$

Our aim, then, is to prove that $\rho(\mathrm{F}) \geqslant \sigma(\mathrm{F})$.

Some additional notation will be required. Let E′, E″ be sets such that E, E′, E″ are pairwise disjoint and $|\mathrm{E}| = |\mathrm{E}'| = |\mathrm{E}''|$. We shall make use of the process of 'substitution' listed among the elementary constructions in § 3.1. Write

$$\mathrm{E} = \{x_1, \ldots, x_m\}_{\neq}, \qquad \mathrm{E}' = \{x_1', \ldots, x_m'\}_{\neq}, \qquad \mathrm{E}'' = \{x_1'', \ldots, x_m''\}_{\neq}.$$

Denote by $\mathscr{E}_1'$ the independence structure on E′ which is isomorphic, in an obvious sense, to $\mathscr{E}_1$; and let ρ_1' be the rank function of $\mathscr{E}_1'$. Similarly, define the independence structure $\mathscr{E}_2''$, with rank function ρ_2'', on E″. Further, denote by ρ^* the rank function of the independence structure $\mathscr{E}_1' + \mathscr{E}_2''$ on $\mathrm{E}' \cup \mathrm{E}''$. It is then plain that, for $\mathrm{X} \subseteq \mathrm{E}' \cup \mathrm{E}''$,

$$\rho^*(\mathrm{X}) = \rho_1'(\mathrm{X} \cap \mathrm{E}') + \rho_2''(\mathrm{X} \cap \mathrm{E}''). \tag{2}$$

Again, let $\mathrm{A}_k = \{x_k', x_k''\}$ $(1 \leqslant k \leqslant m)$, so that $\mathfrak{A} = (\mathrm{A}_1, \ldots, \mathrm{A}_m)$ is a family of subsets of $\mathrm{E}' \cup \mathrm{E}''$.

Let $\mathrm{F} \subseteq \mathrm{E}$ and let k be an integer such that $\sigma(\mathrm{F}) \geqslant k$. Then

$$\rho_1(\mathrm{X}) + \rho_2(\mathrm{X}) + |\mathrm{F} \setminus \mathrm{X}| \geqslant k \qquad (\mathrm{X} \subseteq \mathrm{F}).$$

Write $\mathrm{F} = \{x_1, \ldots, x_p\}$ $(p \leqslant m)$, $\mathrm{X} = \{x_{j_1}, \ldots, x_{j_s}\}$, where

$$1 \leqslant s \leqslant p, \qquad 1 \leqslant j_1 < \ldots < j_s \leqslant p. \tag{3}$$

Then

$$\rho_1(\{x_{j_1}, \ldots, x_{j_s}\}) + \rho_2(\{x_{j_1}, \ldots, x_{j_s}\}) \geqslant s + k - p,$$

i.e.

$$\rho_1'(\{x_{j_1}', \ldots, x_{j_s}'\}) + \rho_2''(\{x_{j_1}'', \ldots, x_{j_s}''\}) \geqslant s + k - p$$

whenever (3) holds. Hence, by (2), we have

$$\rho^*(\mathrm{A}_{j_1} \cup \ldots \cup \mathrm{A}_{j_s}) \geqslant s + k - p$$

whenever (3) holds. By the defect form of Rado's theorem on independent transversals (Theorem 6.2.2), it follows that the family $(\mathrm{A}_1, \ldots, \mathrm{A}_p)$ possesses a PT, say Z, of cardinal k which is a member of $\mathscr{E}_1' + \mathscr{E}_2''$. Write

$$\mathrm{Z} = \{x_{a_1}', \ldots, x_{a_\lambda}', x_{b_1}'', \ldots, x_{b_\mu}''\},$$

where $\lambda + \mu = k$, $\{a_1, \dots, a_\lambda, b_1, \dots, b_\mu\}_{\neq} \subseteq \{1, \dots, p\}$. Then

$$\begin{aligned} k = \rho^*(Z) &= \rho_1'(\{x_{a_1}', \dots, x_{a_\lambda}'\}) + \rho_2''(\{x_{b_1}'', \dots, x_{b_\mu}''\}) \\ &= \rho_1(\{x_{a_1}, \dots, x_{a_\lambda}\}) + \rho_2(\{x_{b_1}, \dots, x_{b_\mu}\}) \\ &= \rho_1(X_1) + \rho_2(X_2), \end{aligned}$$

say. Here $X_1 \cap X_2 = \emptyset$ and $X_1 \cup X_2 \subseteq F$. Let

$$\rho_1(X_1) = |Y_1|, \qquad \rho_2(X_2) = |Y_2|,$$

where $Y_1 \subseteq X_1$, $Y_1 \in \mathscr{E}_1$, $Y_2 \subseteq X_2$, $Y_2 \in \mathscr{E}_2$, so that $Y_1 \cup Y_2 \in \mathscr{E}_1 + \mathscr{E}_2$. Then

$$\rho_1(X_1) + \rho_2(X_2) = |Y_1 \cup Y_2| \leqslant \rho(X_1 \cup X_2) \leqslant \rho(F),$$

and so $\rho(F) \geqslant k$. We have therefore shown that $\sigma(F) \geqslant k$ implies $\rho(F) \geqslant k$. This establishes the inequality $\rho(F) \geqslant \sigma(F)$ for all $F \subseteq E$. The proof of the theorem is now complete.

Exercises 8.1

1. Let $\mathscr{E}_1, \dots, \mathscr{E}_k$ be independence structures on a set E. Show that any basis of $\mathscr{E}_1 + \dots + \mathscr{E}_k$ can be expressed in the form $B_1 \cup \dots \cup B_k$, where $B_1, \dots, B_k$ are bases of $\mathscr{E}_1, \dots, \mathscr{E}_k$ respectively.

2. Let $\mathscr{E}_1, \dots, \mathscr{E}_k$ be independence structures on a finite set E, and let $B_1, \dots, B_k$ be pairwise disjoint bases of $\mathscr{E}_1, \dots, \mathscr{E}_k$ respectively. Show that $B_1 \cup \dots \cup B_k$ is then a basis of $\mathscr{E}_1 + \dots + \mathscr{E}_k$. Also show that this conclusion need not be valid if the qualification 'pairwise disjoint' is omitted.

3. Let $\mathscr{E}_1, \mathscr{E}_2$ be independence structures on a finite set E and denote by ρ_1, ρ_2, ρ the rank functions of $\mathscr{E}_1, \mathscr{E}_2, \mathscr{E}_1 + \mathscr{E}_2$ respectively. Show that, for $F \subseteq E$,

$$\rho(F) = \max \{\rho_1(A_1) + \rho_2(A_2) : A_1 \cup A_2 = F, A_1 \cap A_2 = \emptyset\}.$$

4. Let $\mathscr{E}_1, \dots, \mathscr{E}_k$ be independence structures on a finite set E, and denote by $\rho_1, \dots, \rho_k, \rho$ the rank functions of $\mathscr{E}_1, \dots, \mathscr{E}_k, \mathscr{E}_1 + \dots + \mathscr{E}_k$ respectively. Establish the equivalence of the following statements. (a) No singleton is independent in more than one of $\mathscr{E}_1, \dots, \mathscr{E}_k$. (b) $\rho = \rho_1 + \dots + \rho_k$. (c) For $G \subseteq F \subseteq E$,

$$\rho_1(G) + \dots + \rho_k(G) - |G| \geqslant \rho_1(F) + \dots + \rho_k(F) - |F|.$$

5. Let $(E_1, \dots, E_p)$ be a partition of a finite set E, let $s_1, \dots, s_p$ be non-negative integers, and denote by $\mathscr{E}$ the collection of all subsets X of E which satisfy the relations

$$|X \cap E_j| \leqslant s_j \qquad (1 \leqslant j \leqslant p).$$

Show that $\mathscr{E}$ is an independence structure and that its rank function ρ is given by the equation

$$\rho(F) = \sum_{j=1}^{p} \min(|F \cap E_j|, s_j) \qquad (F \subseteq E).$$

Show further that a family $\mathfrak{A} = (A_1, \ldots, A_n)$ of subsets of E possesses a transversal which is a member of $\mathscr{E}$ if and only if

$$|A(I) \cap E(J)| \geqslant |I| - \sum_{j \notin J} s_j$$

whenever $I \subseteq \{1, \ldots, n\}$, $J \subseteq \{1, \ldots, p\}$.

6. Let $\mathfrak{A} = (A_1, \ldots, A_n)$ be a family of subsets of a finite set E. Let $\mathscr{E}_1, \ldots, \mathscr{E}_k$ be independence structures on E, with rank functions $\rho_1, \ldots, \rho_k$ respectively. Show that $\mathfrak{A}$ possesses a transversal which can be expressed in the form $X_1 \cup \ldots \cup X_k$, where $X_1 \in \mathscr{E}_1, \ldots, X_k \in \mathscr{E}_k$ if and only if

$$\rho_1(F) + \ldots + \rho_k(F) + |A(I)| \geqslant |I| + |F|$$

whenever $I \subseteq \{1, \ldots, n\}$, $F \subseteq A(I)$.

7. Obtain a variant of the proofs of Theorems 8.1.1 and 8.1.2 by applying Theorem 7.4.1 and Corollary 7.4.2 to the deltoid $(E, \Delta, E' \cup E'')$, where

$$\Delta = \{(x_i, x_i'): 1 \leqslant i \leqslant m\} \cup \{(x_i, x_i''): 1 \leqslant i \leqslant m\}.$$

8. Let E, I be arbitrary sets; for each $i \in I$, let $\mathscr{E}_i$ be a collection of subsets of E; and write

$$\mathscr{E} = \left\{ \bigcup_{i \in I} X_i : X_i \in \mathscr{E}_i \text{ for } i \in I \right\}.$$

(i) Show that, if all $\mathscr{E}_i$ are pre-independence structures on E, then $\mathscr{E}$ is also a pre-independence structure on E. (ii) Show, further, that the inference in (i) becomes false if the term 'pre-independence' is replaced by 'independence'.

9. Let $\mathscr{E}_1$ and $\mathscr{E}_2$ be independence structures on an arbitrary set E. Show (e.g. by Rado's selection principle) that $\mathscr{E}_1 + \mathscr{E}_2$, defined as the collection

$$\{X_1 \cup X_2 : X_1 \in \mathscr{E}_1, X_2 \in \mathscr{E}_2\},$$

is again an independence structure. (Cf. the proof of Theorem 4.5.2.)

[Pym & Perfect (**1**)]

8.2 Disjoint independent sets

The remainder of this chapter is largely devoted to applications of the rank formula. We begin with a problem which originated in the following question: under what circumstances can a given subset of a vector space be partitioned into k linearly independent parts? Here we shall treat the more general case of abstract independence.

THEOREM 8.2.1. (Edmonds & Fulkerson) *Let $\mathscr{E}_1, \ldots, \mathscr{E}_k$ be independence structures, with rank functions $\rho_1, \ldots, \rho_k$ respectively, on a finite set* E. *A subset* F *of* E *can then be expressed in the form $F = F_1 \cup \ldots \cup F_k$, where $F_i \in \mathscr{E}_i$ $(1 \leqslant i \leqslant k)$ if and only if*

$$|X| \leqslant \rho_1(X) + \ldots + \rho_k(X)$$

for all $X \subseteq F$.

Denote by ρ the rank function of the independence structure $\mathscr{E}_1 + \ldots + \mathscr{E}_k = \mathscr{E}$ (say). Then F ($\subseteq$ E) can be expressed in the desired form if and only if $F \in \mathscr{E}$, i.e. $\rho(F) = |F|$. In view of Theorem 8.1.2, this requirement is equivalent to

$$\min_{X \subseteq F} \{\rho_1(X) + \ldots + \rho_k(X) + |F \setminus X|\} = |F|.$$

The assertion is therefore proved.

Taking the special case $\mathscr{E}_1 = \ldots = \mathscr{E}_k$, we at once obtain the following consequence.

COROLLARY 8.2.2. (Edmonds) *Let $\mathscr{E}$ be an independence structure, with rank function ρ, on a finite set* E. *A set* $F \subseteq E$ *can be partitioned into k independent subsets if and only if, for all* $X \subseteq F$,

$$|X| \leqslant k.\,\rho(X). \tag{1}$$

It is easy to obtain a transfinite generalization of this result.

COROLLARY 8.2.3. *Let $\mathscr{E}$ be an independence structure, with rank function ρ, on an arbitrary set* E. *A set* $F \subseteq E$ *can then be partitioned into k independent subsets if and only if the inequality* (1) *holds for every finite subset* X *of* F.

This result follows immediately from Corollary 8.2.2 and Theorem 4.5.2 on $(\mathscr{E}, k)$-divisibility. We note, in passing, that Theorem 3.3.2 and its transfinite analogue and also Theorem 4.5.3 are easy consequences of Corollary 8.2.3.

THEOREM 8.2.4. (Edmonds & Fulkerson) *Let $\mathscr{E}_1, \ldots, \mathscr{E}_k$ be independence structures, with rank functions $\rho_1, \ldots, \rho_k$ respectively, on a finite set* E; *and let $s_1, \ldots, s_k$ be non-negative integers. Then there exist pairwise disjoint sets $X_1, \ldots, X_k$ with $X_i \in \mathscr{E}_i$, $|X_i| = s_i$ $(1 \leqslant i \leqslant k)$ if and only if*

$$\sum_{i=1}^{k} (s_i - \rho_i(A))^+ \leqslant |E \setminus A| \text{ for all } A \subseteq E. \tag{2}$$

For any independence structure $\mathscr{F}$ on E and any non-negative integer s, we define

$$\mathscr{F}^{(s)} = \{X \in \mathscr{F} : |X| \leqslant s\}.$$

It is clear (cf. Ex. 6.1.2) that $\mathscr{F}^{(s)}$ is again an independence structure. If σ is the rank function of $\mathscr{F}$, the rank function of $\mathscr{F}^{(s)}$ will be denoted by $\sigma^{(s)}$. We obviously have

$$\sigma^{(s)}(\mathrm{A}) = \min\{\sigma(\mathrm{A}), s\} \qquad (\mathrm{A} \subseteq \mathrm{E}). \tag{3}$$

Next, we note that, for any real number a_i, b_i $(1 \leqslant i \leqslant k)$,

$$\sum_{i=1}^{k} (a_i - b_i)^+ = \sum_{i=1}^{k} a_i - \sum_{i=1}^{k} \min(a_i, b_i).$$

Hence (2) may be rewritten in the form

$$\sum_{i=1}^{k} s_i \leqslant \sum_{i=1}^{k} \min\{\rho_i(\mathrm{A}), s_i\} + |\mathrm{E} \setminus \mathrm{A}| \qquad (\mathrm{A} \subseteq \mathrm{E}).$$

In view of (3), this is equivalent to

$$\sum_{i=1}^{k} s_i \leqslant \sum_{i=1}^{k} \rho_i^{(s_i)}(\mathrm{A}) + |\mathrm{E} \setminus \mathrm{A}| \qquad (\mathrm{A} \subseteq \mathrm{E}),$$

where $\rho_i^{(s_i)}$ is, of course, the rank function of the independence structure $\mathscr{E}_i^{(s_i)}$. By the Nash-Williams rank formula (Theorem 8.1.2), it now follows that condition (2) is equivalent to

$$\sum_{i=1}^{k} s_i \leqslant \tilde{\rho}(\mathrm{E}), \tag{4}$$

where $\tilde{\rho}$ denotes the rank function of the independence structure

$$\tilde{\mathscr{E}} = \mathscr{E}_1^{(s_1)} + \ldots + \mathscr{E}_k^{(s_k)}.$$

If there exist sets $\mathrm{X}_1, \ldots, \mathrm{X}_k$ with the properties specified in the statement of the theorem, then

$$\mathrm{X}_1 \cup \ldots \cup \mathrm{X}_k \in \tilde{\mathscr{E}}, \qquad |\mathrm{X}_1 \cup \ldots \cup \mathrm{X}_k| = \sum_{i=1}^{k} s_i.$$

Hence (4), and so (2), is valid.

Conversely, let (2), and so (4), be satisfied. Then there exist sets $\mathrm{X}_1, \ldots, \mathrm{X}_k$ such that

$$\mathrm{X}_1 \in \mathscr{E}_1^{(s_1)}, \ldots, \mathrm{X}_k \in \mathscr{E}_k^{(s_k)}, \qquad |\mathrm{X}_1 \cup \ldots \cup \mathrm{X}_k| = \sum_{i=1}^{k} s_i.$$

Since $|\mathrm{X}_1| \leqslant s_1, \ldots, |\mathrm{X}_k| \leqslant s_k$, it follows that $\mathrm{X}_1, \ldots, \mathrm{X}_k$ are pairwise disjoint and $|\mathrm{X}_1| = s_1, \ldots, |\mathrm{X}_k| = s_k$. Moreover $\mathrm{X}_i \in \mathscr{E}_i$ $(1 \leqslant i \leqslant k)$, and the proof is complete.

We refer briefly to two specializations of the result just established. Let $\mathfrak{A} = (A_1, ..., A_n)$ be a family of subsets of E, and let each of $\mathscr{E}_1, ..., \mathscr{E}_k$ be identical with the set of all PTs of $\mathfrak{A}$. (This set is an independence structure by Theorem 6.5.2.) Then Theorem 8.2.4 yields necessary and sufficient conditions for the existence of pairwise disjoint PTs with prescribed cardinals. Some calculation now leads to Theorem 5.1.1 (for finite E and I).

Again, let $A_1, ..., A_k$ be any subsets of E and, for $1 \leqslant i \leqslant k$, let $\mathscr{E}_i$ be the collection of all subsets of A_i. With this choice, Theorem 8.2.4 gives necessary and sufficient conditions for the existence of pairwise disjoint subsets of $A_1, ..., A_k$ respectively with prescribed cardinals. In this way, we are led to Theorem 3.3.1 (for finite E).

A further specialization of Theorem 8.2.4 is contained in the next Corollary. Let an independence structure $\mathscr{E}$ be defined on a set E. Then $X(\subseteq E)$ is called a *spanning set* if it contains a basis of $\mathscr{E}$.

COROLLARY 8.2.5. *Let $\mathscr{E}$ be an independence structure, with rank function ρ, on a finite set* E. *Then* E *can be partitioned into k spanning sets if and only if, for every subset* A *of* E,

$$|A| - k.\,\rho(A) \leqslant |E| - k.\,\rho(E).$$

It is clear that E can be partitioned into k spanning sets if and only if $\mathscr{E}$ possesses k pairwise disjoint bases. Taking

$$\mathscr{E}_1 = \ldots = \mathscr{E}_k = \mathscr{E}, \qquad s_1 = \ldots = s_k = \rho(E)$$

in Theorem 8.2.4, we at once obtain our assertion.

Corollaries 8.2.2 and 8.2.5 stand in a 'dual' relation to each other. The former result states into how *few* independent sets, the latter into how *many* spanning sets, the ground set E can be partitioned.

Exercises 8.2

1. Exhibit Corollary 3.3.4 as a consequence of Corollary 8.2.2.

2. Let $\mathfrak{A} = (A_i : i \in I)$ be an infinite family of finite subsets of a set E. Use (i) Theorems 3.3.2 and 4.5.2, (ii) Theorem 3.3.2 and Rado's selection principle, to show that $\mathfrak{A}$ can be partitioned into r subfamilies each of which possesses a transversal if and only if $r|A(J)| \geqslant |J|$ for each $J \subset\subset I$.

3. Deduce Theorem 4.5.3 from Corollary 8.2.3.

4. Show that condition (2) in Theorem 8.2.4 is equivalent to the requirement that, for any $I \subseteq \{1, ..., n\}$,

$$\sum_{i \in I} s_i \leqslant \rho_{(I)}(E),$$

where $\rho_{(I)}$ denotes the rank function of the independence structure $\Sigma\,\{\mathscr{E}_i : i \in I\}$.

8.3 A characterization of transversal structures

In this section we shall pursue some of the questions first raised in Chapters 6 and 7.

Let A be a set. We shall here denote by $\hat{A}$ the collection of sets consisting of Ø and all singletons contained in A. Clearly $\hat{A}$ is an independence structure and, indeed, a transversal structure.

Let $\mathfrak{A} = (A_1, \ldots, A_n)$ be a family of subsets of a finite set E. Then, by Theorem 8.1.1, $\hat{A}_1 + \ldots + \hat{A}_n = \mathscr{E}$ (say) is an independence structure. At the same time, $\mathscr{E}$ is simply the collection of all PTs of $\mathfrak{A}$; and we arrive therefore once again at the familiar conclusion (contained as a special case in Theorem 6.5.2) that a transversal structure defined on a finite set is an independence structure.

Further, with the notation as above, denote by ρ, ρ_k the rank functions of $\mathscr{E}$, $\hat{A}_k$ respectively. It is plain that, for $F \subseteq E$,

$$\rho_k(F) = \begin{cases} 1 & (A_k \cap F \neq \emptyset) \\ 0 & (A_k \cap F = \emptyset) \end{cases}$$

and so

$$\rho_1(F) + \ldots + \rho_n(F) = |\{i : 1 \leqslant i \leqslant n,\ A_i \cap F \neq \emptyset\}|.$$

Therefore, by Theorem 8.1.2,

$$\rho(E) = \min_{F \subseteq E} \left\{ |\{i : 1 \leqslant i \leqslant n,\ A_i \cap F \neq \emptyset\}| + |E \setminus F| \right\}.$$

But $\rho(E)$ is obviously equal to the maximum cardinal of PTs of $\mathfrak{A}$. Hence the transversal index of $\mathfrak{A}$ is given by

$$\min_{F \subseteq E} \left\{ |\{i : 1 \leqslant i \leqslant n,\ A_i \cap F \neq \emptyset\}| + |E \setminus F| \right\}.$$

We recall that a different expression for the transversal index of $\mathfrak{A} = (A_1, \ldots, A_n)$, namely

$$n + \min_{I \subseteq \{1, \ldots, n\}} \{|A(I)| - |I|\},$$

had been derived in §3.2. It is, in fact, not difficult to verify by direct calculation that the two expressions give the same value. This conclusion follows even more readily if we invoke the notion of duality.

Now, an independence structure is not necessarily a transversal structure. An example is ready to hand from the proof of Theorem 7.1.5 (i). Let $E = \{1, 2, \ldots, 6\}$ and let $\mathscr{E}$ be the set of all $X \subseteq E$ such that $|X| \leqslant 2$ with the exception of $\{1, 2\}$, $\{3, 4\}$, $\{5, 6\}$. Then $\mathscr{E}$ is not a transversal structure, but it is an independence structure (and, indeed, a linear structure over **Q**). It is therefore natural to ask for conditions that an independence structure should be a transversal structure. If $\mathscr{E}$ is an independence structure on a finite set E, we sometimes refer to the rank of E in $\mathscr{E}$ as the 'rank of $\mathscr{E}$' and we denote it by $\rho(\mathscr{E})$. We shall call $\mathscr{E}$ *simple* if $\rho(\mathscr{E}) = 1$.

THEOREM 8.3.1. *An independence structure on a finite set is a transversal structure if and only if it is the sum of simple independence structures.*

The proof of this result is implicit in the discussion initiated a few lines earlier.

In the first place, let $\mathscr{E}$ be the transversal structure of the family $(A_1, \ldots, A_n)$ of subsets of a finite set E. Then $\mathscr{E} = \hat{A}_1 + \ldots + \hat{A}_n$, so that the given transversal structure is the sum of simple independence structures.

Next, let $\mathscr{E}_1, \ldots, \mathscr{E}_n$ be given simple independence structures and write $\mathscr{E}_1 + \ldots + \mathscr{E}_n = \mathscr{E}$. Defining

$$A_k = \{x: \{x\} \in \mathscr{E}_k\} \qquad (1 \leqslant k \leqslant n),$$

we see that $\mathscr{E}$ is the transversal structure of the family $(A_1, \ldots, A_n)$.

We can obtain a further refinement of the result just proved by linking the rank of the transversal structure with the cardinal of the index set of an associated family.

LEMMA 8.3.2. *Let $\mathscr{E}_1$, $\mathscr{E}_2$ be independence structures on a finite set* E, *and write $\mathscr{E} = \mathscr{E}_1 + \mathscr{E}_2$. If $\rho(\mathscr{E}) = \rho(\mathscr{E}_1)$, then $\mathscr{E} = \mathscr{E}_1$.*

Let $A \in \mathscr{E}$. Then $A = A_1 \cup A_2$, where $A_1 \in \mathscr{E}_1$, $A_2 \in \mathscr{E}_2$. Let B_1 be any basis of $\mathscr{E}_1$ which contains A_1. We have $B_1 \cup A_2 \in \mathscr{E}$ and, if $A_2 \nsubseteq B_1$, then $|B_1 \cup A_2| > |B_1|$, so that $\rho(\mathscr{E}) > \rho(\mathscr{E}_1)$ contrary to hypothesis. Thus $A_2 \subseteq B_1$ and so $A = A_1 \cup A_2 \subseteq B_1 \in \mathscr{E}_1$. We have therefore shown that $\mathscr{E} \subseteq \mathscr{E}_1$, whence $\mathscr{E} = \mathscr{E}_1$.

THEOREM 8.3.3. *A transversal structure of rank n is the transversal structure of a suitable family of n sets, and so is also the sum of n simple structures.*

Let $\mathscr{E}$ be a transversal structure of rank n. Then there exists an integer $m \geqslant n$ and a family $\mathfrak{A} = (A_1, \ldots, A_m)$ of sets such that $\mathscr{E}$ is the transversal structure of $\mathfrak{A}$. Now $\mathfrak{A}$ has a subfamily of n sets, say $\mathfrak{A}^* = (A_1, \ldots, A_n)$, which possesses a transversal. We know that

$$\mathscr{E} = \hat{A}_1 + \ldots + \hat{A}_m;$$

hence $\mathscr{E} = \mathscr{E}_1 + \mathscr{E}_2$, where

$$\mathscr{E}_1 = \hat{A}_1 + \ldots + \hat{A}_n; \qquad \mathscr{E}_2 = \hat{A}_{n+1} + \ldots + \hat{A}_m.$$

But $\mathscr{E}_1$ is the transversal structure of $\mathfrak{A}^*$ and, since $\mathfrak{A}^*$ possesses a transversal, $\rho(\mathscr{E}_1) = n = \rho(\mathscr{E})$. Hence, by Lemma 8.3.2, $\mathscr{E} = \mathscr{E}_1$, i.e. $\mathscr{E}$ is the transversal structure of the family $(A_1, \ldots, A_n)$.

Exercises 8.3

1. Let $(A_1, \ldots, A_n)$ be a family of subsets of a finite set E. Show (i) by direct calculation, (ii) by duality, that

$$\min_{F \subseteq E} \left\{ |\{i: 1 \leqslant i \leqslant n, A_i \cap F \neq \emptyset\}| + |E \setminus F| \right\}$$

$$= n + \min_{I \subseteq \{1, \ldots, n\}} \{|A(I)| - |I|\}.$$

2. Give an example of a transversal structure expressible as a sum of independence structures which are not themselves transversal structures.

8.4 Symmetrized form of Rado's theorem on independent transversals

The problem to which the finite case of Theorem 6.2.2 (i.e. the finite case of the defect form of Rado's theorem on independent transversals) provides an answer, may be framed thus in the language of deltoids. If (E_1, Δ, E_2) is a finite deltoid and $\mathscr{E}_1$ is an independence structure on E_1, what is the maximum cardinal of independent admissible subsets of E_1? It is natural to extend this problem by imposing independence structures on both E_1 and E_2 and then seeking to determine the maximum cardinal of linked independent sets. At the end of the present section we shall arrive at a solution of this symmetric problem; but first we shall need to discuss a number of preliminary questions, which are also of considerable interest in themselves. We shall continue to operate with finite sets.

Let $\mathscr{E}$ be an independence structure on a finite set E. A subset E′ of E is called a *spanning set* (of $\mathscr{E}$) if it contains a basis of $\mathscr{E}$. With $\mathscr{E}$, we associate a collection $\mathscr{E}^*$ of subsets of E which is specified by the requirement that $X \in \mathscr{E}^*$ if and only if $E \setminus X$ is a spanning set. The collection $\mathscr{E}^*$ will be called the *complementary structure*† of $\mathscr{E}$.

THEOREM 8.4.1. *Let $\mathscr{E}$ be an independence structure on a finite set* E. *Then its complementary structure $\mathscr{E}^*$ is again an independence structure, and the*

† In the literature, the term 'dual' structure is more usual; but we prefer to confine the meaning of 'duality' to situations described in § 2.3.

complementary structure of $\mathscr{E}^*$ is the original structure $\mathscr{E}$. Further, the rank functions ρ, ρ^* of $\mathscr{E}$, $\mathscr{E}^*$ respectively are related by the equation

$$\rho^*(A) = |A| - \rho(E) + \rho(E \setminus A) \qquad (A \subseteq E). \tag{1}$$

Let A, B be spanning sets of $\mathscr{E}$ and suppose that $|A| > |B|$. If D is a maximal independent subset of $A \cap B$, then we can choose sets $A' \subseteq A \setminus B$ and $B' \subseteq B \setminus A$ such that $A' \cup D$, $B' \cup D$ are bases of $\mathscr{E}$ so that, in particular, $|A'| = |B'|$. But $|A| > |B|$ and therefore $|A \setminus B| > |B \setminus A|$. Hence $A' \subset A \setminus B$ and there exists an element $a \in A \setminus B$ such that $A' \subseteq (A \setminus B) \setminus \{a\}$. Thus $A \setminus \{a\}$ contains the basis $A' \cup D$. In other words, we have shown that, if A, B are spanning sets with $|A| > |B|$, then $A \setminus \{a\}$ is a spanning set for some $a \in A \setminus B$.

Next, let $X, Y \in \mathscr{E}^*$, $|Y| = |X| + 1$. Then $E \setminus X$, $E \setminus Y$ are spanning sets and $|E \setminus X| > |E \setminus Y|$. Hence, by the result just proved, there exists an element $y \in (E \setminus X) \setminus (E \setminus Y) = Y \setminus X$ such that $(E \setminus X) \setminus \{y\} = E \setminus (X \cup \{y\})$ is a spanning set, i.e. $X \cup \{y\} \in \mathscr{E}^*$. The collection $\mathscr{E}^*$ therefore satisfies the replacement axiom and so is an independence structure.

By the definition of $\mathscr{E}^*$, the members of $\mathscr{E}^*$ are the subsets of complements of bases of $\mathscr{E}$. Since, as we have shown, $\mathscr{E}^*$ is an independence structure, it follows that the bases of $\mathscr{E}^*$ are precisely the complements of the bases of $\mathscr{E}$. The relation between $\mathscr{E}$, $\mathscr{E}^*$ is thus symmetric; and the complementary structure of $\mathscr{E}^*$ is $\mathscr{E}$. Further, it is now clear that

$$\rho(E) + \rho^*(E) = |E|. \tag{2}$$

To establish (1), let $A \subseteq E$ and denote by B a maximal subset of $E \setminus A$ which is a member of $\mathscr{E}$, so that $\rho(E \setminus A) = |B|$. Then $\mathscr{E}$ possesses a basis of the form $B \cup C$, where $C \subseteq A$; and so

$$\rho(E) = |B| + |C|.$$

Moreover, $E \setminus (A \setminus C)$ contains the basis $B \cup C$; therefore $A \setminus C \in \mathscr{E}^*$ and

$$\begin{aligned} \rho^*(A) &\geqslant |A \setminus C| = |A| - \rho(E) + |B| \\ &= |A| - \rho(E) + \rho(E \setminus A). \end{aligned} \tag{3}$$

Since the relation between $\mathscr{E}$ and $\mathscr{E}^*$ is symmetric, we can now interchange ρ and ρ^* in this inequality, and we also replace A by $E \setminus A$. We then obtain

$$\rho(E \setminus A) \geqslant |E \setminus A| - \rho^*(E) + \rho^*(A).$$

Hence, by (2),

$$\rho^*(A) \leqslant |A| - \rho(E) + \rho(E \setminus A);$$

and (1) now follows by virtue of (3).

THEOREM 8.4.2. (R. A. Brualdi) *Let* $\mathscr{E}_1, \mathscr{E}_2$ *be independence structures, with rank functions* ρ_1, ρ_2 *respectively, on a finite set* E. *Then the maximum cardinal of subsets of* E *which belong to both* $\mathscr{E}_1$ *and* $\mathscr{E}_2$ *is equal to*

$$\min \{\rho_1(X_1) + \rho_2(X_2)\},$$

where the minimum is taken with respect to all partitions (X_1, X_2) *of* E.

We shall denote by M, m the maximum and minimum respectively specified in the statement of the theorem. Let $F \subseteq E$ be a set such that $F \in \mathscr{E}_1$, $F \in \mathscr{E}_2$, $|F| = M$. For any partition (X_1, X_2) of E, we have

$$M = |F| = |F \cap X_1| + |F \cap X_2| \leqslant \rho_1(X_1) + \rho_2(X_2),$$

so that $M \leqslant m$. The crux of the proof lies in establishing the reverse inequality $M \geqslant m$.

We shall write $\tilde{\mathscr{E}} = \mathscr{E}_1 + \mathscr{E}_2^*$, where $\mathscr{E}_2^*$ is the complementary structure of $\mathscr{E}_2$. The rank functions of $\tilde{\mathscr{E}}$, $\mathscr{E}_2^*$ will be denoted by $\tilde{\rho}$, ρ_2^* respectively.

Any basis of $\tilde{\mathscr{E}}$ can be expressed in the form $X \cup Y$, where X, Y are bases of $\mathscr{E}_1, \mathscr{E}_2^*$ respectively. Thus

$$\tilde{\rho}(E) = |X \cup Y| = |Y| + |X \cap (E \setminus Y)|.$$

Further, $E \setminus Y$ is a basis of $\mathscr{E}_2$ and so $|E \setminus Y| = \rho_2(E)$. Hence

$$\tilde{\rho}(E) = |E| - \rho_2(E) + |X \cap (E \setminus Y)|.$$

But $X \in \mathscr{E}_1$, $E \setminus Y \in \mathscr{E}_2$ and so $|X \cap (E \setminus Y)| \leqslant M$. Therefore

$$M \geqslant \tilde{\rho}(E) - |E| + \rho_2(E). \tag{4}$$

Now, by Theorem 8.1.2,

$$\tilde{\rho}(E) = \min_{X \subseteq E} \{\rho_1(X) + \rho_2^*(X) + |E \setminus X|\}$$

and so, by Theorem 8.4.1,

$$\begin{aligned}\tilde{\rho}(E) &= \min_{X \subseteq E} \{\rho_1(X) + |X| - \rho_2(E) + \rho_2(E \setminus X) + |E \setminus X|\} \\ &= |E| - \rho_2(E) + m.\end{aligned}$$

Hence, by (4), $M \geqslant m$, as required.

We are now ready to settle the problem raised at the beginning of the section. Let (E_1, Δ, E_2) be a deltoid. Let $X_1 \subseteq E_1$, $X_2 \subseteq E_2$; then (X_1, X_2) will be called a *disconnecting pair* if

$$\{(E_1 \setminus X_1) \times (E_2 \setminus X_2)\} \cap \Delta = \emptyset. \tag{5}$$

This requirement means simply that the relation $x_1 \leftrightarrow x_2$ cannot hold when $x_1 \in E_1 \setminus X_1$, $x_2 \in E_2 \setminus X_2$. It is clear that (5) is equivalent to each of the relations

$$\Delta(E_1 \setminus X_1) \subseteq X_2, \qquad \Delta(E_2 \setminus X_2) \subseteq X_1.$$

We shall denote by $\mathfrak{D}$ the set of all disconnecting pairs.

THEOREM 8.4.3. (R. A. Brualdi) *Let* (E_1, Δ, E_2) *be a finite deltoid and let independence structures* $\mathscr{E}_1$, $\mathscr{E}_2$, *with rank functions* ρ_1, ρ_2 *respectively, be defined on* E_1, E_2 *respectively. Then the maximum cardinal of members of* $\mathscr{E}_1$ *which are linked to members of* $\mathscr{E}_2$ *is equal to*

$$\min\{\rho_1(X_1) + \rho_2(X_2) : (X_1, X_2) \in \mathfrak{D}\}. \tag{6}$$

Denote by M, m the maximum and minimum respectively specified in the theorem. Let $(X_1, X_2) \in \mathfrak{D}$. Let A_1, A_2 be sets such that $A_1 \in \mathscr{E}_1$, $A_2 \in \mathscr{E}_2$, $A_1 \leftrightarrow A_2$, $|A_1| = |A_2| = M$. Put $A_1 = B_1 \cup C_1$, where $B_1 \subseteq X_1$, $C_1 \subseteq E_1 \setminus X_1$. Since $(X_1, X_2) \in \mathfrak{D}$, C_1 must be linked to a set $C_2 \subseteq X_2$ with $C_2 \in \mathscr{E}_2$. Thus

$$M = |B_1| + |C_1| = |B_1| + |C_2| \leqslant \rho_1(X_1) + \rho_2(X_2),$$

and so $M \leqslant m$.

Next, for $\Gamma \subseteq \Delta$, we denote by $N_\Gamma{}^1$ resp. $N_\Gamma{}^2$ the set of first resp. second components of all elements in Γ. The collections $\mathscr{F}_1$, $\mathscr{F}_2$ of subsets of Δ are defined by the requirement that, for $\Gamma \subseteq \Delta$, $\Gamma \in \mathscr{F}_1$ if and only if $N_\Gamma{}^1 \in \mathscr{E}_1$ and no two elements of Γ have the same first component while $\Gamma \in \mathscr{F}_2$ if and only if $N_\Gamma{}^2 \in \mathscr{E}_2$ and no two elements of Γ have the same second component. It is immediately obvious that $\mathscr{F}_1$, $\mathscr{F}_2$ are independence structures on Δ and that their rank functions σ_1, σ_2 satisfy the relations

$$\sigma_1(\Gamma) = \rho_1(N_\Gamma{}^1), \qquad \sigma_2(\Gamma) = \rho_2(N_\Gamma{}^2) \quad (\Gamma \subseteq \Delta).$$

Now, if $A_1 \in \mathscr{E}_1$, $A_2 \in \mathscr{E}_2$, $A_1 \leftrightarrow A_2$, then there exists $\Gamma \subseteq \Delta$ such that $\Gamma \in \mathscr{F}_1 \cap \mathscr{F}_2$ and $A_1 = N_\Gamma{}^1$, $A_2 = N_\Gamma{}^2$; conversely, if $\Gamma \in \mathscr{F}_1 \cap \mathscr{F}_2$, then $A_1 = N_\Gamma{}^1$ and $A_2 = N_\Gamma{}^2$ are independent linked sets. In either case, we have $|A_1| = |A_2| = |\Gamma|$. Hence

$$M = \max\{|\Gamma| : \Gamma \in \mathscr{F}_1, \cap \mathscr{F}_2\}$$

and so, by Theorem 8.4.2,

$$M = \min_{\Gamma \subseteq \Delta} \{\sigma_1(\Delta \setminus \Gamma) + \sigma_2(\Gamma)\}$$

$$= \min_{\Gamma \subseteq \Delta} \{\rho_1(N^1_{\Delta \setminus \Gamma}) + \rho_2(N_\Gamma{}^2)\}. \tag{7}$$

Let $x_1 \in \mathrm{E}_1$, $x_2 \in \mathrm{E}_2$, $(x_1, x_2) \in \Delta$, $\Gamma \subseteq \Delta$. Then either $(x_1, x_2) \in \Delta \setminus \Gamma$, in which case $x_1 \in N^1_{\Delta \setminus \Gamma}$; or else $(x_1, x_2) \in \Gamma$, in which case $x_2 \in N_\Gamma{}^2$. It follows that

$$(N^1_{\Delta \setminus \Gamma}, N_\Gamma{}^2) \in \mathfrak{D}.$$

Hence, for each $\Gamma \subseteq \Delta$,

$$m \leqslant \rho_1(N^1_{\Delta \setminus \Gamma}) + \rho_2(N_\Gamma{}^2)$$

and so, by (7), $m \leqslant M$.

The proof is now complete but it is worth observing that, in applications of the theorem, an expression formally different from (6), namely

$$\min_{\mathrm{X} \subseteq \mathrm{E}_1} \{\rho_1(\mathrm{E}_1 \setminus \mathrm{X}) + \rho_2(\Delta(\mathrm{X}))\}, \tag{8}$$

is sometimes more convenient. To show that the two expressions, which we shall denote by m and m' respectively, are equal, we first note that, for any $\mathrm{X} \subseteq \mathrm{E}_1$, $(\mathrm{E}_1 \setminus \mathrm{X}, \Delta(\mathrm{X}))$ is a disconnecting pair. Hence

$$m \leqslant \rho_1(\mathrm{E}_1 \setminus \mathrm{X}) + \rho_2(\Delta(\mathrm{X})) \qquad (\mathrm{X} \subseteq \mathrm{E}_1)$$

and so $m \leqslant m'$. Furthermore

$$\begin{aligned} m' &\leqslant \min \{\rho_1(\mathrm{E}_1 \setminus \mathrm{X}_1) + \rho_2(\mathrm{X}_2) : \mathrm{X}_1 \subseteq \mathrm{E}_1, \mathrm{X}_2 \subseteq \mathrm{E}_2, \Delta(\mathrm{X}_1) \subseteq \mathrm{X}_2\} \\ &= \min \{\rho_1(\mathrm{X}_1) + \rho_2(\mathrm{X}_2) : \mathrm{X}_1 \subseteq \mathrm{E}_1, \mathrm{X}_2 \subseteq \mathrm{E}_2, \Delta(\mathrm{E}_1 \setminus \mathrm{X}_1) \subseteq \mathrm{X}_2\} \\ &= m. \end{aligned}$$

This establishes our assertion.

We shall conclude by mentioning briefly three specializations of Theorem 8.4.3.

(i) The proof of Theorem 8.4.3 just exhibited is based on Theorem 8.4.2, and it is of interest to note that we can recover Theorem 8.4.2 by applying Theorem 8.4.3 to the deltoid $(\mathrm{E}, \Delta, \mathrm{E})$, where $\Delta = \{(x, x) : x \in \mathrm{E}\}$, and the given independence structures $\mathscr{E}_1, \mathscr{E}_2$ on E.

(ii) Next, let $\mathfrak{A} = (\mathrm{A}_1, \ldots, \mathrm{A}_n)$ be a family of subsets of a finite set E, and let $\mathscr{E}$ be an independence structure, with rank function ρ, defined on E. Taking $\mathrm{E}_1 = \{1, 2, \ldots, n\}$, $\mathrm{E}_2 = \mathrm{E}$, $\mathscr{E}_1 = \mathscr{P}(\mathrm{E}_1)$, $\mathscr{E}_2 = \mathscr{E}$,

$$\Delta = \{(k, e) : 1 \leqslant k \leqslant n,\ e \in \mathrm{E},\ e \in \mathrm{A}_k\}$$

in Theorem 8.4.3, we are led to the conclusion that the maximum cardinal of independent partial transversals of $\mathfrak{A}$ (i.e. partial transversals of $\mathfrak{A}$ which are members of $\mathscr{E}$) is equal to

$$n + \min_{I \subseteq \{1,\ldots,n\}} \{\rho(A(I)) - |I|\}.$$

This is, in essence, Theorem 6.2.2, i.e. the defect form of Rado's theorem on independent transversals.

(iii) Let G be a finite bipartite graph and denote by E_1, E_2 the two (disjoint) sets of its nodes such that every edge links a node in E_1 and a node in E_2. Further, let Δ be the set of all pairs (e_1, e_2) such that $e_1 \in E_1$, $e_2 \in E_2$, and $\{e_1, e_2\}$ is an edge. Taking $\mathscr{E}_1$, $\mathscr{E}_2$ as the universal structures on E_1, E_2 respectively and using Theorem 8.4.3, we infer at once König's theorem 1.7.1 (for the case of finite graphs). An alternative proof will be offered in §11.2.

Exercises 8.4

1. Write out the proof of König's theorem 1.7.1 (for finite graphs) by specializing the proof of Theorem 8.4.3.

2. Supply the full details of the applications of Theorem 8.4.3 indicated at the end of the section.

3. Let $\mathscr{E}_1, \mathscr{E}_2, \mathscr{E}_3$ be independence structures on a finite set E, and let ρ_1, ρ_2, ρ_3 denote their rank functions. Show that

$$\max\{|X| : X \in \mathscr{E}_1 \cap \mathscr{E}_2 \cap \mathscr{E}_3\} \leqslant \min\{\rho_1(X_1) + \rho_2(X_2) + \rho_3(X_3) : X_1 \cup X_2 \cup X_3 = E\},$$

and that the sign of inequality cannot be replaced by that of equality.

4. Let $\mathscr{E}_1, \mathscr{E}_2$ be independence structures on a finite set E. Using asterisks to denote complementary structures, show that

$$(\mathscr{E}_1^* + \mathscr{E}_2^*)^* \subseteq \mathscr{E}_1 \cap \mathscr{E}_2.$$

Can the sign of inclusion be replaced by that of equality?

5. Let the notation be as in the preceding question, and denote by ρ_1, ρ_2, ρ the rank functions of $\mathscr{E}_1, \mathscr{E}_2, (\mathscr{E}_1^* + \mathscr{E}_2^*)^*$ respectively. Show that, for $A \subseteq E$,

$$\rho(A) = \min_{A \subseteq X \subseteq E} f(X) - \min_{X \subseteq E} f(X),$$

where $f(X) = \rho_1(X) + \rho_2(X) - |X|$.

Notes on Chapter 8

§ 8.1. Theorems 8.1.1 and 8.1.2 both occur in the work of C. St. J. A. Nash-Williams (**1**), although the basic idea seems to be due to J. Edmonds. The proof of Theorem 8.1.1 given above was devised by A. P. Heron, that of Theorem 8.1.2 by

D. J. A. Welsh (**5**). The rank formula contained in Theorem 8.1.2 is a result of crucial importance in the theory of abstract independence since it enables us to treat without difficulty problems which had previously been accessible only to extremely complex arguments. The relation between the Nash-Williams rank formula and Brualdi's symmetrization of Rado's theorem on independent transversals (Theorem 8.4.3) has been investigated by Welsh (**5**). Transfinite analogues of Theorems 8.1.1 and 8.1.2 were discussed by J. S. Pym & Hazel Perfect (**1**).

§ 8.2. All results in this section other than Corollary 8.2.3 are due to Edmonds (**1**) or Edmonds & Fulkerson (**1**). The latter paper contains a wealth of further results of the same general character as those discussed here. For the special case of *linear* independence, Corollary 8.2.3 was originally proved by A. Horn (**1**) and also by R. Rado (**8**). The work of all these authors is difficult: I owe the very transparent treatment offered here to a communication of Dr D. J. A. Welsh; see also Harary & Welsh (**1**) Far-reaching extensions of the findings of this section will be found in Brualdi's paper (**9**). Some of the results are discussed by C. Berge (**2**) in the more general context of 'graphoid' theory.

§ 8.3. The treatment in this section is based largely on Dr Welsh's ideas; in particular, Theorem 8.3.1 is due to him (Welsh (**6**); cf. also (**8**)). A characterization, different from that contained in Theorem 8.3.1, of independence structures which are transversal structures has been given by J. H. Mason (**1**). Theorem 8.3.3 was noted by Professor Brualdi.

§ 8.4. The notion of a complementry structure derives from Whitney's fundamental work (**1**). Theorems 8.4.2 and 8.4.3 made their first appearance in Brualdi's unpublished manuscript (**2**) (see also Aigner & Dowling (**1**)), but the proofs given here are due to D. J. A. Welsh (**5**).

9 Links of Two Finite Families

All the investigations so far have been centred on the existence of transversal-like objects associated with a single family. We shall now extend the scope of the discussion by considering more than one family. In practice, this will amount to a study of pairs of families, since the difficulties of dealing with more than two families have not yet been surmounted.

9.1 The notion of a link

Let $\mathfrak{A}, \mathfrak{B}, \ldots$ be families of subsets of E. If a family $\mathfrak{X}$ of elements of E is a system of representatives of each of these families, then it is called a *common system of representatives* (CSR) of $\mathfrak{A}, \mathfrak{B}, \ldots$. In particular, then, $\mathfrak{X} = (x_k : k \in K)$ is a CSR of $\mathfrak{A} = (A_i : i \in I)$ and $\mathfrak{B} = (B_j : j \in J)$ precisely if there exist bijections $\phi : K \to I, \psi : K \to J$ such that

$$x_k \in A_{\phi(k)} \cap B_{\psi(k)} \qquad (k \in K).$$

Hence $\mathfrak{A}$ and $\mathfrak{B}$ possess a CSR if and only if there exists a bijection $\theta : I \to J$ such that

$$A_i \cap B_{\theta(i)} \neq \emptyset \qquad (i \in I).$$

Again, let a set X be a transversal resp. partial transversal of each of the families $\mathfrak{A}, \mathfrak{B}, \ldots$. Then X is called a *common transversal* (CT) resp. *common partial transversal* (CPT) of these families. It is clear that X is a CT of $\mathfrak{A} = (A_i : i \in I)$ and $\mathfrak{B} = (B_j : j \in J)$ if and only if there exist bijections $\rho : X \to I, \sigma : X \to J$ such that

$$x \in A_{\rho(x)} \cap B_{\sigma(x)} \qquad (x \in X).$$

This statement remains valid if the term 'bijection' is replaced by 'injection' and 'common transversal' is replaced by 'common partial transversal'. We note that, if several families of sets possess a CSR or a CT, then their cardinals must be equal.

The term 'link' will be used as a joint designation for a common system of representatives or a common (partial) transversal.

In the present chapter our discussion will be confined to finite families.

9.2 Common representatives

We begin our examination of pairs of families with a result closely related to Hall's criterion.

THEOREM 9.2.1. *The families* $\mathfrak{A} = (A_1, ..., A_n)$ *and* $\mathfrak{B} = (B_1, ..., B_n)$ *of sets possess a common system of representatives if and only if, for each* k *with* $1 \leqslant k \leqslant n$, *the union of any* k A's *intersects at least* k B's.

The argument here is very similar to that used in the deduction of Theorem 2.2.4 from Theorem 2.2.1 (Hall's criterion). We write

$$C_i = \{j : 1 \leqslant j \leqslant n,\ A_i \cap B_j \neq \emptyset\} \qquad (1 \leqslant i \leqslant n),$$

so that $j \in C_i$ if and only if $A_i \cap B_j \neq \emptyset$. For $1 \leqslant k \leqslant n$ and $1 \leqslant i_1 < ... < i_k \leqslant n$, we therefore have

$$\begin{aligned} C_{i_1} \cup ... \cup C_{i_k} &= \{j : 1 \leqslant j \leqslant n,\ A_{i_1} \cap B_j \neq \emptyset \text{ or } ... \text{ or } A_{i_k} \cap B_j \neq \emptyset\} \\ &= \{j : 1 \leqslant j \leqslant n,\ (A_{i_1} \cup ... \cup A_{i_k}) \cap B_j \neq \emptyset\}. \end{aligned}$$

Thus, if $\mathfrak{A}$ and $\mathfrak{B}$ satisfy the intersection condition stated in the theorem, then $\mathfrak{C} = (C_1, ..., C_n)$ satisfies Hall's condition and so (by Theorem 2.2.1) possesses a transversal, say

$$j_1 \in C_1, ..., j_n \in C_n,$$

where $j_1, ..., j_n$ are the numbers $1, ..., n$ taken in a suitable order. Hence

$$A_1 \cap B_{j_1} \neq \emptyset, ..., A_n \cap B_{j_n} \neq \emptyset$$

and so $\mathfrak{A}$ and $\mathfrak{B}$ possess a CSR. This establishes the sufficiency of the intersection condition; its necessity holds trivially.

The intersection condition in Theorem 9.2.1 is, on the face of it, asymmetric with respect to $\mathfrak{A}$ and $\mathfrak{B}$ (so that it can be restated with the roles of $\mathfrak{A}$ and $\mathfrak{B}$ interchanged). The next result gives a formally symmetric criterion for the existence of a CSR. In formulating it, we shall make use of the function $\chi : \mathbf{Z} \to \{0, 1\}$ (where $\mathbf{Z}$ is the set of integers) which is defined by the equations $\chi(x) = 1\ (x > 0)$, $\chi(x) = 0\ (x \leqslant 0)$.

COROLLARY 9.2.2. *The families* $\mathfrak{A} = (A_1, ..., A_n)$ *and* $\mathfrak{B} = (B_1, ..., B_n)$ *of sets possess a common system of representatives if and only if, for all pairs* I, J *of subsets of* $\{1, ..., n\}$,

$$|A(I) \cap B(J)| \geqslant \chi(|I| + |J| - n). \qquad (1)$$

Suppose, in the first place, that $\mathfrak{A}$ and $\mathfrak{B}$ have a CSR so that, without loss of generality,

$$A_1 \cap B_1 \neq \emptyset, \ldots, A_n \cap B_n \neq \emptyset.$$

Let I, J $\subseteq \{1, \ldots, n\}$. If I $\cap$ J $= \emptyset$, then $|I| + |J| - n \leqslant 0$ and (1) is satisfied. If I $\cap$ J $\neq \emptyset$, then

$$|A(I) \cap B(J)| \geqslant 1 \geqslant \chi(|I| + |J| - n).$$

Suppose, next, that (1) holds for all I, J. Take any subset I of $\{1, \ldots, n\}$ and let $\{1, \ldots, n\} = J_1 \cup J_2$ be a partition such that

$$A(I) \cap B_j \neq \emptyset \quad (j \in J_1), \qquad A(I) \cap B_j = \emptyset \quad (j \in J_2).$$

Then $A(I) \cap B(J_2) = \emptyset$ so that, by (1), $|I| + |J_2| - n \leqslant 0$, i.e. $|I| \leqslant |J_1|$. Thus, for any I, A(I) intersects at least $|I|$ B's. Hence, by Theorem 9.2.1, $\mathfrak{A}$ and $\mathfrak{B}$ have a CSR.

THEOREM 9.2.3. *Let* $A_1 \cup \ldots \cup A_n = B_1 \cup \ldots \cup B_n$ *where the* A's, *and equally the* B's, *are pairwise disjoint and where each* A *and each* B *has the same, finite and non-zero, cardinal. Then the families* $(A_1, \ldots, A_n)$ *and* $(B_1, \ldots, B_n)$ *possess a common system of representatives (and also a common transversal).*

Let the cardinal number of each A and each B be s. Assume that the union of a certain set of k A's (say $A_1 \cup \ldots \cup A_k$) intersects fewer than k B's. Then $A_1 \cup \ldots \cup A_k$ fails to intersect the union of a set of $n - k + 1$ B's, say $B_1 \cup \ldots \cup B_{n-k+1}$. It follows that

$$\begin{aligned} ns &= |A_1 \cup \ldots \cup A_n| = |A_1 \cup \ldots \cup A_n \cup B_1 \cup \ldots \cup B_n| \\ &\geqslant |A_1 \cup \ldots \cup A_k \cup B_1 \cup \ldots \cup B_{n-k+1}| \\ &= |A_1 \cup \ldots \cup A_k| + |B_1 \cup \ldots \cup B_{n-k+1}| = ks + (n - k + 1)\,s \\ &= (n + 1)\,s. \end{aligned}$$

We thus arrive at a contradiction and conclude that, for each k with $1 \leqslant k \leqslant n$, the union of any k A's intersects at least k B's. Hence, by Theorem 9.2.1, the two families possess a CSR. But any two A's (or B's) are disjoint. Hence the range of the CSR is a CT of the two families.

An immediate consequence of Theorem 9.2.3 relating to groups is as follows.

COROLLARY 9.2.4. *Let H be a subgroup of a finite group G. Then the family of left cosets of H and that of right cosets of H possess a common transversal.*

Exercises 9.2

1. Let $(A_1, \ldots, A_n)$, $(B_1, \ldots, B_n)$ be two families of sets. Show that they possess a CSR if and only if $|I| + |J| \leqslant n$ whenever $I, J \subseteq \{1, \ldots, n\}$ and $A(I) \cap B(J) = \emptyset$.

2. Let $\mathfrak{A} = (A_1, \ldots, A_m)$, $\mathfrak{B} = (B_1, \ldots, B_n)$ be two families of sets and suppose that $m \leqslant n$. Establish the equivalence of the following statements. (i) $\mathfrak{A}$ and a subfamily of $\mathfrak{B}$ possess a CSR. (ii) For each k with $1 \leqslant k \leqslant m$, the union of any k A's intersects at least k B's. (iii) For each k with $n - m \leqslant k \leqslant n$, the union of any k B's intersects at least $k - n + m$ A's.
Also deduce Hall's theorem from the above result.

3. Let $\mathfrak{A} = (A_1, \ldots, A_n)$ be a family of pairwise disjoint non-empty sets and let $\mathfrak{B} = (B_1, \ldots, B_n)$ be a family of non-empty subsets of $A_1 \cup \ldots \cup A_n$. Suppose that, for each k with $1 \leqslant k \leqslant n$, the union of any k A's contains at most k B's. Show that $\mathfrak{A}$ and $\mathfrak{B}$ possess a CSR.

[Mann & Ryser (**1**)]

4. Let the assumptions of Theorem 9.2.3 be satisfied, and denote by m the common cardinal of A's and B's. Show that the two families possess m pairwise disjoint CTs.

[van der Waerden (**2**)]

5. Show that, if the three families $(A_1, \ldots, A_n)$, $(B_1, \ldots, B_n)$, $(C_1, \ldots, C_n)$ possess a CSR, then, for all $I, J, K \subseteq \{1, \ldots, n\}$,

$$|A(I) \cap B(J) \cap C(K)| \geqslant \chi(|I| + |J| + |K| - 2n).$$

Also show that the converse inference is false.

6. Let H, H' be subgroups of the same order of a finite group. Show that the family of left cosets of H and the family of left cosets (or of right cosets) of H' possess a CT.

7. Let $\mathfrak{A} = (A_1, \ldots, A_n)$, $\mathfrak{B} = (B_1, \ldots, B_n)$ be two families of sets, and let k be a natural number. Show that, if there exist pairwise disjoint sets $X_1, \ldots, X_n$ with $|X_1| = \ldots = |X_n| = k$ and such that, for a suitable renumbering of A's and B's, $X_i \subseteq A_i \cap B_i$ $(1 \leqslant i \leqslant n)$, then

$$|A(I) \cap B(J)| \geqslant k(|I| + |J| - n)$$

whenever $I, J \subseteq \{1, \ldots, n\}$. Also show that the converse inference is false.

9.3 The criterion of Ford and Fulkerson

Let $\mathfrak{B} = (B_1, \ldots, B_n)$ be a (finite) family of subsets of a set E. We recall (cf. Theorem 6.5.2) that the collection of PTs of $\mathfrak{B}$ is a pre-independence structure on E. Its rank function will be denoted by ρ.

LEMMA 9.3.1. *If* $X \subseteq E$, *then the inequality* $\rho(X) \geqslant k$ *is valid if and only if, for all* $J \subseteq \{1, \ldots, n\}$,

$$|B(J) \cap X| \geqslant |J| + k - n.$$

The inequality $\rho(X) \geqslant k$ means simply that X contains a PT of $\mathfrak{B}$ of cardinal k, i.e. that the family $(B_i \cap X : 1 \leqslant i \leqslant n)$ possesses a PT of cardinal k. The assertion now follows by the Hall–Ore theorem 3.2.1.

THEOREM 9.3.2. *Let* $\mathfrak{A} = (A_1, \ldots, A_m)$, $\mathfrak{B} = (B_1, \ldots, B_n)$ *be finite families of sets, and let k be a natural number. Then* $\mathfrak{A}$ *and* $\mathfrak{B}$ *possess a common partial transversal of cardinal k if and only if*

$$|A(I) \cap B(J)| \geqslant |I| + |J| + k - m - n$$

whenever $I \subseteq \{1, \ldots, m\}$, $J \subseteq \{1, \ldots, n\}$.

As before, we consider the pre-independence structure consisting of all PTs of $\mathfrak{B}$, and we denote its rank function by ρ. Then $\mathfrak{A}$ and $\mathfrak{B}$ possess a CPT of cardinal k if and only if $\mathfrak{A}$ possesses an independent PT of cardinal k. By Theorem 6.2.2, this is the case precisely if

$$\rho(A(I)) \geqslant |I| + k - m \tag{1}$$

whenever $I \subseteq \{1, \ldots, m\}$. Now, by Lemma 9.3.1, (1) holds precisely if

$$|B(J) \cap A(I)| \geqslant |J| + (|I| + k - m) - n$$

for all $J \subseteq \{1, \ldots, n\}$. The assertion is therefore proved.

We record an easy consequence of Theorem 9.3.2. The details of proof may be left to the reader.

COROLLARY 9.3.3. *The maximum cardinal of common partial transversals of the families* $\mathfrak{A} = (A_1, \ldots, A_m)$ *and* $\mathfrak{B} = (B_1, \ldots, B_n)$ *is given by*

$$m + n + \min_{I, J} \{|A(I) \cap B(J)| - |I| - |J|\},$$

where the minimum is taken with respect to all subsets I *of* $\{1, \ldots, m\}$ *and* J *of* $\{1, \ldots, n\}$.

Again, taking $m = n = k$ in Theorem 9.3.2, we obtain at once the following result (which it is instructive to compare with Corollary 9.2.2).

COROLLARY 9.3.4. *The families* $\mathfrak{A} = (A_1, \ldots, A_n)$ *and* $\mathfrak{B} = (B_1, \ldots, B_n)$ *of subsets of* E *possess a common transversal if and only if, for all* $I, J \subseteq \{1, \ldots, n\}$,

$$|A(I) \cap B(J)| \geqslant |I| + |J| - n. \tag{2}$$

We shall refer to both Theorem 9.3.2 and Corollary 9.3.4 as 'the criterion of Ford and Fulkerson'. It will be stated or will be clear from the context which result is meant in each case.

It should be observed that the specialization $B_i = A_1 \cup \ldots \cup A_n$ $(1 \leqslant i \leqslant n)$ reduces Corollary 9.3.4 to Hall's theorem 2.2.1.

We also note briefly a proof of Corollary 9.3.4 which depends on an 'elementary construction' (see § 3.1). Assume, as may be done without loss of generality, that $\{1, \ldots, n\} \cap E = \varnothing$ and consider the family $\mathfrak{C} = (C_k : k \in \{1, \ldots, n\} \cup E)$ of subsets of $\{1, \ldots, n\} \cup E$, where

$$C_k = A_k \qquad (1 \leqslant k \leqslant n),$$
$$C_e = \{e\} \cup \{j : 1 \leqslant j \leqslant n, \quad e \in B_j\} \quad (e \in E).$$

It is not difficult to verify that $\mathfrak{A}$ and $\mathfrak{B}$ possess a CT if and only if $\mathfrak{C}$ possesses a transversal and, applying Hall's theorem, we can show that this is the case if and only if (2) holds whenever $I, J \subseteq \{1, \ldots, n\}$.

We add yet another proof of Corollary 9.3.4, based on Menger's theorem 1.7.2. It will be convenient to write the two families of subsets of E in the form $\mathfrak{A} = (A_i : i \in I)$, $\mathfrak{B} = (B : j \in J)$, where $|I| = |J| = n$ and I, J, E are pairwise disjoint. We have to show that $\mathfrak{A}$ and $\mathfrak{B}$ possess a CT if and only if, for all $I' \subseteq I$, $J' \subseteq J$

$$|A(I') \cap B(J')| \geqslant |I'| + |J'| - n. \tag{3}$$

Let G be the graph $(N, \mathscr{E})$, where $N = I \cup E \cup J$ and

$$\mathscr{E} = \left\{\{i, e\} : i \in I,\ e \in E,\ e \in A_i\right\} \cup \left\{\{j, e\} : j \in J,\ e \in E,\ e \in B_j\right\}.$$

It is then clear that $\mathfrak{A}$ and $\mathfrak{B}$ have a CT if and only if there are n disjoint paths in G which link I and J.

Suppose, in the first place, that I', J' can be so chosen that (3) is false. For these sets I', J', we write $E' = A(I') \cap B(J')$. Then $S = (I \setminus I') \cup E' \cup (J \setminus J')$ separates I and J in G and, furthermore,

$$\begin{aligned} |S| &= |I \setminus I'| + |A(I') \cap B(J')| + |J \setminus J'| \\ &< |I \setminus I'| + (|I'| + |J'| - n) + |J \setminus J'| = n. \end{aligned}$$

Thus the minimum number of elements in a separating set is less than n, and so there cannot exist n disjoint paths linking I and J, i.e. $\mathfrak{A}$ and $\mathfrak{B}$ have no CT.

On the other hand, let (3) be satisfied whenever $I' \subseteq I$, $J' \subseteq J$; and let $S = I' \cup E' \cup J'$ (where $I' \subseteq I$, $E' \subseteq E$, $J' \subseteq J$) be a set separating I and J. Then evidently

$$A(I \setminus I') \cap B(J \setminus J') \subseteq E' \tag{4}$$

and so, by (3) and (4),

$$\begin{aligned}|S| &= |I'| + |E'| + |J'| \\ &\geqslant |I'| + |A(I \setminus I') \cap B(J \setminus J')| + |J'| \\ &\geqslant |I'| + (|I \setminus I'| + |J \setminus J'| - n) + |J'| = n.\end{aligned}$$

The number of elements in any separating set is therefore not less than n; hence there exist n disjoint paths linking I and J, i.e. $\mathfrak{A}$ and $\mathfrak{B}$ have a CT.

Finally, we consider the possibility of extending Corollary 9.3.4 to three families, say $\mathfrak{A} = (A_1, \ldots, A_n)$, $\mathfrak{B} = (B_1, \ldots, B_n)$, $\mathfrak{C} = (C_1, \ldots, C_n)$. It is easy to prove that, if $\mathfrak{A}, \mathfrak{B}, \mathfrak{C}$ possess a CT, then

$$|A(I) \cap B(J) \cap C(K)| \geqslant |I| + |J| + |K| - 2n$$

whenever $I, J, K \subseteq \{1, \ldots, n\}$. However, quite simple counter-examples show that this set of conditions is not sufficient to ensure the existence of a CT of the three families. The reason why an attempt to establish an analogue of Corollary 9.3.4 breaks down for three families is that the collection of CPTs of two families need not be a pre-independence structure. Thus each of the sets $\{1, 2\}$, $\{3\}$ is a CPT of the families

$$(\{1\}, \{2, 3\}), \qquad (\{2\}, \{1, 3\})$$

but neither $\{1, 3\}$ nor $\{2, 3\}$ has this property. The replacement axiom therefore fails to hold.

Exercises 9.3

1. Assuming the validity of Corollary 9.3.4 for a finite E, deduce its validity for an infinite E.

2. Deduce Theorem 9.3.2 from Corollary 9.3.4.

3. Give a proof of Corollary 9.3.3.

4. Let $\mathfrak{A} = (A_1, \ldots, A_m)$, $\mathfrak{B} = (B_1, \ldots, B_n)$ be two families of sets; let $r_1, \ldots, r_m, s_1, \ldots, s_n$ be non-negative integers; and denote by $\mathfrak{A}^*$ the family consisting of r_i copies of A_i, $1 \leqslant i \leqslant m$, and by $\mathfrak{B}^*$ the family consisting of s_j copies of B_j, $1 \leqslant j \leqslant n$. Show that the maximum cardinal of CPTs of $\mathfrak{A}^*$ and $\mathfrak{B}^*$ is equal to

$$\min_{I, J} \left\{ |A(I) \cap B(J)| + \sum_{i \notin I} r_i + \sum_{j \notin J} s_j \right\},$$

where the minimum is taken with respect to all subsets $I \subseteq \{1, \ldots, m\}$, $J \subseteq \{1, \ldots, n\}$.

5. Write out in detail the second proof of Corollary 9.3.4 sketched in the text.

6. Let $\mathfrak{A} = (A_1, \ldots, A_n)$, $B = (B_1, \ldots, B_n)$, $\mathfrak{C} = (C_1, \ldots, C_n)$ be families of finite sets, and consider the following statements relating to these families.

P. $\mathfrak{A}, \mathfrak{B}, \mathfrak{C}$ have a CT.

Q. Each pair of families among $\mathfrak{A}, \mathfrak{B}, \mathfrak{C}$ has a CT.

R. $|A(I) \cap B(J) \cap C(K)| \geqslant |I| + |J| + |K| - 2n$

whenever $I, J, K \subseteq \{1, \ldots, n\}$.

S. $$|A(I) \cap B(J)| + |A(I) \cap C(K)| + |B(J) \cap C(K)| - |A(I) \cap B(J) \cap C(K)| \geqslant |I| + |J| + |K| - n$$

whenever $I, J, K \subseteq \{1, \ldots, n\}$.

Show that the only valid implications between pairs of the above statements are $P \Rightarrow Q$, $P \Rightarrow R$, $P \Rightarrow S$, $R \Rightarrow Q$, $S \Rightarrow Q$.

7. Let $\mathfrak{A} = (A_1, \ldots, A_n)$ be a family of subsets of E, and let $M \subseteq E$. Show that $\mathfrak{A}$ possesses a transversal which contains M if and only if $\mathfrak{A}$ has a CT with the family consisting of $|M|$ copies of M and $n - |M|$ copies of $E \setminus M$. Hence use Corollary 9.3.4 to deduce Theorem 3.3.6.

8. Let $\mathfrak{A} = (A_1, \ldots, A_n)$, $\mathfrak{B} = (B_1, \ldots, B_n)$ be two families of subsets of E and let r be a natural number. Prove that $\mathfrak{A}$ and $\mathfrak{B}$ possess a CSR in which no element occurs more than r times if and only if, for all $I, J \subseteq \{1, \ldots, n\}$,

$$r|A(I) \cap B(J)| \geqslant |I| + |J| - n.$$

9.4 Common representatives with restricted frequencies

Let $\mathfrak{X} = (x_i : i \in I)$ be a family of elements of a set E. For each $x \in E$, we shall write

$$f(\mathfrak{X}; x) = |\{i \in I : x_i = x\}|.$$

Thus $f(\mathfrak{X}; x)$ is simply the 'frequency of occurrence' of x in $\mathfrak{X}$.

The main problem to be considered in this section is as follows: under what circumstances do two given families possess a CSR in which the frequencies of elements lie between prescribed bounds? Before we can solve this problem, we need to settle some preliminary questions.

LEMMA 9.4.1. *Let $\mathscr{E}$ be a pre-independence structure on a set* E; *let* M *be a finite independent subset of* E; *and write*

$$\mathscr{E}^* = \{X \subseteq E : X \cup M \in \mathscr{E}\}.$$

Then $\mathscr{E}^*$ is again a pre-independence structure.

It suffices to show that the replacement axiom is valid. We shall denote by A, B finite members of $\mathscr{E}^*$ such that $|B| = |A| + 1$.

Suppose, in the first place, that $|B \cup M| > |A \cup M|$. Then, since $A \cup M$, $B \cup M \in \mathscr{E}$, there exists an element $x \in B \cup M$ such that $x \notin A \cup M$ and $A \cup M \cup \{x\} \in \mathscr{E}$. Therefore $x \in B \setminus A$ and $A \cup \{x\} \in \mathscr{E}^*$.

If, on the other hand, $|B \cup M| \leqslant |A \cup M|$, then $|B \cap M| > |A \cap M|$ and there exists an element x such that $x \in B \cap M$ and $x \notin A \cap M$ so that $x \in B \setminus A$. Further $(A \cup \{x\}) \cup M = A \cup M \in \mathscr{E}$, and so $A \cup \{x\} \in \mathscr{E}^*$. The proof is therefore complete.

LEMMA 9.4.2. *Let the notation be as in Lemma* 9.4.1, *and denote by* ρ, ρ^* *the rank functions of* $\mathscr{E}$, $\mathscr{E}^*$ *respectively. Then, for any subset* F *of* E,

$$\rho^*(F) = \rho(F \cup M) - |M| + |F \cap M|.$$

Let n be any integer such that $\rho(F \cup M) \geqslant n \geqslant |M|$. Applying Lemma 6.1.1, we see that there exists a set $X \subseteq F \setminus M$ such that $X \cup M \in \mathscr{E}$ and $|X \cup M| = n$. Therefore

$$\rho^*(F) \geqslant |X| + |F \cap M|$$

and so

$$\rho^*(F) \geqslant n - |M| + |F \cap M|. \tag{1}$$

If $\rho(F \cup M) < \infty$, we take $n = \rho(F \cup M)$ and obtain the relation

$$\rho^*(F) \geqslant \rho(F \cup M) - |M| + |F \cap M|. \tag{2}$$

If $\rho(F \cup M) = \infty$, then (1) is true for every n, so that $\rho^*(F) = \infty$ and (2) is again valid.

To obtain the reverse inequality, let n be any non-negative integer such that $\rho^*(F) \geqslant n$ and let Y be a set such that $Y \subseteq F$, $|Y| = n$, $Y \cup M \in \mathscr{E}$. Then

$$\begin{aligned} \rho(F \cup M) &\geqslant |Y \cup M| = |Y| + |M| - |Y \cap M| \\ &\geqslant n + |M| - |F \cap M|. \end{aligned}$$

Arguing as above, we conclude that

$$\rho(F \cup M) \geqslant \rho^*(F) + |M| - |F \cap M|;$$

and the assertion follows.

We next settle the question of 'marginal' elements for a CT of two families. (Cf. the discussion in § 6.6.)

THEOREM 9.4.3. *Let* $\mathfrak{A} = (A_1, \ldots, A_n)$, $\mathfrak{B} = (B_1, \ldots, B_n)$ *be families of subsets of an arbitrary set* E, *and let* $M \subseteq E$. *Then* $\mathfrak{A}$ *and* $\mathfrak{B}$ *possess a common transversal which contains* M *if and only if, for all* $I, J \subseteq \{1, \ldots, n\}$,

$$|A(I) \cap B(J)| + |\{A(I) \cup B(J)\} \cap M| \geqslant |I| + |J| + |M| - n. \tag{3}$$

For $M = \emptyset$, this result reduces to Corollary 9.3.4.

The set of all PTs of $\mathfrak{B}$ will be denoted by $\mathscr{E}$. We may assume that $M \in \mathscr{E}$ (so that, in particular, M is finite). For, if $\mathfrak{A}$, $\mathfrak{B}$ possess a CT containing M, this is obvious. And if (3) holds for all I, J, then, taking $I = \emptyset$, we obtain

$$|B(J) \cap M| \geqslant |J| + |M| - n \qquad \text{(all J)}.$$

In view of Corollary 3.2.2, this means precisely that $M \in \mathscr{E}$.

By Theorem 6.5.2, $\mathscr{E}$ is a pre-independence structure on E. Let $\mathscr{E}^*$ be defined as in Lemma 9.4.1 and ρ, ρ^* as in Lemma 9.4.2. Suppose that $\mathfrak{A}$, $\mathfrak{B}$ possess a CT, say X, which contains M. Then $X \cup M$ (= X) is a transversal of $\mathfrak{B}$ and so $X \cup M \in \mathscr{E}$, i.e. $X \in \mathscr{E}^*$. Thus $\mathfrak{A}$ has a transversal which is a member of $\mathscr{E}^*$. Conversely, suppose that some $X \in \mathscr{E}^*$ is a transversal of $\mathfrak{A}$. We then have $X \cup M \in \mathscr{E}$, i.e. $X \cup M$ is a PT of $\mathfrak{B}$. Consequently X is also a PT of $\mathfrak{B}$ and, since $|X| = n$, X is a transversal of $\mathfrak{B}$. But $X \cup M$ is a PT of $\mathfrak{B}$ and so $M \subseteq X$.

We have thus shown that $\mathfrak{A}$ and $\mathfrak{B}$ possess a CT which contains M if and only if $\mathfrak{A}$ possesses a transversal which is a member of the pre-independence structure $\mathscr{E}^*$. In view of Rado's theorem 6.2.1 on independent transversals, this is the case if and only if $\rho^*(A(I)) \geqslant |I|$ for all I; and by Lemma 9.4.2 this means that

$$\rho(A(I) \cup M) \geqslant |I| + |M| - |A(I) \cap M| \qquad \text{(all I)}.$$

Further, by Lemma 9.3.1, these inequalities hold if and only if

$$|B(J) \cap \{A(I) \cup M\}| \geqslant |J| + |I| + |M| - |A(I) \cap M| - n \qquad \text{(all I, J)}.$$

Finally, it is easily verified that these conditions are equivalent to the validity of (3) for all I and J.

We are now able to deal with the problem formulated at the beginning of the section.

THEOREM 9.4.4. (Ford & Fulkerson) *Let* $\mathfrak{A} = (A_1, \ldots, A_n)$ *and* $\mathfrak{B} = (B_1, \ldots, B_n)$ *be families of subsets of* E. *For each* $x \in E$, *let* r_x, s_x *be integers with* $0 \leqslant r_x \leqslant s_x$ *and suppose that* $r_x \neq 0$ *for only finitely many* x*'s. Then* $\mathfrak{A}$ *and*

$\mathfrak{B}$ possess a common system of representatives $\mathfrak{X}$ with

$$r_x \leqslant f(\mathfrak{X}; x) \leqslant s_x \qquad (x \in \mathrm{E}) \tag{4}$$

if and only if, for each pair I, J *of subsets of* $\{1, \ldots, n\}$,

$$\sum_{x \in \mathrm{A(I)} \cap \mathrm{B(J)}} s_x - \sum_{x \notin \mathrm{A(I)} \cup \mathrm{B(J)}} r_x \geqslant |\mathrm{I}| + |\mathrm{J}| - n.$$

To establish this assertion, we write

$$\begin{aligned}
\hat{\mathrm{E}} &= \{(x, k) : x \in \mathrm{E},\ 1 \leqslant k \leqslant s_x\}, \\
\hat{\mathrm{M}} &= \{(x, k) : x \in \mathrm{E},\ 1 \leqslant k \leqslant r_x\}, \\
\hat{\mathrm{A}}_i &= \{(x, k) : x \in \mathrm{A}_i,\ 1 \leqslant k \leqslant s_x\} \qquad (1 \leqslant i \leqslant n), \\
\hat{\mathrm{B}}_i &= \{(x, k) : x \in \mathrm{B}_i,\ 1 \leqslant k \leqslant s_x\} \qquad (1 \leqslant i \leqslant n), \\
\hat{\mathfrak{A}} &= (\hat{\mathrm{A}}_1, \ldots, \hat{\mathrm{A}}_n),\ \hat{\mathfrak{B}} = (\hat{\mathrm{B}}_1, \ldots, \hat{\mathrm{B}}_n).
\end{aligned}$$

The set $\hat{\mathrm{M}}$ is finite; and the following statements are then plainly equivalent.

(i) $\mathfrak{A}$ and $\mathfrak{B}$ possess a CSR $\mathfrak{X}$ which satisfies (4).

(ii) $\hat{\mathfrak{A}}$ and $\hat{\mathfrak{B}}$ possess a CT containing $\hat{\mathrm{M}}$.

Now, by Theorem 9.4.3, statement (ii)—and hence also statement (i)—is valid if and only if, for all $\mathrm{I}, \mathrm{J} \subseteq \{1, \ldots, n\}$,

$$|\hat{\mathrm{A}}(\mathrm{I}) \cap \hat{\mathrm{B}}(\mathrm{J})| + |\{\hat{\mathrm{A}}(\mathrm{I}) \cup \hat{\mathrm{B}}(\mathrm{J})\} \cap \hat{\mathrm{M}}| \geqslant |\mathrm{I}| + |\mathrm{J}| + |\hat{\mathrm{M}}| - n.$$

A simple verification shows that

$$|\hat{\mathrm{A}}(\mathrm{I}) \cap \hat{\mathrm{B}}(\mathrm{J})| = \sum_{x \in \mathrm{A(I)} \cap \mathrm{B(J)}} s_x,$$

$$|\{\hat{\mathrm{A}}(\mathrm{I}) \cup \hat{\mathrm{B}}(\mathrm{J})\} \cap \hat{\mathrm{M}}| = \sum_{x \in \mathrm{A(I)} \cup \mathrm{B(J)}} r_x,$$

$$|\hat{\mathrm{M}}| = \sum_{x \in \mathrm{E}} r_x;$$

and Theorem 9.4.4 therefore follows.

Exercises 9.4

1. Exhibit Theorem 9.4.3 as a consequence of Theorem 9.4.4.
2. Deduce Theorem 3.3.6 from Theorem 9.4.3.
3. Deduce Corollary 9.3.4 from Theorem 9.4.4.

4. Let $\mathfrak{A} = (A_1, \ldots, A_n)$ be a family of subsets of $\{x_1, \ldots, x_m\}_{\neq}$ and let r_k, s_k $(1 \leqslant k \leqslant m)$ be integers such that $0 \leqslant r_k \leqslant s_k$ $(1 \leqslant k \leqslant m)$. Use Theorem 9.4.4 to show that $\mathfrak{A}$ possesses a system $\mathfrak{X}$ of representatives with $r_k \leqslant f(\mathfrak{X}; x_k) \leqslant s_k$ $(1 \leqslant k \leqslant m)$ if and only if, for all $I \subseteq \{1, \ldots, n\}$,

$$|I| \leqslant \min \left\{ n - \sum_{x_k \notin A(I)} r_k, \sum_{x_k \in A(I)} s_k \right\}.$$

Deduce Theorem 3.3.6.

5. Let $\mathscr{E}$ be a pre-independence structure on a set E, and let M be an infinite member of $\mathscr{E}$. Show that the collection

$$\{X \subseteq E : X \cup M \in \mathscr{E}\}$$

need not be a pre-independence structure. (Cf. Lemma 9.4.1.)

6. Let $\mathscr{E}$ be an independence structure on a set E, and let M be an arbitrary independent set. Use Lemma 9.4.1 and Rado's selection principle to prove that the collection

$$\{X \subseteq E : X \cup M \in \mathscr{E}\}$$

is again an independence structure.

7. Let $\mathscr{E}$ be an independence structure, with rank function ρ, on a finite set E; let $M \subseteq E$, $M \in \mathscr{E}$; let n be a natural number; and write

$$\tilde{\mathscr{E}} = \{X \subseteq E : X \cup M \in \mathscr{E}, |X \cup M| \leqslant n\}.$$

By means of Lemmas 9.4.1 and 9.4.2, show that $\tilde{\mathscr{E}}$ is an independence structure, whose rank function $\tilde{\rho}$ is given by

$$\tilde{\rho}(X) = \min \{\rho(X \cup M), n\} - |M| + |X \cap M| \qquad (X \subseteq E).$$

Further, let $\mathfrak{A} = (A_1, \ldots, A_n)$ be a family of subsets of E. Verify that $\mathfrak{A}$ possesses a transversal in $\mathscr{E}$ which contains M if and only if $\mathfrak{A}$ possesses a transversal in $\tilde{\mathscr{E}}$. Using Rado's theorem 6.2.1, prove that $\mathfrak{A}$ possesses a transversal in $\mathscr{E}$ which contains M if and only if, for all $I \subseteq \{1, \ldots, n\}$,

$$\min \{\rho(A(I) \cup M), n\} \geqslant |I| + |M| - |A(I) \cap M|.$$

Hence deduce Theorem 9.4.3.

[H. Perfect]

9.5 An insertion theorem for common transversals

In the present section, we shall obtain a general structural principle for common transversals. Later, a number of further results will be derived as consequences of this principle.

LEMMA 9.5.1. *Let $\mathfrak{A} = (A_1, \ldots, A_m)$, $\mathfrak{B} = (B_1, \ldots, B_n)$ be two families of sets. Then one of these families and a subfamily of the other possess a common transversal if and only if*

$$|A(I) \cap B(J)| \geqslant |I| + |J| - \max(m, n)$$

whenever $I \subseteq \{1, \ldots, m\}$, $J \subseteq \{1, \ldots, n\}$.

This is simply the case $k = \min(m, n)$ of Theorem 9.3.2.

THEOREM 9.5.2. *Let* $\mathfrak{A}$, $\mathfrak{B}$ *be two finite families of subsets of a set* E, *and let* $\mathfrak{A}' \subseteq \mathfrak{A}$, $\mathfrak{B}' \subseteq \mathfrak{B}$. *Then the following statements are equivalent.*

(i) *There exist families* $\mathfrak{A}_0$, $\mathfrak{B}_0$ *with* $\mathfrak{A}' \subseteq \mathfrak{A}_0 \subseteq \mathfrak{A}$, $\mathfrak{B}' \subseteq \mathfrak{B}_0 \subseteq \mathfrak{B}$ *which possess a common transversal.*

(ii) (a) $\mathfrak{A}'$ *and a subfamily of* $\mathfrak{B}$ *possess a common transversal.*

(b) $\mathfrak{B}'$ *and a subfamily of* $\mathfrak{A}$ *possess a common transversal.*

The term 'insertion theorem' in the heading refers to the fact that the families $\mathfrak{A}_0$, $\mathfrak{B}_0$ are inserted between $\mathfrak{A}'$, $\mathfrak{A}$ and $\mathfrak{B}'$, $\mathfrak{B}$ respectively. Theorem 2.3.1 is another result of this type, and so is Theorem 10.1.7 for CSRs.

We note that statement (i) in the theorem obviously implies (ii). To show that (ii) implies (i), we write

$$\mathfrak{A} = (A_1, \ldots, A_m), \qquad \mathfrak{A}' = (A_1, \ldots, A_\mu) \qquad (\mu \leqslant m),$$

$$\mathfrak{B} = (B_1, \ldots, B_n), \qquad \mathfrak{B}' = (B_1, \ldots, B_\nu) \qquad (\nu \leqslant n).$$

We note that, in view of (a) and (b), $\mu \leqslant n$, $\nu \leqslant m$. By Lemma 9.5.1 and (a), we infer that

$$|A(I) \cap B(J)| \geqslant |I| + |J| - n \text{ when } I \subseteq \{1, \ldots, \mu\},\ J \subseteq \{1, \ldots, n\}. \qquad (1)$$

Again, by Lemma 9.5.1 and (b),

$$|A(I) \cap B(J)| \geqslant |I| + |J| - m \text{ when } I \subseteq \{1, \ldots, m\},\ J \subseteq \{1, \ldots, \nu\}. \qquad (2)$$

We shall assume, as may be done without loss of generality, that $m \leqslant n$. Let D be a set such that $D \cap E = \emptyset$, and let $\mathfrak{A}^*$, $\mathfrak{B}^*$ be families consisting of the following sets.

$$\mathfrak{A}^*: A_1, \ldots, A_\mu, \quad A_{\mu+1} \cup D, \ldots, A_m \cup D, \quad A_{m+1} = D, \ldots, A_n = D;$$

$$\mathfrak{B}^*: B_1, \ldots, B_\nu, \quad B_{\nu+1} \cup D, \quad \ldots \quad \ldots \quad , B_n \cup D.$$

Next, we propose to show that, if D contains enough elements, then $\mathfrak{A}^*$ and $\mathfrak{B}^*$ possess a CT. In view of Corollary 9.3.4, we need to verify that the relations

$$I_1 \subseteq \{1, \ldots, \mu\}, \qquad I_2 \subseteq \{\mu + 1, \ldots, m\}, \qquad I_3 \subseteq \{m + 1, \ldots, n\},$$

$$J_1 \subseteq \{1, \ldots, \nu\}, \qquad J_2 \subseteq \{\nu + 1, \ldots, n\}$$

imply the inequality†

$$\left|[A(I_1) \cup \{A(I_2) \cup |I_2|\, D\} \cup |I_3|\, D] \cap [B(J_1) \cup \{B(J_2) \cup |J_2|\, D\}]\right| \\ \geqslant |I_1| + |I_2| + |I_3| + |J_1| + |J_2| - n,$$

that is

$$\left|\{A(I_1 \cup I_2) \cup |I_2 \cup I_3| D\} \cap \{B(J_1 \cup J_2) \cup |J_2| D\}\right| \\ \geqslant |I_1 \cup I_2| + |I_3| + |J_1 \cup J_2| - n. \tag{3}$$

If $I_2 \cup I_3 = \emptyset$, then (3) reduces to the statement

$$|A(I_1) \cap B(J_1 \cup J_2)| \geqslant |I_1| + |J_1 \cup J_2| - n,$$

and this holds by (1). If $I_2 \cup I_3 \neq \emptyset$, $J_2 = \emptyset$, then (3) becomes

$$|A(I_1 \cup I_2) \cap B(J_1)| \geqslant |I_1 \cup I_2| + |I_3| + |J_1| - n.$$

Now the greatest value that $|I_3|$ can take is $n - m$. We have therefore to show that

$$|A(I_1 \cup I_2) \cap B(J_1)| \geqslant |I_1 \cup I_2| + |J_1| - m,$$

and this holds by virtue of (2). Finally, if $I_2 \cup I_3 \neq \emptyset$, $J_2 \neq \emptyset$, then (3) assumes the form

$$|D| + |A(I_1 \cup I_2) \cap B(J_1 \cup J_2)| \geqslant |I_1 \cup I_2| + |I_3| + |J_1 \cup J_2| - n;$$

and this is clearly valid if $|D|$ is sufficiently large.

We now examine how the CT, whose existence we have established, links the sets in $\mathfrak{A}^*$ and $\mathfrak{B}^*$. Each of the sets $A_{m+1}, \ldots, A_n$ is identical to D and so cannot be linked to any set among $B_1, \ldots, B_\nu$. Hence $A_{m+1}, \ldots, A_n$ must be linked to $n - m$ sets among $B_{\nu+1} \cup D, \ldots, B_n \cup D$; and, for notational convenience, we may assume that $A_{m+1}, \ldots, A_n$ are linked to $B_{m+1} \cup D, \ldots, B_n \cup D$ respectively. Then the residual families, namely

$$A_1, \ldots, A_\mu, A_{\mu+1} \cup D, \ldots, A_m \cup D,$$
$$B_1, \ldots, B_\nu, B_{\nu+1} \cup D, \ldots, B_m \cup D$$

have a CT. The elements in this CT which belong to D (if there are such) provide links between certain pairs of sets of the form $A_i \cup D$, $B_j \cup D$. Let us suppose that the sets linked in this way are

$$A_{k+1} \cup D, \ldots, A_m \cup D; \qquad B_{k+1} \cup D, \ldots, B_m \cup D,$$

† We recall (cf. §3.3) that, when k is a non-negative integer and X a set, kX is defined as X or $\emptyset$ according as $k > 0$ or $k = 0$.

where max $(\mu, \nu) \leqslant k \leqslant m$. Then the residual families, namely

$$A_1, \ldots, A_\mu, A_{\mu+1} \cup D, \ldots, A_k \cup D,$$
$$B_1, \ldots, B_\nu, B_{\nu+1} \cup D, \ldots, B_k \cup D$$

have a CT consisting of elements of E only. Thus

$$\mathfrak{A}_0 = (A_1, \ldots, A_\mu, A_{\mu+1}, \ldots, A_k), \qquad \mathfrak{B}_0 = (B_1, \ldots, B_\nu, B_{\nu+1}, \ldots, B_k)$$

have a CT.

As an application of the result just proved, we shall give an alternative treatment of Theorem 7.4.3 for the case of finite families.

Let, then, $\mathfrak{A} = (A_i : i \in I)$ and $\mathfrak{B} = (B_j : j \in J)$ be two finite families of sets and denote by $\mathscr{J}$ the collection of all subsets I′ of I such that $\mathfrak{A}(I')$ has a CT with some subfamily of $\mathfrak{B}$. It is required to show that $\mathscr{J}$ *is an independence structure on* I, and to do this we need only verify the replacement axiom. For convenience of notation, we shall write $I' \leftrightarrow J'$ whenever I′, J′ are subsets of I, J respectively such that $\mathfrak{A}(I')$ and $\mathfrak{B}(J')$ possess a CT.

Now let $I_1, I_2 \in \mathscr{J}$ and $|I_1| = r$, $|I_2| = r + 1$. Denote by J_1, J_2 subsets of J such that $I_1 \leftrightarrow J_1$, $I_2 \leftrightarrow J_2$. Then $\mathfrak{A}(I_1)$ and a subfamily of $\mathfrak{B}(J_1 \cup J_2)$ have a CT; equally, $\mathfrak{B}(J_2)$ and a subfamily of $\mathfrak{A}(I_1 \cup I_2)$ have a CT. By Theorem 9.5.2 there exist, therefore, sets I_0, J_0 such that $I_0 \leftrightarrow J_0$ and

$$I_1 \subseteq I_0 \subseteq I_1 \cup I_2, \qquad J_2 \subseteq J_0 \subseteq J_1 \cup J_2.$$

Hence

$$|I_0| = |J_0| \geqslant |J_2| = |I_2| = r + 1.$$

But $I_0 \in \mathscr{J}$ and so there exists a set $I' \in \mathscr{J}$ such that $I_1 \subseteq I' \subseteq I_1 \cup I_2$ and $|I'| = r + 1$. This establishes the replacement property.

Exercises 9.5

1. Deduce Theorem 3.4.1 from Theorem 9.5.2.
2. Deduce Theorem 9.5.2 from Menger's theorem 1.7.2.

[H. Perfect (**2**)]

3. Use Theorem 7.4.3 to derive the case $\mathfrak{B}' = \mathfrak{B}$ of Theorem 9.5.2.

9.6 Harder results for a single family

We now return to the study of transversals and similar objects associated with a single family. However, all results established below are based on the insertion theorem of the preceding section.

THEOREM 9.6.1. *Let* $(A_1, \ldots, A_n)$ *be a family of subsets of* E, *and let* $(E_1, \ldots, E_p)$ *be a partition of* E. *Further, let* $0 \leqslant r_i' \leqslant r_i$ $(1 \leqslant i \leqslant n)$, $0 \leqslant s_j' \leqslant s_j$ $(1 \leqslant j \leqslant p)$ *be integers. Then the following statements are equivalent.*

(i) *There exist pairwise disjoint sets* $X_1, \ldots, X_n$ *with* $X_i \subseteq A_i$, $r_i' \leqslant |X_i| \leqslant r_i$ $(1 \leqslant i \leqslant n)$ *such that the set* $X = X_1 \cup \ldots \cup X_n$ *satisfies the inequalities*

$$s_j' \leqslant |X \cap E_j| \leqslant s_j \qquad (1 \leqslant j \leqslant p).$$

(ii) *Whenever* $I \subseteq \{1, \ldots, n\}$, $J \subseteq \{1, \ldots, p\}$, *we have*

$$|A(I) \cap E(J)| \geqslant \max\left\{ \sum_{i \in I} r_i' - \sum_{j \notin J} s_j,\ \sum_{j \in J} s_j' - \sum_{i \notin I} r_i \right\}.$$

We shall denote by $\mathfrak{A}$ resp. $\mathfrak{A}'$ the family consisting of r_i resp. r_i' copies of A_i, $1 \leqslant i \leqslant n$; and by $\mathfrak{E}$ resp. $\mathfrak{E}'$ the family consisting of s_j resp. s_j' copies of E_j, $1 \leqslant j \leqslant p$. Statement (i) in the theorem is evidently valid if and only if there exist families $\mathfrak{A}_0$, $\mathfrak{E}_0$ with a CT for which $\mathfrak{A}' \subseteq \mathfrak{A}_0 \subseteq \mathfrak{A}$, $\mathfrak{E}' \subseteq \mathfrak{E}_0 \subseteq \mathfrak{E}$. By Theorem 9.5.2, this is the case precisely if both the following conditions are satisfied: (a) $\mathfrak{A}'$ and a subfamily of $\mathfrak{E}$ have a CT; (b) $\mathfrak{E}'$ and a subfamily of $\mathfrak{A}$ have a CT. Now, by the case $k = m$ of Theorem 9.3.2, condition (a) holds if and only if

$$\left| \bigcup_{i=1}^{n} \rho_i' A_i \cap \bigcup_{j=1}^{p} \sigma_j E_j \right| \geqslant \sum_{i=1}^{n} \rho_i' + \sum_{j=1}^{p} \sigma_j - \sum_{j=1}^{p} s_j$$

whenever $0 \leqslant \rho_i' \leqslant r_i'$ $(1 \leqslant i \leqslant n)$, $0 \leqslant \sigma_j \leqslant s_j$ $(1 \leqslant j \leqslant p)$. This means that, for $I \subseteq \{1, \ldots, n\}$, $J \subseteq \{1, \ldots, p\}$,

$$\left| \bigcup_{i \in I} A_i \cap \bigcup_{j \in J} E_j \right| \geqslant \sum_{i \in I} r_i' + \sum_{j \in J} s_j - \sum_{j=1}^{p} s_j$$
$$= \sum_{i \in I} r_i' - \sum_{j \notin J} s_j.$$

By analogous reasoning it follows that (b) holds if and only if, for all $I \subseteq \{1, \ldots, n\}$, $J \subseteq \{1, \ldots, p\}$,

$$\left| \bigcup_{i \in I} A_i \cap \bigcup_{j \in J} E_j \right| \geqslant \sum_{j \in J} s_j' - \sum_{i \notin I} r_i.$$

This completes the proof.

COROLLARY 9.6.2. (Hoffman & Kuhn) *Let* $\mathfrak{A} = (A_1, \ldots, A_n)$ *be a family of subsets of* E, *and let* $(E_1, \ldots, E_p)$ *be a partition of* E. *Further, let* $0 \leqslant r_j \leqslant s_j$

$(1 \leqslant j \leqslant p)$ be integers. Then $\mathfrak{A}$ possesses a transversal X *such that*

$$r_j \leqslant |X \cap E_j| \leqslant s_j \qquad (1 \leqslant j \leqslant p) \tag{1}$$

if and only if, for all $I \subseteq \{1, \ldots, n\}$, $J \subseteq \{1, \ldots, p\}$,

$$|A(I) \cap E(J)| \geqslant |I| - \min\left\{n - \sum_{j \in J} r_j, \sum_{j \notin J} s_j\right\}. \tag{2}$$

This result follows at once if we take $r_i' = r_i = 1$ $(1 \leqslant i \leqslant n)$ in Theorem 9.6.1 and replace the symbol s_j' by r_j.

COROLLARY 9.6.3. *The preceding result remains valid if the term 'transversal' is replaced by 'partial transversal' and inequality* (2) *is replaced by*

$$|A(I) \cap E(J)| \geqslant |I| - n + \sum_{j \in J} r_j.$$

This is again a special case of Theorem 9.6.1. To obtain it, we take $r_i' = 0$, $r_i = 1$ $(1 \leqslant i \leqslant n)$ and once more replace the symbol s_j' by r_j.

A feature of Corollary 9.6.3 which, at first sight, may seem surprising is the absence of the numbers s_j from the necessary and sufficient conditions for the existence of the desired PT. The reader may like to investigate for himself why this is the case.

If we specify that X should be a partial transversal of prescribed cardinal, then the problem becomes harder. The solution is contained in the next theorem.

THEOREM 9.6.4. *Let the notation be as in Corollary* 9.6.2. *Then $\mathfrak{A}$ possesses a partial transversal of cardinal t and satisfying* (1) *if and only if*

$$\sum_{j=1}^{p} r_j \leqslant t \tag{3}$$

and, for all $I \subseteq \{1, \ldots, n\}$, $J \subseteq \{1, \ldots, p\}$,

$$|A(I) \cap E(J)| \geqslant |I| - n + \max\left\{\sum_{j \in J} r_j, t - \sum_{j \notin J} s_j\right\}. \tag{4}$$

For $t = n$, this result reduces to Corollary 9.6.2. In this connection it should be noted that, for $I = \emptyset$, $J = \{1, \ldots, p\}$, condition (2) becomes

$$\sum_{1}^{p} r_j \leqslant n.$$

We shall prove the assertion by applying Corollary 9.6.2 to a family derived from $\mathfrak{A}$ by an elementary construction. We first note that $t \leqslant n$; for if a PT of the desired kind exists, this is obvious; and equally it follows from (4). Let E_0 be any set such that $E_0 \cap E = \emptyset$, $|E_0| = n - t$, and write $\mathfrak{A}^* = (A_1 \cup E_0, \ldots, A_n \cup E_0)$. Also put $r_0 = s_0 = n - t$. We propose to establish the equivalence of the following two statements.

(i) $\mathfrak{A}$ possesses a PT X of cardinal t which satisfies (1).

(ii) $\mathfrak{A}^*$ possesses a transversal X^* such that

$$r_k \leqslant |X^* \cap E_k| \leqslant s_k \qquad (0 \leqslant k \leqslant p). \tag{5}$$

Let (i) be satisfied and take $X^* = X \cup E_0$. Then (5) is valid. Further, let X be a transversal of $(A_1, \ldots, A_t)$, and so of $(A_1 \cup E_0, \ldots, A_t \cup E_0)$. Since E_0 is plainly a transversal of $(A_{t+1} \cup E_0, \ldots, A_n \cup E_0)$, it follows that X^* is a transversal of $\mathfrak{A}^*$. Conversely, let (ii) be satisfied; and write $X = X^* \cap E$. Then X is a PT of $\mathfrak{A}$ and (1) clearly holds. Moreover $r_0 \leqslant |X^* \cap E_0| \leqslant s_0$, i.e. $|X^* \cap E_0| = n - t$ and so $|X| = t$.

The equivalence of (i) and (ii) has now been established. Now, by Corollary 9.6.2, statement (ii)—and consequently also (i)—holds if and only if, for all $I \subseteq \{1, \ldots, n\}$, $J \subseteq \{1, \ldots, p\}$, $\lambda = 0, 1$, we have

$$\Big|\{A(I) \cup |I|\, E_0\} \cap \{E(J) \cup \lambda E_0\}\Big|$$

$$\geqslant \max\left\{|I| - n + \lambda r_0 + \sum_{j \in J} r_j,\ |I| - \sum_{0}^{p} s_j + \lambda s_0 + \sum_{j \in J} s_j\right\}$$

$$= \max\left\{|I| - n + \lambda r_0 + \sum_{j \in J} r_j,\ |I| - \sum_{1}^{p} s_j - s_0 + \lambda s_0 + \sum_{j \in J} s_j\right\}.$$

Thus the required condition is:

$$|A(I) \cap E(J)| + \chi(\lambda|I|)\,.\,(n - t)$$

$$\geqslant \max\left\{|I| - n + \sum_{j \in J} r_j,\ |I| - n + t - \sum_{j \notin J} s_j\right\} + \lambda(n - t)$$

(where the function χ was defined in § 9.2).

For $\lambda = 0$ and also for $\lambda = 1$, $I \neq \emptyset$, this condition simply reduces to (4) while for $\lambda = 1$, $I = \emptyset$, it reduces to (3). The proof is therefore complete.

THEOREM 9.6.5. *Let* $(E_1, \ldots, E_m)$ *and* $(F_1, \ldots, F_n)$ *be two partitions of a set* E. *Further, let* $0 \leqslant r_i' \leqslant r_i$ $(1 \leqslant i \leqslant m)$, $0 \leqslant s_j' \leqslant s_j$ $(1 \leqslant j \leqslant n)$ *be integers. Then there exists a set* $X \subseteq E$ *satisfying the relations*

$$r_i' \leqslant |X \cap E_i| \leqslant r_i \quad (1 \leqslant i \leqslant m), \qquad s_j' \leqslant |X \cap F_j| \leqslant s_j \quad (1 \leqslant j \leqslant n) \tag{6}$$

if and only if, for all $I \subseteq \{1, ..., m\}$, $J \subseteq \{1, ..., n\}$,

$$\sum_{\substack{i \in I \\ j \in J}} |E_i \cap F_j| \geqslant \max \left\{ \sum_{i \in I} r_i' - \sum_{j \notin J} s_j,\ \sum_{j \in J} s_j' - \sum_{i \notin I} r_i \right\}.$$

Taking $(A_1, ..., A_n)$ to be a partition of E and suitably changing the notation in Theorem 9.6.1, we infer at once that a set X which satisfies (6) exists if and only if, for all $I \subseteq \{1, ..., m\}$, $J \subseteq \{1, ..., n\}$,

$$\left| \bigcup_{i \in I} E_i \cap \bigcup_{j \in J} F_j \right| \geqslant \max \left\{ \sum_{i \in I} r_i' - \sum_{j \notin J} s_j,\ \sum_{j \in J} s_j' - \sum_{i \notin I} r_i \right\}.$$

But

$$\left| \bigcup_{i \in I} E_i \cap \bigcup_{j \in J} F_j \right| = \sum_{\substack{i \in I \\ j \in J}} |E_i \cap F_j|,$$

and the assertion follows.

THEOREM 9.6.6. *Let* $A_1, ..., A_n$ *be subsets of* $E = \{x_1, ..., x_m\}_{\neq}$. *Further, let* $0 \leqslant s_i' \leqslant s_i$ $(1 \leqslant i \leqslant n)$, $0 \leqslant r_j' \leqslant r_j$ $(1 \leqslant j \leqslant m)$ *be integers. The following statements are then equivalent.*

(i) *There exist sets* $X_1, ..., X_n$ *with* $X_i \subseteq A_i$, $s_i' \leqslant |X_i| \leqslant s_i$ $(1 \leqslant i \leqslant n)$ *such that, for* $1 \leqslant j \leqslant m$, x_j *belongs to at least* r_j' *and at most* r_j *X's.*

(ii) *The inequality*

$$\sum_{i \in I} |A_i \cap F| \geqslant \max \left\{ \sum_{i \in I} s_i' - \sum_{x_j \notin F} r_j,\ \sum_{x_j \in F} r_j' - \sum_{i \notin I} s_i \right\}$$

holds whenever $I \subseteq \{1, ..., n\}$, $F \subseteq E$.

To establish this result, we write

$$\hat{E} = \{(i,j) : 1 \leqslant i \leqslant n,\ 1 \leqslant j \leqslant m,\ x_j \in A_i\},$$
$$\hat{A}_i = \{(i,j) : 1 \leqslant j \leqslant m,\ x_j \in A_i\} \qquad (1 \leqslant i \leqslant n),$$
$$\hat{B}_j = \{(i,j) : 1 \leqslant i \leqslant n,\ x_j \in A_i\} \qquad (1 \leqslant j \leqslant m).$$

Then $(\hat{A}_1, ..., \hat{A}_n)$ and $(\hat{B}_1, ..., \hat{B}_m)$ are partitions of $\hat{E}$. Statement (i) in the theorem plainly holds if and only if there exists a set $Y \subseteq \hat{E}$ such that

$$s_i' \leqslant |Y \cap \hat{A}_i| \leqslant s_i \quad (1 \leqslant i \leqslant n), \qquad r_j' \leqslant |Y \cap \hat{B}_j| \leqslant r_j \quad (1 \leqslant j \leqslant m);$$

and, by Theorem 9.6.5, this is plainly the case if and only if

$$\sum_{\substack{i \in I \\ j \in J}} |\hat{A}_i \cap \hat{B}_j| \geqslant \max \left\{ \sum_{i \in I} s_i' - \sum_{j \notin J} r_i,\ \sum_{j \in J} r_j' - \sum_{i \notin I} s_i \right\}$$

whenever $I \subseteq \{1, \ldots, n\}$, $J \subseteq \{1, \ldots, m\}$. Now write $F = \{x_j : j \in J\}$. We have $|\hat{A}_i \cap \hat{B}_j| = 1$ or 0 according as $x_j \in A_i$ or $x_j \notin A_i$; and so

$$\sum_{\substack{i \in I \\ j \in J}} |\hat{A}_i \cap \hat{B}_j| = \sum_{i \in I} |A_i \cap F|.$$

The proof is therefore complete.

Exercises 9.6

1. Exhibit Theorem 3.3.6 as a special case of Corollary 9.6.2.

2. Let $A_1, \ldots, A_n$ be subsets of $E = \{x_1, \ldots, x_m\}_{\neq}$. Further, let s_i $(1 \leqslant i \leqslant n)$ and r_j $(1 \leqslant j \leqslant m)$ be non-negative integers. Show that there exist sets $X_1, \ldots, X_n$ with $X_i \subseteq A_i$, $s_i \leqslant |X_i|$ $(1 \leqslant i \leqslant n)$ such that, for $1 \leqslant j \leqslant m$, x_j belongs to at most r_j X's if and only if

$$\sum_{i \in I} |A_i \cap F| \geqslant \sum_{i \in I} s_i - \sum_{x_j \notin F} r_j$$

whenever $I \subseteq \{1, \ldots, n\}$ and $F \subseteq E$.

3. By specializing Theorem 9.6.6, show that the family $(A_1, \ldots, A_n)$ of subsets of a finite set E possesses a transversal if and only if, for all $I \subseteq \{1, \ldots, n\}$, $F \subseteq E$,

$$\sum_{i \in I} |A_i \cap F| \geqslant |I| + |F| - |E|.$$

(Cf. Ex. 2.2.1.)

4. Let $\mathfrak{A} = (A_1, \ldots, A_n)$ be a family of subsets of $E = \{x_1, \ldots, x_m\}_{\neq}$, and let $0 \leqslant r_k \leqslant s_k$ $(1 \leqslant k \leqslant m)$ be integers. Writing $s = s_1 + \ldots + s_m$, $N = \{1, 2, \ldots, s\}$,

$$A_i^* = A_i \times N, \qquad \mathfrak{A}^* = (A_1^*, \ldots, A_n^*), \qquad E^* = E \times N, \qquad E_k^* = \{x_k\} \times N,$$

and applying the Hoffman-Kuhn theorem (Corollary 9.6.2) to the family $\mathfrak{A}^*$ of subsets of E^* and the partition $(E_1^*, \ldots, E_m^*)$ of E^*, show that $\mathfrak{A}$ possesses a system $\mathfrak{X}$ of representatives with

$$r_k \leqslant f(\mathfrak{X}; x_k) \leqslant s_k \qquad (1 \leqslant k \leqslant m)$$

if and only if, for all $I \subseteq \{1, \ldots, n\}$,

$$|I| \leqslant \min\left\{n - \sum_{x_k \notin A(I)} r_k, \sum_{x_k \in A(I)} s_k\right\}.$$

(Cf. Ex. 9.4.4.)

[Ford & Fulkerson]

5. Let $\mathfrak{A} = (A_1, \ldots, A_n)$ be a family of subsets of E, and let $(E_1, \ldots, E_p)$ be a partition of E. Further, let $0 \leqslant r_j \leqslant s_j$ $(1 \leqslant j \leqslant p)$ be integers. Suppose that $\mathfrak{A}$ possesses transversals T', T'' which satisfy the conditions

$$r_j \leqslant |T' \cap E_j|, \qquad |T'' \cap E_j| \leqslant s_j \qquad (1 \leqslant j \leqslant p).$$

Show that $\mathfrak{A}$ possesses a transversal T such that

$$r_j \leqslant |T \cap E_j| \leqslant s_j \qquad (1 \leqslant j \leqslant p).$$

6. Let $(E_1, \ldots, E_m)$ and $(F_1, \ldots, F_n)$ be two partitions of a set E. Further, let $0 \leqslant r_i' \leqslant r_i\ (1 \leqslant i \leqslant m)$, $0 \leqslant s_j' \leqslant s_j\ (1 \leqslant j \leqslant n)$ be integers. Suppose there exist subsets X^*, X^{**} of E such that

$$r_i' \leqslant |X^* \cap E_i| \quad (1 \leqslant i \leqslant m), \qquad |X^* \cap F_j| \leqslant s_j \quad (1 \leqslant j \leqslant n),$$
$$|X^{**} \cap E_i| \leqslant r_i \quad (1 \leqslant i \leqslant m), \qquad s_j' \leqslant |X^{**} \cap F_j| \quad (1 \leqslant j \leqslant n).$$

Show that there exists a subset X of E such that

$$r_i' \leqslant |X \cap E_i| \leqslant r_i \quad (1 \leqslant i \leqslant m), \qquad s_j' \leqslant |X \cap F_j| \leqslant s_j \quad (1 \leqslant j \leqslant n).$$

Notes on Chapter 9

§ 9.2. Theorem 9.2.1 occurs in P. Hall's classical paper (**1**), and we have followed here his very simple and natural argument. W. Maak (**1**, **2**) discovered the theorem independently and used it in his account (**1**) of almost periodic functions. Theorem 9.2.3 was first established by van der Waerden (**2**) in the form stated here, and another proof was subsequently given by Sperner (**1**). Here we adopt quite a different approach. It may be noted that Theorem 9.2.3 is equivalent to a much older result in the theory of graphs due to D. König (**1**; **2**; **7**, chap. 11, Satz 13).
Corollary 9.2.4 is another venerable result. It was proved by G. A. Miller (**1**) in 1910 and rediscovered a few years later by H. W. Chapman (**1**). For a variant, see G. Scorza (**1**). All these authors used group-theoretic arguments, but the discussion was shifted to combinatorial ground when van der Waerden pointed out that the result in question is a trivial consequence of Theorem 9.2.3.

§ 9.3. The most significant special case of Theorem 9.3.2 is contained in Corollary 9.3.4; this last result was established by Ford and Fulkerson in the context of their broad theory of 'flows in networks' (**1**; **2**, 67–75). The treatment of Theorem 9.3.2 given here is modelled on the discussion of Mirsky & Perfect (**2**). The proof of Corollary 9.3.4 indicated in outline only is due to Dr Hazel Perfect (**4**), who has also pointed out to me that a slight refinement of her argument leads to the much stronger Theorem 9.4.3. The deduction of Corollary 9.3.4 from Menger's theorem is also due to Hazel Perfect; indeed, in her paper (**2**) she succeeded in exhibiting a large part of transversal theory as a series of consequences of Menger's theorem.

§ 9.4. The principal result in this section (Theorem 9.4.4) is due to Ford & Fulkerson (**1**; **2**, 67–75), but in deriving it from Theorem 9.4.3, we follow Mirsky & Perfect (**3**). Theorem 9.4.3, which deals with marginal elements of common transversals, is a special case of the principal theorem; the proof given here is due to D. J. A. Welsh (**1**), For an alternative proof, depending on the process of 'substitution' (cf. § 3.1), see Mirsky & Perfect (**3**). R. A. Brualdi (**7**) discussed the problem of marginal elements for common transversals of arbitrary families.

§ 9.5. The insertion theorem for common transversals (Theorem 9.5.2) is due to Mirsky (**4**). A transfinite generalization has been given by J. S. Pym (**3**, **4**). An analogous insertion theorem for common systems of representatives (of not necessarily finite families) will be found in §10.1. The case of common transversals of two arbitrary families is discussed in §10.4.

§ 9.6. Corollary 9.6.2 is a well-known result of Hoffman & Kuhn (**2**), established originally with the aid of linear programming. Another proof, due to D. J. A. Welsh (**1**), makes use of Rado's theorem 6.2.1. The generalization of the Hoffman–Kuhn result contained in Theorem 9.6.1 is due to Mirsky, as are also Theorems

9.6.5 and 9.6.6 (Mirsky (**5**)). A result slightly more special than Theorem 9.6.6 had been proved earlier by W. Vogel (**1**). A radical generalization of the Hoffman–Kuhn theorem (Corollary 9.6.2) has been given by R. A. Brualdi (**5**), who made the statement symmetric with respect to elements and sets and at the same time extended the theorem to the case of arbitrary families. His final conclusion is very comprehensive and contains as special cases virtually all known results on transversals. Theorem 9.6.4 is due to Brualdi (**1**), though the treatment offered here depends on quite different ideas.

10 Links of Two Arbitrary Families

We shall now widen the scope of the discussion initiated in the preceding chapter and consider links between pairs of *arbitrary* families.

10.1 The theorem of Mendelsohn and Dulmage and its interpretations

We begin with an obvious consequence of Theorem 4.2.2.

LEMMA 10.1.1. *Let* (X, Δ, Y) *be a deltoid; let* $X' \subseteq X$; *and suppose that* $\Delta(x)$ *is finite for each* $x \in X'$. *Then* X' *is admissible if and only if* $|\Delta(A)| \geqslant |A|$ *for every finite subset* A *of* X'.

Write $\mathscr{D} = (X, \Delta, Y)$, $\mathscr{D}' = (X', \Delta', Y)$, where

$$\Delta' = \{(x, y) \in \Delta : x \in X',\ y \in Y\}.$$

Then $\mathscr{D}'$ is locally right-finite, and a subset of X' is admissible with respect to $\mathscr{D}$ if and only if it is admissible with respect to $\mathscr{D}'$. Applying the equivalence of (a) and (c) in Theorem 4.2.2 to the deltoid $\mathscr{D}'$, we obtain at once the desired result.

THEOREM 10.1.2. (Transfinite form of the Mendelsohn–Dulmage theorem) *Let* (X, Δ, Y) *be a deltoid and let* X', Y' *be subsets of* X, Y *respectively. Suppose that* $\Delta(x)$ *is finite for each* $x \in X'$ *and* $\Delta(y)$ *for each* $y \in Y'$. *The following statements are then equivalent.*

(i) *There exist linked sets* X_0, Y_0 *such that* $X' \subseteq X_0 \subseteq X$, $Y' \subseteq Y_0 \subseteq Y$.

(ii) (a) $|\Delta(A)| \geqslant |A|$ *for every finite subset* A *of* X'.

(b) $|\Delta(B)| \geqslant |B|$ *for every finite subset* B *of* Y'.

This powerful result derives from the fusion of (the deltoid form of) Hall's theorem and Theorem 2.3.1.

Let (i) be given and denote by $\phi : X_0 \to Y_0$ an admissible bijection. Let A be any finite subset of X'. Now $(x, \phi(x)) \in \Delta$ for all $x \in X_0$ and, a fortiori, for all $x \in X'$. Then $\phi(x) \in \Delta(x)$ for all $x \in X'$, and (ii a) follows since ϕ is injective. By considering the bijection $\phi^{-1} : Y_0 \to X_0$, we infer, in the same way, the validity of (ii b).

Next, let (ii) be given. By (a) and Lemma 10.1.1, we see that X′ is admissible; by (b) and the dual of Lemma 10.1.1, we see that Y′ is admissible. Statement (i) now follows by Theorem 2.3.1.

On the face of it, Theorem 10.1.2 may appear to be a somewhat superficial result of only limited interest. This impression is misleading, and the force of the theorem is seen in the interpretations to which it gives rise.

THEOREM 10.1.3. *Let* I, I′, E, E′ *be sets such that* $I' \subseteq I$, $E' \subseteq E$, *and let* $\mathfrak{A} = (A_i : i \in I)$ *be a family of subsets of* E. *Suppose that* A_i *is finite for each* $i \in I'$ *and that each element of* E′ *belongs to at most a finite number of* A's. *Then there exist sets* I_0, E_0 *with* $I' \subseteq I_0 \subseteq I$, $E' \subseteq E_0 \subseteq E$ *such that* E_0 *is a transversal of the subfamily* $\mathfrak{A}(I_0)$ *if and only if both the following conditions are satisfied.* (i) *For every finite subset* I* *of* I′, A(I*) *contains at least* |I*| *elements.* (ii) *Every finite subset* E* *of* E′ *intersects at least* |E*| *sets in* $\mathfrak{A}$.

Theorem 10.1.3 (which is essentially the transfinite analogue of Theorem 3.4.1) is, of course, equivalent to Theorem 10.1.2. To obtain it, we use Theorem 10.1.2 with X = E, X′ = E′, Y = I, Y′ = I′ and with Δ defined by the requirement that $(e, i) \in \Delta$ precisely when $e \in A_i$.

It should be noted that the specialization I′ = I in Theorem 10.1.3 yields at once the Hoffman–Kuhn–Rado theorem 6.6.3.

THEOREM 10.1.4. *Let* $\mathfrak{A} = (A_i : i \in I)$, $\mathfrak{B} = (B_j : j \in J)$ *be two families of subsets of a set* E; *and let* $I' \subseteq I$, $J' \subseteq J$. *Suppose that each* A_i *with* $i \in I'$ *intersects at most a finite number of* B's *and that each* B_j *with* $j \in J'$ *intersects at most a finite number of* A's. *Then there exist sets* I_0, J_0 *with* $I' \subseteq I_0 \subseteq I$, $J' \subseteq J_0 \subseteq J$ *such that the subfamilies* $\mathfrak{A}(I_0)$ *and* $\mathfrak{B}(J_0)$ *possess a common system of representatives if and only if both the following conditions are satisfied.*

(i) *For each finite subset* I* *of* I′, *the set* A(I*) *intersects at least* |I*| B*'s.*

(ii) *For each finite subset* J* *of* J′, *the set* B(J*) *intersects at least* |J*| A*'s.*

We have here, once again, a result equivalent to Theorem 10.1.2. To prove it, we take X = I, X′ = I′, Y = J, Y′ = J′ in that theorem and specify Δ by the requirement that, for $i \in I$, $j \in J$, the relation $(i, j) \in \Delta$ holds if and only if $A_i \cap B_j \neq \emptyset$.

Two families $\mathfrak{A} = (A_i : i \in I)$ and $\mathfrak{B} = (B_j : j \in J)$ will be said to be *relatively finite* if each A resp. B intersects only a finite number of B's resp. A's.

THEOREM 10.1.5. *Let* $\mathfrak{A} = (A_i : i \in I)$, $\mathfrak{B} = (B_j : j \in J)$ *be two infinite, relatively finite families of sets. The following statements are then equivalent.*

(i) $\mathfrak{A}$ *and* $\mathfrak{B}$ *possess a common system of representatives.*

(ii) *For each natural number k, the union of any k* A's *resp.* B's *intersects at least k* B's *resp.* A's.

(iii) *Each finite subfamily of* $\mathfrak{A}$ *resp.* $\mathfrak{B}$ *has a common system of representatives with some subfamily of* $\mathfrak{B}$ *resp.* $\mathfrak{A}$.

The equivalence of (i) and (ii) is simply the case $I' = I$, $J' = J$ of Theorem 10.1.4. Further, the implications (i) $\Rightarrow$ (iii) and (iii) $\Rightarrow$ (ii) hold trivially. The assertion is therefore valid. It would, of course, have been an easy matter to obtain an alternative proof by invoking Theorem 1.3.4 or Theorem 2.3.1.

When we compare Theorem 10.1.5 with its finite analogue (Theorem 9.2.1), we recognize at once the essential point of difference between the two cases: in the transition from the finite to the infinite version, the 'one-sided' statement is superseded by one that is 'two-sided'. Indeed, the two-sided form of clauses (ii) and (iii) in Theorem 10.1.5 is essential, as is easily demonstrated by a counter-example (cf. Ex. 10.1.2).

We now resume the discussion of an algebraic problem raised in § 9.2.

THEOREM 10.1.6. *Let H be a finite subgroup of an arbitrary group G. Then the family of left cosets of H and that of right cosets of H possess a common transversal.*

When G is finite, the result is already known from Corollary 9.2.4. When G is infinite, let $\mathfrak{A} = (A_i : i \in I)$, $\mathfrak{B} = (B_j : j \in J)$ denote the (infinite) families of left cosets and right cosets respectively. Obviously, $\mathfrak{A}$ and $\mathfrak{B}$ are relatively finite. Moreover, the union of any k A's contains precisely $k|H|$ elements and so intersects at least k B's. Similarly, the union of any k B's intersects at least k A's. Hence, by Theorem 10.1.5, $\mathfrak{A}$ and $\mathfrak{B}$ possess a CSR and consequently also a CT.

The conclusion of Theorem 10.1.6 is still valid if H is a subgroup of finite index; but in that case the proof is group-theoretic rather than combinatorial, and we omit the details. Finally, if both the order and the index of the subgroup H of G are infinite, then the theorem ceases to be valid. This can be demonstrated by the following example (in which **Z** and **Q** denote the set of integers and the set of rational numbers respectively). For $a, b \in \mathbf{Q}$, let $\theta_b{}^a$ be the mapping $x \to ax + b$ of **Q** into itself; let G be the group, with respect to the composition of mappings, of all $\theta_b{}^a$ with $a, b \in \mathbf{Q}$, $a \neq 0$; and let H be the subgroup $\{\theta_n{}^1 : n \in \mathbf{Z}\}$ of G. Then

$$\theta_0{}^2 H = \{\theta_{2n}{}^2 : n \in \mathbf{Z}\}, \qquad \theta_1{}^2 H = \{\theta_{2n+1}{}^2 : n \in \mathbf{Z}\},$$
$$H\theta_0{}^2 = \{\theta_n{}^2 : n \in \mathbf{Z}\}.$$

Hence $\theta_0^2 H \cup \theta_1^2 H = H\theta_0^2$. If therefore X is a transversal of the family of right cosets, then it fails to intersect at least one of the two disjoint left cosets $\theta_0^2 H, \theta_1^2 H$. Thus the two families do not possess a CT.

Next, we return to general combinatorial theory and formulate a 'qualitative' variant of Theorem 10.1.4, in which there is no restriction on cardinals.

THEOREM 10.1.7. *Let $\mathfrak{A}$, $\mathfrak{B}$ be two arbitrary families of sets, and let $\mathfrak{A}' \subseteq \mathfrak{A}$, $\mathfrak{B}' \subseteq \mathfrak{B}$. The following statements are then equivalent.*

(i) *There exist families $\mathfrak{A}_0$, $\mathfrak{B}_0$ with $\mathfrak{A}' \subseteq \mathfrak{A}_0 \subseteq \mathfrak{A}$, $\mathfrak{B}' \subseteq \mathfrak{B}_0 \subseteq \mathfrak{B}$ which possess a common system of representatives.*

(ii) (a) *$\mathfrak{A}'$ and a subfamily of $\mathfrak{B}$ possess a common system of representatives.*

(b) *$\mathfrak{B}'$ and a subfamily of $\mathfrak{A}$ possess a common system of representatives.*

The implication (i) ⇒ (ii) is trivial. Let, then, (ii) be given and write $\mathfrak{A} = (A_i : i \in I)$, $\mathfrak{A}' = (A_i : i \in I')$, $\mathfrak{B} = (B_j : j \in J)$, $\mathfrak{B}' = (B_j : j \in J')$, where $I' \subseteq I$, $J' \subseteq J$. We shall consider the deltoid (I, Δ, J), where

$$\Delta = \{(i, j) : i \in I,\ j \in J,\ A_i \cap B_j \neq \emptyset\}.$$

By (a), I′ is an admissible subset of I; by (b), J′ is an admissible subset of J. Hence, by Theorem 2.3.1, there exist linked sets I_0, J_0 such that $I' \subseteq I_0 \subseteq I$, $J' \subseteq J_0 \subseteq J$; and this means that the families $\mathfrak{A}_0 = (A_i : i \in I_0)$, $\mathfrak{B}_0 = (B_j : j \in J_0)$ possess a CSR.

We recall that, for the case of finite families, an analogous result (Theorem 9.5.2) had been proved earlier in relation to common transversals. It is therefore natural to inquire whether Theorem 10.1.7 remains valid if the term 'common system of representatives' is replaced by 'common transversal' throughout. This question will be examined in §10.4.

Exercises 10.1

1. Let $\mathfrak{A} = (A_i : i \in I)$ be a family of subsets of E; let $M \subseteq E$; and suppose that no element of M occurs in infinitely many A_i. Use Lemma 10.1.1 to show that M is a PT of $\mathfrak{A}$ if and only if every $N \subset\subset M$ intersects at least $|N|$ A's.

2. Let $\mathfrak{A} = (A_i : i \in I)$, $\mathfrak{B} = (B_i : i \in I)$ be two infinite, relatively finite families of sets with a common index set I. Suppose that, for each natural number k, the union of any k A's intersects at least k B's. Show that $\mathfrak{A}$ and $\mathfrak{B}$ need not have a CSR.

3. Show that two families $\mathfrak{A}$, $\mathfrak{B}$ of sets possess a CSR if and only if $\mathfrak{A}$ resp. $\mathfrak{B}$ has a CSR with some subfamily of $\mathfrak{B}$ resp. $\mathfrak{A}$.

10.2 Systems of representatives with repetition

We begin with a preliminary result which may be regarded as a 'one-sided' version of Theorem 10.1.4.

LEMMA 10.2.1. *Let* $\mathfrak{A} = (A_i : i \in I)$, $\mathfrak{B} = (B_j : j \in J)$ *be two families of sets and suppose that each* A *intersects only a finite number of* B's. *Then* $\mathfrak{A}$ *and a subfamily of* $\mathfrak{B}$ *possess a common system of representatives if and only if, for each natural number* $k \leqslant |I|$, *the union of any* k A's *intersects at least* k B's.

The necessity of the stated condition holds trivially. To establish its sufficiency, we shall use a type of argument which is by now familiar (cf. e.g. Theorems 2.2.4 and 9.2.1). Write

$$C_i = \{j \in J : A_i \cap B_j \neq \varnothing\} \qquad (i \in I).$$

Then, for $I^* \subseteq I$,

$$\bigcup_{i \in I^*} C_i = \{j \in J : A(I^*) \cap B_j \neq \varnothing\}$$

and so, for $I^* \subset\subset I$,

$$\left|\bigcup_{i \in I^*} C_i\right| \geqslant |I^*|.$$

Now each C_i is finite and so, by the transfinite form of Hall's theorem (Theorem 4.2.1), the family $(C_i : i \in I)$ possesses a transversal. Thus there exists an injective mapping $\theta : I \to J$ such that $\theta(i) \in C_i$ $(i \in I)$, i.e. $A_i \cap B_{\theta(i)} \neq \varnothing$ $(i \in I)$; and this means that $\mathfrak{A}$ and a subfamily of $\mathfrak{B}$ have a CSR.

The symbol $f(\mathfrak{X}; x)$ used in the next theorem has already been defined in §9.4.

THEOREM 10.2.2. *Let* E *be a set; and for each* $x \in E$, *let* r_x, s_x *be non-negative integers satisfying* $r_x \leqslant s_x$. *Write* $E^* = \{x \in E : r_x > 0\}$. *Let* $\mathfrak{A} = (A_i : i \in I)$ *be a family of finite subsets of* E *and suppose that no element of* E^* *occurs in infinitely many* A's. *Then* $\mathfrak{A}$ *possesses a system of representatives* $\mathfrak{X}$ *such that*

$$r_x \leqslant f(\mathfrak{X}; x) \leqslant s_x \qquad (x \in E)$$

if and only if both the following conditions are satisfied.

(a) *For each finite subset* J *of* I,

$$\sum_{x \in A(J)} s_x \geqslant |J|.$$

(b) *For each finite subset* F *of* E* (*or, equivalently, for each finite subset* F *of* E),

$$\sum_{x \in F} r_x \leqslant |\{i \in I : A_i \cap F \neq \emptyset\}|.$$

We shall denote by $\mathfrak{B}$ the family consisting of s_x copies of $\{x\}$, $x \in E$, and by $\mathfrak{B}'$ the subfamily of $\mathfrak{B}$ consisting of r_x copies of $\{x\}$, $x \in E$.† It follows by Theorem 10.1.7 that $\mathfrak{A}$ possesses a system of representatives $\mathfrak{X}$ subject to the stated restrictions if and only if both the following conditions are satisfied.

(a*) $\mathfrak{A}$ and a subfamily of $\mathfrak{B}$ have a CSR.

(b*) $\mathfrak{B}'$ and a subfamily of $\mathfrak{A}$ have a CSR.

Since each A is finite, we see by Lemma 10.2.1, that (a*) is equivalent to the following statement.

(a**) For each $J \subset\subset I$, A(J) intersects at least |J| sets in $\mathfrak{B}$.

Now write $A(J) = \{x_1, \ldots, x_p\}_{\neq}$. Then A(J) intersects exactly $s_{x_1} + \ldots + s_{x_p}$ sets in $\mathfrak{B}$. Consequently (a**), and so also (a*), is equivalent to (a).

Again, no element of E* occurs in infinitely many A's, i.e. each set in $\mathfrak{B}'$ intersects only a finite number of A's. Hence, by Lemma 10.2.1, (b*) is equivalent to the following statement.

(b**) For each natural number‡ k, the union of any k sets in $\mathfrak{B}'$ intersects at least k A's.

Now consider an arbitrary collection of k sets in $\mathfrak{B}'$, and let it consist of α_1 copies of $\{x_1\}$, ..., α_p copies of $\{x_p\}$, where $x_1, \ldots, x_p$ are distinct elements in E* and

$$\alpha_1 + \ldots + \alpha_p = k,$$
$$0 < \alpha_1 \leqslant r_{x_1}, \ldots, 0 < \alpha_p \leqslant r_{x_p}.$$

Write $F = \{x_1, \ldots, x_p\}$. Then the number of A's intersected by the union of the k sets of $\mathfrak{B}'$ specified above is $|\{i \in I : A_i \cap F \neq \emptyset\}|$, and so

$$|\{i \in I : A_i \cap F \neq \emptyset\}| \geqslant k = \alpha_1 + \ldots + \alpha_p.$$

It follows that (b**) is equivalent to the statement: for each finite subset $F = \{x_1, \ldots, x_p\}_{\neq}$ of E*,

$$|\{i \in I : A_i \cap F \neq \emptyset\}| \geqslant r_{x_1} + \ldots + r_{x_p}.$$

† If it is desired to put the matter more formally, denote by S the set of all pairs (x, n), where $x \in E$ and n is an integer such that $0 < n \leqslant s_x$. Then

$$\mathfrak{B} = (\{x\} : (x, n) \in S);$$

and if s_x is replaced by r_x, we have the definition of $\mathfrak{B}'$.

‡ If E* is finite, then of course we only consider $k \leqslant \Sigma\{r_x : x \in E^*\}$.

Thus (b**), and so also (b*), is equivalent to (b).

We have thus demonstrated the implications (a) $\Leftrightarrow$ (a*) and (b) $\Leftrightarrow$ (b*). Now a system $\mathfrak{X}$ of the required kind exists if and only if (a*) and (b*) are satisfied, and so if and only if (a) and (b) are satisfied. The theorem is therefore proved.

In conclusion, we note a specialization of the above result. If we take $s_x = 1$ (all $x \in \mathrm{E}$) and $r_x = 1$ or 0 according as $x \in \mathrm{E}^*$ or $x \in \mathrm{E} \setminus \mathrm{E}^*$, then Theorem 10.2.2 reduces to the Hoffman–Kuhn–Rado theorem 6.6.3.

Exercises 10.2

1. Let $\mathfrak{A} = (\mathrm{A}_i : i \in \mathrm{I})$ be a family of non-empty subsets of E; let $\mathrm{M} \subseteq \mathrm{E}$; and suppose that no element of M occurs in infinitely many A's. By specializing Theorem 10.2.2, show that $\mathfrak{A}$ possesses a system of representatives whose range contains M if and only if each $\mathrm{N} \subset\subset \mathrm{M}$ intersects at least $|\mathrm{N}|$ A's.

2. Let $\mathfrak{A} = (\mathrm{A}_1, \ldots, \mathrm{A}_n)$ be a family of subsets of $\{x_1, \ldots, x_m\}_{\neq}$, and let $0 \leqslant r_k \leqslant s_k$ $(1 \leqslant k \leqslant m)$ be integers. Show that, if $\mathfrak{A}$ possesses systems of representatives $\mathfrak{X}'$ and $\mathfrak{X}''$ such that

$$r_k \leqslant f(\mathfrak{X}'; x_k), \qquad f(\mathfrak{X}''; x_k) \leqslant s_k \qquad (1 \leqslant k \leqslant m),$$

then it possesses a system of representatives $\mathfrak{X}$ such that

$$r_k \leqslant f(\mathfrak{X}; x_k) \leqslant s_k \qquad (1 \leqslant k \leqslant m).$$

10.3 Common systems of representatives with defect

We shall now deduce a generalization of the inference (ii) $\Rightarrow$ (i) in Theorem 10.1.5. We recall the abbreviation $z^+ = \max(z, 0)$.

THEOREM 10.3.1. *Let d, e be non-negative integers, and let* $\mathfrak{A} = (\mathrm{A}_i : i \in \mathrm{I})$, $\mathfrak{B} = (\mathrm{B}_j : j \in \mathrm{J})$ *be relatively finite families of subsets of* E. *Suppose that, for each natural number* $k \leqslant |\mathrm{I}|$, *the union of any k* A*'s intersects at least* $(k - d)^+$ B*'s and that, for any natural number* $k \leqslant |\mathrm{J}|$, *the union of any k* B*'s intersects at least* $(k - e)^+$ A*'s. Then there exist sets* $\mathrm{I}_0 \subseteq \mathrm{I}$, $\mathrm{J}_0 \subseteq \mathrm{J}$ *with* $|\mathrm{I} \setminus \mathrm{I}_0| \leqslant d$, $|\mathrm{J} \setminus \mathrm{J}_0| \leqslant e$ *such that* $\mathfrak{A}(\mathrm{I}_0)$ *and* $\mathfrak{B}(\mathrm{J}_0)$ *possess a common system of representatives.*

To prove this result, we shall make use of Theorem 2.3.1, M. Hall's theorem 4.2.1, and the standard procedure for dealing with questions involving defect (namely the introduction of dummy elements).

We shall consider the deltoid $\mathscr{D} = (\mathrm{I}, \Delta, \mathrm{J})$, where

$$\Delta = \{(i, j) : i \in \mathrm{I}, j \in \mathrm{J}, \mathrm{A}_i \cap \mathrm{B}_j \neq \varnothing\}.$$

By hypothesis, $\Delta(i)$ is finite for each $i \in \mathrm{I}$. Also, for $\mathrm{I}^* \subseteq \mathrm{I}$, $|\mathrm{I}^*| = k$, we have

$$\left| \bigcup_{i \in \mathrm{I}^*} \Delta(i) \right| \geqslant (k - d)^+.$$

Denote by D any set such that $\mathrm{J} \cap \mathrm{D} = \emptyset$, $|\mathrm{D}| = d$ and put

$$\mathrm{F}_i = \Delta(i) \cup \mathrm{D} \qquad (i \in \mathrm{I}).$$

Then, for any $k \leqslant |\mathrm{I}|$, the union of any k F's contains at least $(k - d)^+ + d \geqslant k$ elements. The family $(\Delta(i) \cup \mathrm{D} : i \in \mathrm{I})$ of finite subsets of $\mathrm{J} \cup \mathrm{D}$ thus satisfies Hall's condition and therefore, by Theorem 4.2.1, there exists an injective mapping $\theta : \mathrm{I} \rightarrow \mathrm{J} \cup \mathrm{D}$ such that $\theta(i) \in \mathrm{F}_i$ for all $i \in \mathrm{I}$. Hence, for a certain set $\mathrm{I}' \subseteq \mathrm{I}$ with $|\mathrm{I} \setminus \mathrm{I}'| \leqslant d$, we have $\theta(i) \in \Delta(i)$ for all $i \in \mathrm{I}'$; and this means that I' is admissible in $\mathscr{D}$.

The same argument, with the roles of A's and B's reversed, shows that there exists an admissible set $\mathrm{J}' \subseteq \mathrm{J}$ with $|\mathrm{J} \setminus \mathrm{J}'| \leqslant e$. By Theorem 2.3.1 there exist, therefore, linked sets I_0, J_0 such that $\mathrm{I}' \subseteq \mathrm{I}_0 \subseteq \mathrm{I}$ and $\mathrm{J}' \subseteq \mathrm{J}_0 \subseteq \mathrm{J}$. Hence $|\mathrm{I} \setminus \mathrm{I}_0| \leqslant d$, $|\mathrm{J} \setminus \mathrm{J}_0| \leqslant e$, and $\mathfrak{A}(\mathrm{I}_0)$, $\mathfrak{B}(\mathrm{J}_0)$ possess a CSR.

Exercises 10.3

1. Let d, e be minimal values for which the conditions of Theorem 10.3.1 are satisfied. Show that the conclusion of the theorem holds with sets I_0, J_0 such that $|\mathrm{I} \setminus \mathrm{I}_0| = d$, $|\mathrm{J} \setminus \mathrm{J}_0| = e$.

2. Let $\mathrm{A}_1, \ldots, \mathrm{A}_m, \mathrm{B}_1, \ldots, \mathrm{B}_n$ be sets and let d be the least non-negative integer with the property that, for $1 \leqslant k \leqslant m$, the union of any k A's intersects at least $(k - d)^+$ B's. Further, let s be the greatest integer such that some family of s A's and some family of s B's possess a common transversal. Show that $s = m - d$.

3. Let $(\mathrm{A}_1, \ldots, \mathrm{A}_n)$ and $(\mathrm{B}_1, \ldots, \mathrm{B}_n)$ be two families of non-empty, pairwise disjoint sets. Denote by d the least number such that, for each k with $1 \leqslant k \leqslant n$, the union of any k A's intersects at least $(k - d)^+$ B's; and denote by r the least number such that the two families possess a common representing set of $n + r$ elements. Show that $d = r$.

[G. Kreweras (**1**)]

10.4 Common transversals of two families

So far in the present chapter we have been mainly concerned with common systems of representatives. We shall now turn to common transversals and shall ultimately obtain a transfinite analogue of the Ford–Fulkerson criterion (Corollary 9.3.4).

THEOREM 10.4.1. *Let* $\mathfrak{A} = (\mathrm{A}_i : i \in \mathrm{I})$ *and* $\mathfrak{B} = (\mathrm{B}_j : j \in \mathrm{J})$ *be two arbitrary families of sets, and let k be a natural number. Then* $\mathfrak{A}$ *and* $\mathfrak{B}$ *possess a common partial transversal of cardinal k if and only if, for all cofinite subsets* I*, J* *of* I, J *respectively*

$$|A(I^*) \cap B(J^*)| \geqslant k - |I \setminus I^*| - |J \setminus J^*|.$$

This generalization of Theorem 9.3.2 is deduced from Theorem 6.4.1 in much the same way as Theorem 9.3.2 is deduced from Theorem 6.2.2.

By Theorem 6.5.2, the set of all PTs of $\mathfrak{B}$ is a pre-independence structure, whose rank function we shall denote by ρ. Now $\mathfrak{A}$ and $\mathfrak{B}$ have a CPT of cardinal k if and only if $\mathfrak{A}$ has an independent PT of cardinal k and, by Theorem 6.4.1, this is precisely the case if, for all cofinite $I^* \subseteq I$,

$$\rho(A(I^*)) \geqslant k - |I \setminus I^*|.$$

Further, the statement $\rho(X) \geqslant h$ means that the family $(B_j \cap X : j \in J)$ possesses a PT of cardinal h. By Corollary 6.4.2, this holds if and only if

$$|B(J^*) \cap X| \geqslant h - |J \setminus J^*|$$

for all cofinite $J^* \subseteq J$. The assertion therefore follows.

THEOREM 10.4.2. *Let* $\mathfrak{A} = (A_i : i \in I)$ *be a family of finite subsets of* E, *and* $\mathfrak{B} = (B_j : j \in J)$ *a restricted family of subsets of* E. *The following statements are then equivalent.*

(i) $\mathfrak{A}$ *has a common transversal with a subfamily of* $\mathfrak{B}$.

(ii) *Every finite subfamily of* $\mathfrak{A}$ *has a common transversal with some subfamily of* $\mathfrak{B}$.

(iii) *For all finite subsets* I*, J* *of* I, J *respectively, we have*

$$|A(I^*) \cap B(J \setminus J^*)| \geqslant |I^*| - |J^*|. \tag{1}$$

Since $\mathfrak{B}$ is restricted, the set of all its PTs is, by Theorem 6.5.3, an independence structure on E. Thus (i) means that $\mathfrak{A}$ has an independent transversal. Now all A_i are finite and so, by Theorem 6.2.4, this is the case if and only if every finite subfamily of $\mathfrak{A}$ has an independent transversal. This establishes the equivalence of (i) and (ii).

Again, by Theorem 10.4.1, the finite subfamily $\mathfrak{A}(\bar{I})$ of $\mathfrak{A}$ has a CT with some subfamily of $\mathfrak{B}$ if and only if, for all $I^* \subseteq \bar{I}$ and all cofinite $J^* \subseteq J$, we have

$$|A(I^*) \cap B(J^*)| \geqslant |\bar{I}| - |\bar{I} \setminus I^*| - |J \setminus J^*|,$$

i.e.

$$|A(I^*) \cap B(J^*)| \geqslant |I^*| - |J \setminus J^*|. \tag{2}$$

Thus (ii) holds if and only if (2) is satisfied for all $\bar{\mathrm{I}} \subset\subset \mathrm{I}$, all $\mathrm{I}^* \subseteq \bar{\mathrm{I}}$, and all cofinite $\mathrm{J}^* \subseteq \mathrm{J}$. This is plainly equivalent to the requirement that (2) should be satisfied for all $\mathrm{I}^* \subset\subset \mathrm{I}$ and all cofinite $\mathrm{J}^* \subseteq \mathrm{J}$; and this, in turn, means that (1) is satisfied for all $\mathrm{I}^* \subset\subset \mathrm{I}$ and all $\mathrm{J}^* \subset\subset \mathrm{J}$. Thus (ii) and (iii) are equivalent.

Next, we recall the insertion theorem 9.5.2. It can be shown that this theorem remains valid for arbitrary families but the proof falls outside the scope of the present treatment, and we shall therefore content ourselves with establishing the transfinite analogue of the special case $\mathfrak{A}' = \mathfrak{A}$, $\mathfrak{B}' = \mathfrak{B}$. It will be useful to record first an almost obvious preliminary result.

LEMMA 10.4.3. *Let* H, K, E *be sets. Let* $(\mathrm{C}_h : h \in \mathrm{H})$ *be a family of subsets of* E, *and let* $\rho: \mathrm{K} \to \mathrm{H}$, $\sigma: \mathrm{K} \to \mathrm{E}$ *be mappings. Suppose that* (i) *for any* $k, k' \in \mathrm{K}$, *the relations* $\rho(k) = \rho(k')$ *and* $\sigma(k) = \sigma(k')$ *imply each other*; (ii) $\sigma(k) \in \mathrm{C}_{\rho(k)}$ *for all* $k \in \mathrm{K}$. *Then* $\sigma(\mathrm{K})$ *is a transversal of* $(\mathrm{C}_h : h \in \rho(\mathrm{K}))$.

For $h \in \rho(\mathrm{K})$, we shall write

$$\rho^*(h) = \{k \in \mathrm{K} : \rho(k) = h\} \quad (\neq \emptyset).$$

If $k, k' \in \rho^*(h)$, then $\rho(k) = h = \rho(k')$ and so, by (i), $\sigma(k) = \sigma(k')$. It follows that $\sigma(\rho^*(h))$ is a single element in $\sigma(\mathrm{K})$. We shall denote it by $\tau(h)$, so that τ is a mapping of $\rho(\mathrm{K})$ into $\sigma(\mathrm{K})$. It is a matter of immediate verification that τ is bijective.

Moreover, let $h \in \rho(\mathrm{K})$ and $k \in \rho^*(h)$, so that $\rho(k) = h$. Then $\tau(h) = \sigma(k)$ and, by (ii), we infer that $\tau(h) \in \mathrm{C}_h$ for all $h \in \rho(\mathrm{K})$.

THEOREM 10.4.4. *Let* $\mathfrak{A}$ *and* $\mathfrak{B}$ *be two arbitrary families of subsets of an arbitrary set* E. *If* $\mathfrak{A}$ *resp.* $\mathfrak{B}$ *has a common transversal with a subfamily of* $\mathfrak{B}$ *resp.* $\mathfrak{A}$, *then* $\mathfrak{A}$ *and* $\mathfrak{B}$ *have a common transversal.*

We shall write $\mathfrak{A} = (\mathrm{A}_i : i \in \mathrm{I})$, $\mathfrak{B} = (\mathrm{B}_j : j \in \mathrm{J})$ and shall assume, as may be done without loss of generality, that $\mathrm{I} \cap \mathrm{J} = \emptyset$. By hypothesis, $\mathfrak{A}$ and a subfamily of $\mathfrak{B}$ have a CT, i.e. there exists a family $(x_i : i \in \mathrm{I})$ of distinct elements of E and an injection $\phi: \mathrm{I} \to \mathrm{J}$ such that $x_i \in \mathrm{A}_i \cap \mathrm{B}_{\phi(i)}$ $(i \in \mathrm{I})$. Further, $\mathfrak{B}$ and a subfamily of $\mathfrak{A}$ have a CT, i.e. there exists a family $(y_j : j \in \mathrm{J})$ of distinct elements of E and an injection $\psi: \mathrm{J} \to \mathrm{I}$ such that $y_j \in \mathrm{A}_{\psi(j)} \cap \mathrm{B}_j$ $(j \in \mathrm{J})$.

We define

$$\bar{\mathrm{A}}_i = \begin{cases} \{x_i, y_j\} & \text{if } i = \psi(j) \\ \{x_i\} & \text{if } i \notin \psi(\mathrm{J}) \end{cases} \qquad (i \in \mathrm{I}),$$

$$\overline{\mathrm{B}}_j = \begin{cases} \{y_j, x_i\} & \text{if } j = \phi(i) \\ \{y_j\} & \text{if } j \notin \phi(\mathrm{I}) \end{cases} \qquad (j \in \mathrm{J}),$$

$$\overline{\mathfrak{A}} = (\overline{\mathrm{A}}_i : i \in \mathrm{I}), \qquad \overline{\mathfrak{B}} = (\overline{\mathrm{B}}_j : j \in \mathrm{J}).$$

It is plain that

$$\overline{\mathrm{A}}_i \subseteq \mathrm{A}_i \quad (i \in \mathrm{I}), \qquad \overline{\mathrm{B}}_j \subseteq \mathrm{B}_j \quad (j \in \mathrm{J}). \tag{3}$$

Further, we define

$$\begin{aligned} \mathrm{F} &= \{(i, e, j) \in \mathrm{I} \times \mathrm{E} \times \mathrm{J} \ : \ e \in \overline{\mathrm{A}}_i \cap \overline{\mathrm{B}}_j\}, \\ \mathrm{F}_{i_0} &= \{(i, e, j) \in \mathrm{F} : i = i_0\} \quad (i_0 \in \mathrm{I}), \\ \mathrm{F}_{j_0} &= \{(i, e, j) \in \mathrm{F} : j = j_0\} \quad (j_0 \in \mathrm{J}), \\ \mathfrak{F} &= (\mathrm{F}_k : k \in \mathrm{I} \cup \mathrm{J}). \end{aligned}$$

We note that every set F_k is finite. For let $i \in \mathrm{I}$. Then $e \in \overline{\mathrm{A}}_i$ for only finitely many $e \in \mathrm{E}$. Further, if $e \in \mathrm{E}$, then $e \in \overline{\mathrm{B}}_j$ for only finitely many $j \in \mathrm{J}$. Therefore $e \in \overline{\mathrm{A}}_i \cap \overline{\mathrm{B}}_j$ for only finitely many pairs (e, j), and this means that F_i is finite. Similarly, F_j is finite for each $j \in \mathrm{J}$.

Let $\mathrm{I}' \subseteq \mathrm{I}$, $\mathrm{J}' \subseteq \mathrm{J}$, and denote by η a choice function of the subfamily $\mathfrak{F}(\mathrm{I}' \cup \mathrm{J}')$ of $\mathfrak{F}$. We shall then write

$$\eta(k) = \big(\eta^1(k), \eta^2(k), \eta^3(k)\big) \in \mathrm{F}_k \qquad (k \in \mathrm{I}' \cup \mathrm{J}').$$

The choice function η will be said to be 'coherent' if, whenever $h, k \in \mathrm{I}' \cup \mathrm{J}'$ and $\eta^r(h) = \eta^r(k)$ for *some* r with $1 \leqslant r \leqslant 3$, then $\eta^r(h) = \eta^r(k)$ for *all* r with $1 \leqslant r \leqslant 3$.

Let $\mathrm{I}' \subset\subset \mathrm{I}$, $\mathrm{J}' \subset\subset \mathrm{J}$ and put

$$\mathrm{I}'' = \mathrm{I}' \cup \psi(\mathrm{J}'), \qquad \mathrm{J}'' = \phi(\mathrm{I}') \cup \mathrm{J}'.$$

Then I'', J'' are finite. Now

$$x_i \in \overline{\mathrm{A}}_i \cap \overline{\mathrm{B}}_{\phi(i)} \qquad (i \in \mathrm{I})$$

and so the subfamily $\overline{\mathfrak{A}}(\mathrm{I}')$ of the finite family $\overline{\mathfrak{A}}(\mathrm{I}'')$ has a CT with a certain subfamily, namely $\overline{\mathfrak{B}}(\phi(\mathrm{I}'))$, of the finite family $\overline{\mathfrak{B}}(\mathrm{J}'')$. Similarly, the subfamily $\overline{\mathfrak{B}}(\mathrm{J}')$ of $\overline{\mathfrak{B}}(\mathrm{J}'')$ has a CT with the subfamily $\overline{\mathfrak{A}}(\psi(\mathrm{J}'))$ of $\overline{\mathfrak{A}}(\mathrm{I}'')$. It follows by Theorem 9.5.2 that there exist (finite) sets I_0, J_0 with $\mathrm{I}' \subseteq \mathrm{I}_0 \subseteq \mathrm{I}''$, $\mathrm{J}' \subseteq \mathrm{J}_0 \subseteq \mathrm{J}''$ such that $\overline{\mathfrak{A}}(\mathrm{I}_0)$ and $\overline{\mathfrak{B}}(\mathrm{J}_0)$ possess a CT. Thus there exists a family $(e_i : i \in \mathrm{I}_0)$ of distinct elements of E and a bijection $\pi : \mathrm{I}_0 \to \mathrm{J}_0$ such that

$$e_i \in \overline{\mathrm{A}}_i \cap \overline{\mathrm{B}}_{\pi(i)} \qquad (i \in \mathrm{I}_0).$$

From this it is clear that

$$(i, e_i, \pi(i)) \in \mathrm{F} \qquad (i \in \mathrm{I}_0). \tag{4}$$

We now define the mapping $\eta: \mathrm{I}' \cup \mathrm{J}' \to \mathrm{F}$ by the equations

$$\eta(i) = (i, e_i, \pi(i)) \qquad (i \in \mathrm{I}'),$$
$$\eta(j) = (\pi^{-1}(j), e_{\pi^{-1}(j)}, j) \qquad (j \in \mathrm{J}').$$

By (4) and the definition of F_k, we see that η is a choice function of $\mathfrak{F}(\mathrm{I}' \cup \mathrm{J}')$. Moreover, since π is bijective and $(e_i: i \in \mathrm{I}_0)$ is a family of distinct elements, it follows that η is coherent.

We have thus shown that, whenever $\mathrm{I}' \subset\subset \mathrm{I}$, $\mathrm{J}' \subset\subset \mathrm{J}$, the subfamily $\mathfrak{F}(\mathrm{I}' \cup \mathrm{J}')$ of the family $\mathfrak{F}$ of finite sets possesses a coherent choice function, say $\theta_{\mathrm{I}' \cup \mathrm{J}'}$. Denote by θ a corresponding Rado choice function of $\mathfrak{F}$. Let $\{h, k\} \subseteq \mathrm{I} \cup \mathrm{J}$. Then there exist sets $\mathrm{I}' \subset\subset \mathrm{I}$, $\mathrm{J}' \subset\subset \mathrm{J}$ with $\{h, k\} \subseteq \mathrm{I}' \cup \mathrm{J}'$ such that $\theta | \{h, k\} = \theta_{\mathrm{I}' \cup \mathrm{J}'} | \{h, k\}$. Thus

$$\theta^1(m) = \theta_{\mathrm{I}' \cup \mathrm{J}'}{}^1(m), \qquad \theta^2(m) = \theta_{\mathrm{I}' \cup \mathrm{J}'}{}^2(m), \qquad \theta^3(m) = \theta_{\mathrm{I}' \cup \mathrm{J}'}{}^3(m)$$

for $m = h$ and $m = k$. Hence, since $\theta_{\mathrm{I}' \cup \mathrm{J}'}$ is coherent, θ is also coherent.

Now θ takes its values in F and therefore, in particular, the mappings $\theta^1: \mathrm{I} \cup \mathrm{J} \to \mathrm{I}$, $\theta^2: \mathrm{I} \cup \mathrm{J} \to \mathrm{J}$ satisfy the relations

$$\theta^2(k) \in \overline{\mathrm{A}}_{\theta^1(k)} \qquad (k \in \mathrm{I} \cup \mathrm{J}).$$

Moreover, θ is coherent and so, for $h, k \in \mathrm{I} \cup \mathrm{J}$, the relations $\theta^1(h) = \theta^1(k)$ and $\theta^2(h) = \theta^2(k)$ imply each other. Hence, by Lemma 10.4.3, $\theta^2(\mathrm{I} \cup \mathrm{J})$ is a transversal of

$$(\overline{\mathrm{A}}_i: i \in \theta^1(\mathrm{I} \cup \mathrm{J})).$$

But, since θ is a choice function of $\mathfrak{F}$, $\theta^1(k) \in \mathrm{I}$ $(k \in \mathrm{I} \cup \mathrm{J})$ and $\theta^1(i) = i$ $(i \in \mathrm{I})$. Thus $\theta^1(\mathrm{I} \cup \mathrm{J}) = \mathrm{I}$ and so $\theta^2(\mathrm{I} \cup \mathrm{J})$ is a transversal of $\overline{\mathfrak{A}}$, and similarly of $\overline{\mathfrak{B}}$. Thus $\overline{\mathfrak{A}}$ and $\overline{\mathfrak{B}}$ possess a CT and, by (3), so do the families $\mathfrak{A}$ and $\mathfrak{B}$.

To obtain a result which contains the transfinite analogue of the Ford–Fulkerson criterion, we only need to gather up the threads.

THEOREM 10.4.5. *Let* $\mathfrak{A} = (\mathrm{A}_i: i \in \mathrm{I})$ *and* $\mathfrak{B} = (\mathrm{B}_j: j \in \mathrm{J})$ *be two restricted families of finite sets. The following statements are then equivalent.*

(i) $\mathfrak{A}$ *and* $\mathfrak{B}$ *possess a common transversal.*

(ii) (a) *Every finite subfamily of* $\mathfrak{A}$ *has a common transversal with some subfamily of* $\mathfrak{B}$.

(b) *Every finite subfamily of* $\mathfrak{B}$ *has a common transversal with some subfamily of* $\mathfrak{A}$.

(iii) (a) *For all* $I^* \subset\subset I$, $J^* \subset\subset J$, *we have*

$$|A(I^*) \cap B(J \setminus J^*)| \geqslant |I^*| - |J^*|.$$

(b) *For all* $I^* \subset\subset I$, $J^* \subset\subset J$, *we have*

$$|A(I \setminus I^*) \cap B(J^*)| \geqslant |J^*| - |I^*|.$$

Since all A_i are finite while $\mathfrak{B}$ is restricted, we know by Theorem 10.4.2 that (ii a) and (iii a) are equivalent. Similarly, since all B_j are finite while $\mathfrak{A}$ is restricted, (ii b) and (iii b) are equivalent. Hence (ii) and (iii) are equivalent. Moreover, again by Theorem 10.4.2, either of these statements is equivalent to the requirement that $\mathfrak{A}$ resp. $\mathfrak{B}$ should have a CT with a subfamily of $\mathfrak{B}$ resp. $\mathfrak{A}$; and this, in turn, is equivalent to (i) by Theorem 10.4.4.

10.5 Common transversals of maximal subfamilies

We possess no general method for discussing links of more than two families, and are only able to handle rather special cases. It is simplest to frame the argument in terms of deltoids.

THEOREM 10.5.1. *Let* (X, Δ, Y) *be a deltoid, and suppose that a certain maximal admissible subset of* X *can be linked to only a finite number of subsets of* Y. *Then all maximal admissible subsets of* X *can be linked to a single subset of* Y.

Let X_0 be a maximal admissible subset of X which is linked to the (different) subsets $Y_0, Y_1, \ldots, Y_m$ of Y and to no others. We shall assume, as may be done without loss of generality, that

$$Y_0 \nsubseteq Y_1, \ldots, Y_0 \nsubseteq Y_m. \tag{1}$$

Let Z be an admissible subset of Y such that $Y_0 \subseteq Z$. Then, by Theorem 2.3.1, there exist linked sets $\overline{X}$, $\overline{Z}$ with $X_0 \subseteq \overline{X} \subseteq X$, $Z \subseteq \overline{Z} \subseteq Y$. But X_0 is maximal and so $X_0 \leftrightarrow \overline{Z}$. Hence $\overline{Z}$ is identical with one of the sets $Y_0, Y_1, \ldots, Y_m$ and, in view of (1) and the relation $Y_0 \subseteq \overline{Z}$, it follows that $\overline{Z} = Y_0$. Consequently, $Z = Y_0$ and thus Y_0 is maximal. Corollary 2.3.2 now shows that every maximal admissible subset of X is linked to Y_0.

The standard interpretation of Theorem 10.5.1 yields at once the following result.

THEOREM 10.5.2. *Let* $\mathfrak{A}$ *be a family of sets and suppose that some maximal subfamily of* $\mathfrak{A}$ *possesses only a finite number of transversals. Then all maximal subfamilies of* $\mathfrak{A}$ *possess a common transversal.*

Notes on Chapter 10

§ 10.1. The theorem of Mendelsohn and Dulmage (**1**) was, in its original form, essentially the finite case of Theorem 2.3.1. (The standard interpretation of that result yields Theorem 3.4.1). Here we are concerned with its transfinite analogue, which is embodied in Theorem 10.1.2; this result and its interpretations are taken from the work of Mirsky & Perfect (**2**). A conclusion closely related to Theorem 10.1.5 was first established by Shmushkovitch (**1**); subsequently it was rediscovered by de Bruijn (**1**) and also by Everett & Whaples (**1**).

An alternative derivation of Theorem 10.1.6 about representation of cosets can be based on the work of König & Valkó (**1**) in the theory of graphs. As has been mentioned, the conclusion of the theorem remains valid if it is assumed that H is a subgroup of finite index of the group G; see van der Waerden (**2**). The counter-example after the proof of Theorem 10.1.6 is also due to van der Waerden. Our discussion does not, of course, exhaust the problem since we can ask under what additional conditions the left cosets and the right cosets of an infinite subgroup with infinite index possess a common transversal. However, investigations beyond this point cannot be carried out solely in terms of combinatorial ideas and must take account of the group structure. For further results in this field, we refer to papers of S. Shü (**1**), O. Ore (**2**), and R. A. Rankin (**1**).

§ 10.2. In this section, we follow closely the argument of Mirsky's paper (**3**). The case of Theorem 10.2.2 relating to finite families had been established previously by Ford & Fulkerson (**2**, Theorem 10.5). For a far-reaching generalization of Theorem 10.2.2, see Folkman & Fulkerson (**1**).

§ 10.3. Theorem 10.3.1 was discovered by O. Ore (**1**) in the context of a broad investigation in the theory of graphs. The proof presented here is due to Perfect & Pym (**1**).

§ 10.4. Theorems 10.4.1 and 10.4.2 are due to Dr Hazel Perfect, Theorem 10.4.4 and, indeed, more general results were found by J. S. Pym (**3**, **4**) and, independently, by R. A. Brualdi (**4**); see also Brualdi & Pym (**1**). The proof of Theorem 10.4.4 offered here has been shown to me by Dr Pym: it is based on the argument in his paper (**4**).

§ 10.5. Theorem 10.5.1 is due to J. S. Pym (see Mirsky & Perfect (**2**)). Pym (**1**) has also obtained a series of other results on deltoids. It should be noted that Theorem 10.5.1, and consequently Theorem 10.5.2, assert much less than the full truth. Indeed, a stronger result of Brualdi & Scrimger (**1**) states that any transversal of a maximal subfamily of $\mathfrak{A}$ is a common transversal of all maximal subfamilies of $\mathfrak{A}$.

11 Combinatorial Properties of Matrices

Combinatorial aspects of the theory of (finite and infinite) matrices have been studied extensively. Here we do not attempt a comprehensive discussion of the subject but confine ourselves to such parts of the theory as are closely linked to or suggested by results in general transversal tneory. We shall begin by laying down some points of terminology.

A *line* is a common designation for a row or a column of a matrix. A *line-sum* is the sum of all elements on a line. A *place* is the position of an element in a matrix (fixed by the specification of its row and column). Thus, for example, we can speak of 'the places occupied by non-zero elements'. In practice, we usually adopt a looser mode of expression and simply say 'the non-zero elements'.

A set S of elements (or, more precisely, of places!) is said to be *incident* with a set L of lines if each line in L contains at least one element of S.

A set S of elements is said to be *scattered* if no two elements of S lie on the same line. A set S of elements in a square matrix Q is called a *diagonal* if every line of Q contains precisely one element of S. We say that a diagonal is positive (non-zero) if all its constituent elements are positive (non-zero).

A *Z-matrix* is a (rectangular) matrix with at most one 1 in each line and with all its other elements equal to 0. A *permutation matrix* is a square matrix with precisely one 1 in each line and with all other elements equal to 0.

A (finite or infinite) square matrix is said to be *doubly-stochastic* (d.s.) if its elements are real non-negative numbers and all its line-sums are equal to 1.

The *width* $w(Q)$ of the $m \times n$ matrix Q is defined as $w(Q) = m + n$. Further, we shall denote by $w_0(Q)$ the maximum width of all zero submatrices of Q.

11.1 The language of matrix theory

It is a fairly immediate observation that much of transversal theory relating to finite families of finite sets can be rewritten in the language of matrices. In the present section we shall compile the appropriate dictionary and give some specimens of translation.

A family $\mathfrak{A} = (A_1, \ldots, A_n)$ of subsets of $E = \{x_1, \ldots, x_m\}_{\neq}$ and an $m \times n$ matrix $Q = \|q_{ij}\|$ are said to be *associated* if $q_{ij} \neq 0$ precisely when $x_i \in A_j$. The matrix associated with $\mathfrak{A}$ all of whose non-zero elements are equal to 1 is called the *incidence matrix* of $\mathfrak{A}$.† (When the positive elements in the incidence matrix of $\mathfrak{A}$ are replaced by independent indeterminates, we obtain a 'formal incidence matrix' of $\mathfrak{A}$; cf. § 6.5.) Indeed, we shall find it convenient to call any matrix all of whose elements are 0 or 1 an incidence matrix: it is, of course, the incidence matrix of a suitable family. For if $Q = \|q_{ij}\|$ is an $m \times n$ matrix and if we write $\mathfrak{A} = (A_1, \ldots, A_n)$, where

$$A_j = \{i : 1 \leqslant i \leqslant m,\ q_{ij} \neq 0\} \qquad (1 \leqslant j \leqslant n),$$

then the family $\mathfrak{A}$ is associated with Q.

The correspondence we have established between families and matrices is not bijective, but this deficiency is not serious. For two matrices associated with the same family have their non-zero elements in precisely the same places,‡ and two families associated with the same matrix only differ in the notation used for their elements. (The correspondence could therefore be made bijective if we cared to introduce suitable equivalence classes.)

It should also be noted that in our correspondence the elements of the ground set are associated with the rows of the matrix while the sets of the family are associated with the columns. Thus, if the matrix Q is associated with a family $\mathfrak{A}$, then its transposed matrix Q^T is associated with the dual family $\tilde{\mathfrak{A}}$ (defined in § 2.3). The process of dualization for families is therefore reflected in the transposition of associated matrices.

Let the family $\mathfrak{A}$ and the matrix Q be associated. Then $\mathfrak{A}$ possesses a transversal precisely if Q has a scattered set of non-zero elements incident with all columns. More generally, $\mathfrak{A}$ has a PT of cardinal r precisely if Q has a scattered set of r non-zero elements. Again, suppose $\mathfrak{A}$ consists of n sets (so that Q has n columns); then $\mathfrak{A}$ satisfies Hall's condition precisely if, for each k with $1 \leqslant k \leqslant n$, the non-zero elements in any k columns of Q belong to at least k rows.

When we study the structure of overlaps of sets in two families, say $\mathfrak{A} = (A_1, \ldots, A_m)$, $\mathfrak{B} = (B_1, \ldots, B_n)$, we may also find it useful to introduce matrices into the discussion. An obvious matrix that can be defined in this case is an $m \times n$ matrix $Q = \|q_{ij}\|$ such that $q_{ij} \neq 0$ precisely when $A_i \cap B_j \neq \emptyset$.

We shall begin our discussion of results by stating the matrix equivalent of the Hall–Ore theorem 3.2.1.

THEOREM 11.1.1. *An $m \times n$ matrix Q possesses a scattered set of r non-zero*

† To be quite precise, we ought to speak of the incidence matrix of $\mathfrak{A}$ *and* E, since $\mathfrak{A}$ remains unchanged if the ground set is enlarged while the incidence matrix is augmented by the adjunction of zero rows.

‡ In the terminology of §11.4 below, they have the same 'pattern'.

elements if and only if, for each k with $1 \leqslant k \leqslant n$, *the non-zero elements in any k columns belong to at least* $(k + r - n)^+$ *rows.*†

It is useful to record the special case $r = m = n$ of this result.

COROLLARY 11.1.2. *An* $n \times n$ *matrix Q possesses a non-zero diagonal if and only if, for each k with* $1 \leqslant k \leqslant n$, *the non-zero elements in any k columns of Q belong to at least k rows.*

The terms 'rows' and 'columns' in this statement can, of course, be interchanged. An interesting deduction from the Corollary runs as follows.

THEOREM 11.1.3. *Every finite doubly-stochastic matrix possesses a positive diagonal.*

Let D be a d.s. $n \times n$ matrix. Assume that, for some value of k in the range $1 \leqslant k \leqslant n$, all positive elements in a certain set of k columns of D (say the first k columns) are contained in a set of $k - 1$ rows (say the first $k - 1$ rows). Then the sum of all elements in the $(k - 1) \times k$ top left-hand submatrix of D is k if reckoned by columns and at most $k - 1$ if reckoned by rows. We thus arrive at a contradiction. The assertion now follows by Corollary 11.1.2.

We shall next consider the generalization of Hall's theorem due to Mendelsohn and Dulmage. It is useful to introduce the following, almost self-explanatory, terminology. Let Q be a finite matrix, and let R be a set of rows in Q. We shall say that R satisfies condition $\mathscr{H}$ (Hall's condition) if, for each k with $1 \leqslant k \leqslant |\mathrm{R}|$, the non-zero elements in any k rows of R belong to at least k columns. Condition $\mathscr{H}$ for sets of columns is defined analogously.

THEOREM 11.1.4. *Let Q be a rectangular matrix; let* R *be a set of rows and* C *a set of columns in Q. Then Q possesses a scattered set of non-zero elements incident with both* R *and* C *if and only if both* R *and* C *satisfy* $\mathscr{H}$.

This is, in essence, the matrix equivalent of Theorem 3.4.1. We denote by $\mathfrak{A}$ the family associated with Q and by E the ground set of $\mathfrak{A}$. Let $\mathfrak{A}'$ be the subfamily of $\mathfrak{A}$ specified by C, and E′ the subset of E specified by R. Now Q possesses a scattered set of non-zero elements incident with both R and C precisely if there exists a family $\mathfrak{A}_0$ with transversal E_0 such that $\mathfrak{A}' \subseteq \mathfrak{A}_0 \subseteq \mathfrak{A}$ and $\mathrm{E}' \subseteq \mathrm{E}_0 \subseteq \mathrm{E}$. By Theorem 3.4.1, this is the case if and only if both the following conditions are satisfied: (i) E′ is a PT of $\mathfrak{A}$; (ii) $\mathfrak{A}'$ possesses a trans-

† We recall the notation $x^+ = \max(x, 0)$.

versal. Now (i) means that there exists a scattered set of non-zero elements in Q incident with R and, by Theorem 11.1.1, this is the case if and only if R satisfies $\mathscr{H}$. Again, (ii) means that there exists a scattered set of non-zero elements incident with C, and this is the case precisely if C satisfies $\mathscr{H}$.

We shall conclude this section by discussing some consequences of the results obtained so far.

THEOREM 11.1.5. *Let Q be a square matrix whose elements are non-negative integers. If every line-sum in Q is equal to k, then Q can be expressed as the sum of k permutation matrices.*

We may assume that $k \neq 0$. The matrix $k^{-1}Q$ is d.s. and so, by Theorem 11.1.3, possesses a positive diagonal, say Δ. Let P be the permutation matrix whose positive elements occupy the places of Δ.. Then $Q - P$ is a square matrix whose elements are non-negative integers and all of whose line-sums are equal to $k - 1$. The assertion therefore follows by induction with respect to k.

THEOREM 11.1.6. *Let Q be a rectangular matrix whose elements are non-negative integers, and let $s(Q)$ denote the maximum line-sum of Q. Then Q can be expressed as the sum of $s(Q)$, but not fewer, Z-matrices.*

Suppose that $Q = Z_1 + \ldots + Z_k$, where $Z_1, \ldots, Z_k$ are Z-matrices. The sum of elements on at least one line of Q is $s(Q)$; the sum of elements on every line of each Z_i is at most 1. Hence $s(Q) \leqslant k$, i.e. we require at least $s(Q)$ summands.

To prove the main part of the theorem, we shall call a line in Q *critical* if its line-sum is equal to $s = s(Q)$. Let R denote the set of critical rows in Q. Consider any k rows in R and let them, for simplicity, be the first k rows in Q. Let the positive elements in these k rows be contained in h columns, which we take to be the first h columns. Then

$$Q = \begin{Vmatrix} Q_1 & O \\ Q_2 & Q_3 \end{Vmatrix},$$

where Q_1 is of type $k \times h$. Now the sum of elements in each row of Q_1 is s and so the sum of all elements in Q_1 is sk. Again, every column-sum in Q is at most s and so the sum of all elements in Q_1 is at most sh. Thus $k \leqslant h$, i.e. R satisfies condition $\mathscr{H}$. Similarly, the set of all critical columns of Q satisfies $\mathscr{H}$.

Therefore, by Theorem 11.1.4, there exists a scattered set S of positive elements in Q which is incident with every critical line. Let Z be a Z-matrix of the same type as Q and having a 1 in each place of S. Then $Q - Z$ has non-negative integral elements and $s(Q - Z) = s(Q) - 1$. The assertion now follows by induction with respect to $s(Q)$.

Exercises 11.1

1. Let Q be a rectangular matrix. Let R be a set of rows and C a set of columns in Q. Suppose that there exists a scattered set of non-zero elements in Q incident with R and another such set incident with C. Deduce that there exists a scattered set of non-zero elements incident with both R and C.

2. In a certain college, there are m lecturers, say $L_1, \ldots, L_m$, and n classes, say $C_1, \ldots, C_n$. The number of one-hour lectures to be given by L_i to C_j is denoted by a_{ij}. Let h be the least number of hours into which the time-table can be fitted (so that no lecturer has to give more than one lecture in any one hour and no class has to attend more than one lecture in any one hour). Use Theorem 11.1.6 to show that h is equal to the maximum line-sum in the $m \times n$ matrix $||a_{ij}||$.

3. At a dance, m boys and n girls take part. Each boy is acquainted with precisely k girls and each girl is acquainted with precisely k boys. Show that $m = n$ and that it is possible to arrange partners for a dance in such a way that everyone takes part and that everyone dances with a partner with whom he or she is acquainted.

4. Give a proof of Theorem 11.1.6 which depends on the use of Theorem 11.1.5.
[Mirsky (**8**)]

5. Let $1 \leqslant r < n$, $1 \leqslant k \leqslant n$ and let A be an $r \times n$ incidence matrix all of whose row-sums are equal to k. Show that the following two statements are equivalent. (i) It is possible to adjoin to A $n - r$ rows consisting of 1's and 0's such that every line-sum in the resulting $n \times n$ matrix is equal to k. (ii) We have $k - n + r \leqslant N(i) \leqslant k$ for $1 \leqslant i \leqslant n$, where $N(i)$ denotes the number of 1's in the i-th column of A.
[H. J. Ryser (**1**)]

6. An $r \times n$ incidence matrix (where $1 \leqslant r < n$) with exactly one 1 in each row and at most one 1 in each column will be called a 'truncated permutation matrix'.
Let $1 \leqslant r < n$, $1 \leqslant k \leqslant n$ and let A be an $r \times n$ incidence matrix all of whose row-sums are equal to k and which satisfies condition (ii) in Ex. 11.1.5. Use Theorem 11.1.5 to prove that A can be expressed as the sum of k truncated permutation matrices.
[H. J. Ryser (**1**)]

11.2 Theorems of König, Frobenius, and Rado

We have already met at least three statements (Theorems 3.2.4 and 3.2.6, Corollary 3.2.7) which are equivalent† to the defect form of Hall's theorem

† The term 'equivalent' used here is mathematically empty though it has a methodological content: when we say that two theorems are equivalent, we simply mean that it is much easier to derive each result from the other than to give a self-contained proof of either.

(Theorem 3.2.1). We shall now add to their number by obtaining some matrix versions of the same result.

We shall continue to use the notation and terminology introduced in §3.2. In particular, we associate the numbers t^*, t_*, $\bar{t}$ with the family $\mathfrak{A} = (A_1, \ldots, A_n)$ of subsets of $E = \{x_1, \ldots, x_m\}_{\neq}$.

THEOREM 11.2.1. (König) *In any rectangular matrix, the maximum number of non-zero elements of a scattered set is equal to the minimum number of lines which contain all non-zero elements.*

Let Q be the given matrix, and let it be of type $m \times n$. Denote by M^* and M_* the maximum and minimum specified in the theorem. Let $\mathfrak{A} = (A_1, \ldots, A_n)$ be a family associated with Q. Then M^* is clearly equal to the transversal index of $\mathfrak{A}$, and M_* to the minimum number of objects in an incidence-bound collection. It follows, by Theorem 3.2.4, that $M^* = M_*$.

Conversely, Theorem 3.2.4 is an immediate consequence of König's theorem. It should also be noted that the result just proved and Theorem 1.7.1 (for finite G) are identical in substance and differ only in their terminology.

LEMMA 11.2.2. *Let Q be an $m \times n$ matrix and let M^*, M_* be the maximum and minimum respectively specified in Theorem 11.2.1. If $M_* < \min(m, n)$, then $M_* + w_0(Q) = m + n$.*

Consider a set of M_* lines which contain all non-zero elements of Q. Since $M_* < \min(m, n)$, the elements of Q outside these lines form a zero submatrix of width $m + n - M_* \leqslant w_0(Q)$.

Again, consider a zero submatrix of Q having width $w_0(Q) \geqslant 0$. Then the $m + n - w_0(Q)$ lines not incident with this submatrix contain all non-zero elements of Q, i.e. $M_* \leqslant m + n - w_0(Q)$; and the assertion follows.

LEMMA 11.2.3. *For any $m \times n$ matrix Q, $M^* = \min(m, n)$ or $m + n - w_0(Q)$ according as $w_0(Q) \leqslant \max(m, n)$ or $w_0(Q) \geqslant \max(m, n)$.*

Adjoin a zero row and a zero column to Q, and denote the resulting matrix by Q'. The value of M_* then remains unchanged and, obviously, $M_* < \min(m+1, n+1)$. Hence, by Lemma 11.2.2,

$$M_* = (m+1) + (n+1) - w_0(Q').$$

But $w_0(Q') = \max\{\max(m+2, n+2), w_0(Q) + 2\}$, and the assertion now follows at once in view of Theorem 11.2.1.

THEOREM 11.2.4. *Let Q be an $m \times n$ matrix and let* $1 \leqslant r \leqslant \min(m, n)$. *Then every term in the determinantal expansion of every $r \times r$ submatrix of Q vanishes if and only if Q possesses a zero submatrix of width $m + n - r + 1$.*

Every term in the determinantal expansion of every $r \times r$ submatrix of Q vanishes precisely if $M^* < r$. Now, if $M^* < r$, then $M^* < \min(m, n)$ and so, by Lemma 11.2.2 and Theorem 11.2.1, $M^* = m + n - w_0(Q)$. Hence $w_0(Q) \geqslant m + n - r + 1$. If, on the other hand, $w_0(Q) \geqslant m + n - r + 1$, then

$$w_0(Q) \geqslant m + n - \min(m, n) + 1 > \max(m, n);$$

and so, by Lemma 11.2.3, $M^* = m + n - w_0(Q) < r$.

For a subsequent application, it is convenient to record a variant of the result just proved.

COROLLARY 11.2.5. *Let Q be an $m \times n$ matrix and let* $1 \leqslant k \leqslant \min(m, n)$. *Then every scattered set of* $\min(m, n)$ *elements of Q contains at least k zeros if and only if Q has a zero submatrix of width* $\max(m, n) + k$.

We obtain this result by taking $r = \min(m, n) - k + 1$ in Theorem 11.2.4.

The special case $m = n, k = 1$ of Corollary 11.2.5 (or, equivalently, the case $m = n = r$ of Theorem 11.2.4) is worth stating separately.

COROLLARY 11.2.6. (Frobenius) *Every diagonal of an $n \times n$ matrix Q contains at least one zero element if and only if Q possesses a submatrix of width $n + 1$.*

We shall now use this result to give another proof of Theorem 9.2.1. Suppose that, for any k with $1 \leqslant k \leqslant n$, the union of any k A's intersects at least k B's. Let $Q = \|q_{ij}\|$ be the $n \times n$ matrix such that $q_{ij} = 1$ or 0 according as A_i and B_j do or do not intersect. If all terms in the determinantal expansion of Q vanish, then, by Corollary 11.2.6, Q possesses a zero submatrix (situated, say, in the top left-hand corner) of type $r \times s$, where $r + s = n + 1$. Thus $A_i \cap B_j = \emptyset$ whenever $1 \leqslant i \leqslant r$, $1 \leqslant j \leqslant s$; and it follows that, contrary to our hypothesis, $A_1 \cup \ldots \cup A_r$ intersects at most $n - s = r - 1$ B's. We conclude that at least one term in the determinantal expansion of Q is non-zero – a fact which is equivalent to the existence of a CSR.

Next, we deduce a theorem on 'representing sets' and exhibit in matrix form certain of its consequences. Let $\mathfrak{A} = (A_i : i \in I)$ be a family of subsets of E. We recall (cf. § 2.1) that a set $R \subseteq E$ is called a representing set of $\mathfrak{A}$ if $A_i \cap R \neq \emptyset$ $(i \in I)$. The family $\mathfrak{A}$ is said to be *bipartite* if there exists a partition $I = I_1 \cup I_2$ such that the sets in the subfamily $\mathfrak{A}(I_1)$, and equally those in $\mathfrak{A}(I_2)$, are pairwise disjoint.

Let E, I be finite. Below, we shall denote by M^* the maximum number of pairwise disjoint sets in $\mathfrak{A}$† and by M_* the minimum number of elements in a representing set of $\mathfrak{A}$. It is then trivial that $M^* \leqslant M_*$; but in general we do not have $M^* = M_*$, as is seen by the case

$$\mathfrak{A} = (A_1, A_2, A_3),\ A_1 = \{2, 3\},\ A_2 = \{1, 3\},\ A_3 = \{1, 2\},\ E = \{1, 2, 3\}.$$

The next result furnishes a *sufficient* condition for the validity of the equation $M^* = M_*$.

THEOREM 11.2.7. (R. Rado) *Let $\mathfrak{A}$ be a finite, bipartite family of finite, non-empty sets. Then the maximum number of pairwise disjoint sets in $\mathfrak{A}$ is equal to the minimum number of elements in a representing set of $\mathfrak{A}$.*

That the condition of bipartiteness in this theorem is not a necessary condition is demonstrated by the case of the family

$$\mathfrak{A} = (A_1, A_2, A_3), \qquad A_1 = A_2 = A_3 = E = \{1, 2\}.$$

To prove the theorem, let us denote by

$$A_1, \ldots, A_m, \quad B_1, \ldots, B_n$$

the sets of $\mathfrak{A}$; and suppose that any two A's, and equally any two B's, are disjoint. We may assume that $m > 0$, $n > 0$; for otherwise the assertion holds trivially. Let P be the set of distinct objects

$$x_1, \ldots, x_m, \quad y_1, \ldots, y_n.$$

We introduce a partial order in P by declaring that $x_i < y_j$ if and only if $A_i \cap B_j \neq \emptyset$ and that there are no other order relations. We shall denote by μ^* the maximum number of elements in an antichain of P, and by μ_* the minimum number of chains into which P can be decomposed. Let M^*, M_* be the maximum and minimum specified in the theorem. Further, let

$$\{x_1, y_1\}, \ldots, \{x_i, y_i\}, \quad \{x_{i+1}\}, \ldots, \{x_m\}, \quad \{y_{i+1}\}, \ldots, \{y_n\}$$

be a decomposition of P into μ_* chains‡ (so that $\mu_* = m + n - i$). Let $e_1, \ldots, e_{\mu_*}$ be any elements contained in

$$A_1 \cap B_1, \ldots, A_i \cap B_i, \quad A_{i+1}, \ldots, A_m, \quad B_{i+1}, \ldots, B_n$$

respectively. Then we have a representing set of $\mathfrak{A}$ consisting of at most μ_* elements, and so $M_* \leqslant \mu_*$.

† If no two sets in $\mathfrak{A}$ are disjoint, then M^* is defined as 1.

‡ It is assumed that the A's and B's have, if necessary, been renumbered.

Again, let R = $\{e_1, \ldots, e_{M_*}\}_{\neq}$ be a representing set of $\mathfrak{A}$. For $e \in$ R, we define $f(e)$ as $\{x_i\}$ if e belongs to A_i but to no B; as $\{y_j\}$ if e belongs to B_j but to no A; and as $\{x_i, y_j\}$ if $e \in A_i \cap B_j$. Then $f(e_1), \ldots, f(e_{M_*})$ is a set of chains whose union is P, and so $\mu_* \leqslant M_*$. It follows that $M_* = \mu_*$.

Furthermore, we obviously have $M^* = \mu^*$. Since, by Dilworth's decomposition theorem 4.4.1, $\mu^* = \mu_*$, we infer that $M^* = M_*$.

We now resume our discussion of matrices. A set L of lines in a matrix will be called *non-intersecting* if any two lines in L are either parallel or else their common place is occupied by a zero element of the matrix.

THEOREM 11.2.8. *Let Q be a rectangular matrix without zero lines. Then the maximum number of non-intersecting lines in Q is equal to the minimum number of non-zero elements incident with all lines.*

Let Q be of type $m \times n$. Define A_i resp. B_j as the set of places in the i-th row resp. j-th column of Q occupied by non-zero elements. Then $\mathfrak{A} = (A_1, \ldots, A_m, B_1, \ldots, B_n)$ is a bipartite family of non-empty sets. The maximum number of non-intersecting lines in Q is equal to the maximum number of pairwise disjoint sets in $\mathfrak{A}$. Again, a representing set of $\mathfrak{A}$ is a set of places in Q occupied by non-zero elements and incident with all lines. Hence the minimum number of non-zero elements incident with all lines of Q is equal to the minimum number of elements in a representing set of $\mathfrak{A}$. The assertion therefore follows by Theorem 11.2.7.

It should be noted that Theorem 11.2.8 is a special case and not a matrix equivalent of Rado's theorem 11.2.7. For to establish Theorem 11.2.8, we consider a bipartite family in which any two sets have at most one element in common.

The reader's attention is also drawn to the remarkable formal resemblance between König's theorem 11.2.1 and Theorem 11.2.8—a resemblance which becomes even more arresting when certain changes of nomenclature are made in the statement of the theorems. This naturally raises the question of the logical relation between the two results. It is, in fact, possible to deduce either of them from the other; but, perhaps surprisingly, these deductions are not particularly short or illuminating.

Exercises 11.2

1. Deduce König's theorem 11.2.1 from Lemma 11.2.2 and Corollary 11.2.5.

2. From Theorem 3.2.6 it is known that a finite family of subsets of a finite set E possesses a transversal if and only if no collection of $|E| + 1$ 'objects' is incident-free. Use this result to deduce Frobenius's theorem (Corollary 11.2.6).

3. Deduce Theorem 11.2.8 from Dilworth's theorem 4.4.1.

4. Deduce Frobenius's theorem (Corollary 11.2.6) from (a) Hall's theorem 2.2.1; (b) König's theorem 11.2.1.

5. Deduce König's theorem 11.2.1 from Dilworth's theorem 4.4.1.

6. Let E, I be finite sets; let $\mathfrak{A} = (A_i : i \in I)$ be a family of subsets of E; and suppose that $E = \cup \{A_i : i \in I\}$. Suppose, further, that there exists a partition $E = E' \cup E''$ such that $|A_i \cap E'| \leqslant 1$, $|A_i \cap E''| \leqslant 1$ for all $i \in I$. Show that the maximum number of elements in E such that no two belong to the same A is equal to the minimum number of A's whose unions is E.

7. Frame König's theorem 11.2.1 and Rado's theorem 11.2.7 in graph-theoretic terms.

8. Use Frobenius's theorem (Corollary 11.2.6) to derive the result stated in Ex. 9.2.1.

11.3 Diagonals of doubly-stochastic matrices

We recall from Theorem 11.1.3 that every finite d.s. matrix possesses a positive diagonal. The main purpose of the present section is to study quantitative refinements of this statement. We shall begin with an important result which shows that every d.s. matrix can be built up from the simplest d.s. matrices, namely the permutation matrices.

THEOREM 11.3.1. (G. Birkhoff) *An $n \times n$ matrix belongs to the convex hull of the $n \times n$ permutation matrices if and only if it is doubly-stochastic.*

We shall denote by $\nu(A)$ the number of positive elements in the d.s. $n \times n$ matrix A. Then $\nu(A) \geqslant n$. If $\nu(A) = n$, then A is a permutation matrix and so lies, of course, in the convex hull of permutation matrices. Let $r > n$ and suppose that every d.s. matrix A with $\nu(A) < r$ lies in the convex hull of permutation matrices.

Next, consider a d.s. matrix A with $\nu(A) = r$. By Theorem 11.1.3, A has a positive diagonal, say Δ. Denote by P the permutation matrix whose positive elements are situated in the places of Δ, and let λ be the least (positive) element of A on Δ. Now $\lambda = 1$ would imply that $A = P$, so that $\nu(A) = n$. Thus $\lambda < 1$ and the d.s. matrix $(A - \lambda P)/(1 - \lambda)$ satisfies the relation

$$\nu\left(\frac{A - \lambda P}{1 - \lambda}\right) < \nu(A) = r.$$

Hence, by the induction hypothesis, there exist positive numbers $\lambda_1, \ldots, \lambda_k$ with sum 1 and permutation matrices $P_1, \ldots, P_k$ such that

$$\frac{A - \lambda P}{1 - \lambda} = \sum_{i=1}^{k} \lambda_i P_i,$$

i.e.

$$A = \lambda P + \sum_{i=1}^{k} (1-\lambda)\lambda_i P_i.$$

This completes the induction proof of the proposition that every d.s. matrix is a convex combination of permutation matrices. The converse inference holds, of course, trivially.

In the subsequent discussion, we shall denote by $\mathscr{D}_n$ the set of all d.s. $n \times n$ matrices, and by $\mathfrak{S}_n$ the set of the $n!$ permutations of $\{1, 2, \ldots, n\}$.

COROLLARY 11.3.2. *Let k_{rs} $(1 \leqslant r, s \leqslant n)$ be given real numbers; and, for any $n \times n$ matrix $A = \|a_{rs}\|$, define*

$$f(A) = \sum_{r,s=1}^{n} k_{rs} a_{rs}.$$

Then $f(A)$ attains its maximum on $\mathscr{D}_n$ at a permutation matrix.

Write $N = n!$ and denote by $P_1, \ldots, P_N$ the N permutation matrices of type $n \times n$. If $A \in \mathscr{D}_n$, then, by Theorem 11.3.1,

$$A = \sum_{r=1}^{N} \lambda_r P_r,$$

where the λ's are non-negative numbers with sum 1. Hence

$$f(A) = \sum_{r=1}^{N} \lambda_r f(P_r);$$

and therefore

$$\max_{A \in \mathscr{D}_n} f(A) = \max \left\{ \sum_{r=1}^{N} \lambda_r f(P_r) : \lambda_1, \ldots, \lambda_N \geqslant 0,\ \lambda_1 + \ldots + \lambda_N = 1 \right\}$$

$$= \max_{1 \leqslant r \leqslant N} f(P_r),$$

as required.

The remaining results in this section state that all (or 'many') elements on some diagonal of every d.s. matrix are not too 'small'. The study of such questions has received a powerful impetus from attempts to prove a conjecture enunciated by van der Waerden. Let $A = \|a_{ik}\|$ be an $n \times n$ matrix. Its *permanent*, per A, is defined by the formula

$$\text{per } A = \sum_{\pi \in \mathfrak{S}_n} a_{1,\pi(1)} \cdots a_{n,\pi(n)}.$$

We shall, throughout, denote by J_n the (d.s.) $n \times n$ matrix all of whose elements are equal to n^{-1}. Van der Waerden conjectured that, for every $A \in \mathscr{D}_n$, per $A \geqslant$ per $J_n = n!\,n^{-n}$, with equality only for $A = J_n$. This relation has not been proved (for general n); but it is plain that its truth would imply that, for each $A \in \mathscr{D}_n$, there exists some $\pi \in \mathfrak{S}_n$ with

$$a_{1,\,\pi(1)} \cdots a_{n,\,\pi(n)} \geqslant n^{-n}. \tag{1}$$

Again, in view of the inequality of the arithmetic and geometric means, the validity of (1) would imply that

$$a_{1,\,\pi(1)} + \ldots + a_{n,\,\pi(n)} \geqslant 1; \qquad a_{1,\,\pi(1)} > 0, \ldots, a_{n,\,\pi(n)} > 0. \tag{2}$$

Both these results will be established below. We shall deal with the inequalities in (2) first for, though an immediate consequence of (1), they admit of a much easier proof.

THEOREM 11.3.3. *Let* $A = \|a_{kj}\|$ *be a doubly-stochastic* $n \times n$ *matrix. The relations* (2) *are then valid for some permutation* $\pi \in \mathfrak{S}_n$.

Let $\sigma \in \mathfrak{S}_n$. We shall denote by P_σ the $n \times n$ permutation matrix associated with σ, so that

$$(P_\sigma)_{kj} = \delta_{\sigma(k),\,j}$$

(where δ stands for the Kronecker delta).

In view of Birkhoff's theorem 11.3.1, we can write

$$A = \sum_{\sigma \in S} \lambda_\sigma P_\sigma, \tag{3}$$

where S is a suitable subset of $\mathfrak{S}_n$, and the λ_σ are *positive* numbers with sum 1. Hence

$$\sum_{k,\,j=1}^{n} a_{kj}^2 = \operatorname{tr}(AA^T) = \sum_{\sigma \in S} \lambda_\sigma \operatorname{tr}(AP_\sigma^{\,T})$$

$$\leqslant \max_{\sigma \in S} \operatorname{tr}(AP_\sigma^{\,T}) = \operatorname{tr}(AP_\pi^{\,T}),$$

where π is some permutation in S. Thus

$$\sum_{k,\,j=1}^{n} a_{kj}^2 \leqslant \sum_{k=1}^{n} a_{k,\,\pi(k)}. \tag{4}$$

But

$$\left(\sum_{k,j=1}^{n} a_{kj}\right)^2 \leqslant \left(\sum_{k,j=1}^{n} a_{kj}^2\right)\left(\sum_{k,j=1}^{n} 1\right)$$

and so

$$1 \leqslant \sum_{k,j=1}^{n} a_{kj}^2.$$

Hence, by (4), $a_{1,\pi(1)} + \ldots + a_{n,\pi(n)} \geqslant 1$. Moreover, by (3),

$$a_{kj} = \sum_{\sigma \in S} \lambda_\sigma \delta_{\sigma(k),j} \geqslant \lambda_\pi \delta_{\pi(k),j} \qquad (1 \leqslant k, j \leqslant n)$$

and therefore

$$a_{k,\pi(k)} \geqslant \lambda_\pi > 0 \qquad (1 \leqslant k \leqslant n).$$

THEOREM 11.3.4. *If $A = \|a_{kj}\|$ is a doubly-stochastic $n \times n$ matrix, then* (1) *is valid for some permutation $\pi \in \mathfrak{S}_n$.*

This result is clearly best possible since, for $A = J_n$ and every $\pi \in \mathfrak{S}_n$, relation (1) reduces to an equality. An analogous remark applies, of course, to Theorem 11.3.3.

We begin by recalling that a real-valued function f is said to be *convex* on the interval I if

$$f(\alpha_1 x_1 + \alpha_2 x_2) \leqslant \alpha_1 f(x_1) + \alpha_2 f(x_2)$$

whenever $x_1, x_2 \in I$, $0 < \alpha_1, \alpha_2 < 1$, $\alpha_1 + \alpha_2 = 1$. It is well-known† that, if f'' exists and is non-negative in I, then f is convex in I. In particular, then, $f(x) = x \log x$ is convex in (0, 1].

We shall denote by $\mathscr{D}_n^+$ the set of all d.s. $n \times n$ matrices without zero elements. For $B = \|b_{kj}\| \in \mathscr{D}_n^+$, we define

$$\phi(B) = \sum_{k,j=1}^{n} b_{kj} \log b_{kj}.$$

We then have

$$\phi(\alpha_1 B_1 + \alpha_2 B_2) \leqslant \alpha_1 \phi(B_1) + \alpha_2 \phi(B_2)$$

whenever $B_1, B_2 \in \mathscr{D}_n^+$ and $\alpha_1, \alpha_2 > 0$, $\alpha_1 + \alpha_2 = 1$. Hence, by obvious induction,

$$\phi\left(\sum_{i=1}^{r} \alpha_i B_i\right) \leqslant \sum_{i=1}^{r} \alpha_i \phi(B_i) \tag{5}$$

† See e.g. Titchmarsh (**1**, 172).

whenever $B_1, \ldots, B_r \in \mathscr{D}_n^+$ and $\alpha_1, \ldots, \alpha_r > 0$, $\alpha_1 + \ldots + \alpha_r = 1$.

Now for any permutation matrix P and any $B \in \mathscr{D}_n^+$, we have $\phi(PB) = \phi(B)$. Hence, taking

$$P = \begin{Vmatrix} 0 & 1 & 0 & 0 & \ldots & 0 \\ 0 & 0 & 1 & 0 & \ldots & 0 \\ \ldots & \ldots & \ldots & \ldots & \ldots & \ldots \\ 0 & 0 & 0 & 0 & \ldots & 1 \\ 1 & 0 & 0 & 0 & \ldots & 0 \end{Vmatrix}$$

and using (5), we obtain

$$\phi(B) = \frac{1}{n} \sum_{k=0}^{n-1} \phi(P^k B) \geqslant \phi\left(\sum_{k=0}^{n-1} \frac{1}{n} P^k B\right)$$

$$= \phi(J_n B) = \phi(J_n) = n \log \frac{1}{n}.$$

Thus

$$\phi(B) \geqslant n \log \frac{1}{n} \qquad (B \in \mathscr{D}_n^+). \tag{6}$$

Next, let $A \in \mathscr{D}_n$ and put $B = (A + \varepsilon J_n)/(1 + \varepsilon)$, where $\varepsilon > 0$. Then $B \in \mathscr{D}_n^+$. Denote by π a permutation of $\{1, \ldots, n\}$ which satisfies

$$\sum_{k=1}^{n} \log b_{k,\pi(k)} = \max_{\sigma \in \mathfrak{S}_n} \sum_{k=1}^{n} \log b_{k,\sigma(k)}.$$

Then

$$\sum_{k=1}^{n} \log b_{k,\pi(k)} = \max_{P} \sum_{k,j=1}^{n} p_{kj} \log b_{kj},$$

where $P = \|p_{kj}\|$ denotes a typical permutation matrix. In view of Corollary 11.3.2 and inequality (6), we infer that

$$\max_{P} \sum_{k,j=1}^{n} p_{kj} \log b_{kj} = \max_{D \in \mathscr{D}_n} \sum_{k,j=1}^{n} d_{kj} \log b_{kj}$$

$$\geqslant \sum_{k,j=1}^{n} b_{kj} \log b_{kj} \geqslant n \log \frac{1}{n}.$$

Thus

$$\sum_{k=1}^{n} \log b_{k,\,\pi(k)} \geqslant n \log \frac{1}{n},$$

i.e.

$$\prod_{k=1}^{n} \left(\frac{a_{k,\,\pi(k)} + \varepsilon n^{-1}}{1+\varepsilon} \right) \geqslant n^{-n}.$$

Letting $\varepsilon \to 0$, we obtain the desired result.

We next turn to a somewhat different kind of statement.

THEOREM 11.3.5. *Let $1 \leqslant k \leqslant n$ and write*

$$\mu_{nk} = \begin{cases} \dfrac{4k}{(n+k)^2} & \text{if } n+k \text{ is even,} \\[2ex] \dfrac{4k}{(n+k)^2 - 1} & \text{if } n+k \text{ is odd.} \end{cases}$$

Then every doubly-stochastic $n \times n$ matrix possesses a diagonal on which at least $n - k + 1$ elements are greater than or equal to μ_{nk}. Moreover, this result is best possible in the sense that the phrase 'or equal to' cannot be omitted.

Assume that every diagonal of the d.s. $n \times n$ matrix A has fewer than $n - k + 1$ elements $\geqslant \mu_{nk}$. Thus, every diagonal contains at least k elements $< \mu_{nk}$. Hence, by Corollary 11.2.5,† A possesses a submatrix B of width $n + k$ in which all elements are $< \mu_{nk}$. Without loss of generality, we can write

$$A = \begin{Vmatrix} B & C \\ D & E \end{Vmatrix},$$

where B is of type $p \times q$ and $p + q = n + k$. Let b, c, d, e denote the sum of all elements in B, C, D, E respectively. Then $b + c = p$, $b + d = q$ and so

$$2b + c + d = p + q = n + k.$$

But $b + c + d + e = n$ and so $b - e = k$. Hence

$$k \leqslant b < pq\, \mu_{nk} \leqslant \mu_{nk} \,.\, \max\{xy:\ x + y = n + k\} = \mu_{nk}\, m$$

† In applying Corollary 11.2.5, we distinguish not between zero and non-zero elements but between those $\geqslant \mu_{nk}$ and $< \mu_{nk}$.

say (where x, y take positive integral values). Now it is an easy matter to verify that $m = k/\mu_{nk}$, and we thus arrive at a contradiction. It follows that some diagonal has at least $n - k + 1$ elements $\geqslant \mu_{nk}$.

To show that this result is best possible, we shall denote by U_{pq} the $p \times q$ matrix all of whose elements are equal to 1.

Let $n + k$ be even, and write $(n + k)/2 = r$, $(n - k)/2 = s$. Then

$$A_0 = \left\| \begin{matrix} kr^{-2}U_{rr} & r^{-1}U_{rs} \\ r^{-1}\ U_{sr} & O \end{matrix} \right\|$$

is a d.s. $n \times n$ matrix. Its submatrix $kr^{-2}U_{rr}$ is of width $n + k$ and all its elements are equal to μ_{nk}. Hence, by Corollary 11.2.5, every diagonal of A_0 contains at least k elements equal to μ_{nk} and so no diagonal contains $n - k + 1$ elements greater than μ_{nk}.

The case when $n + k$ is odd is dealt with similarly. We write

$$r = (n + k + 1)/2, \ \rho = (n + k - 1)/2, \ s = (n - k + 1)/2, \ \sigma = (n - k - 1)/2.$$

The $n \times n$ matrix

$$A_1 = \left\| \begin{matrix} k(r\rho)^{-1}U_{\rho r} & \rho^{-1}U_{\rho\sigma} \\ r^{-1}U_{sr} & O \end{matrix} \right\|$$

is then d.s. Its submatrix $k(r\rho)^{-1}U_{\rho r}$ is of width $n + k$ and has all its elements equal to μ_{nk}. Hence no diagonal of A_1 contains $n - k + 1$ elements greater than μ_{nk}.

The case $k = 1$ of the theorem just proved is worth stating separately.

Corollary 11.3.6. *Let μ_n be defined as $4/n(n + 2)$ or $4/(n + 1)^2$ according as n is even or odd. Then every doubly-stochastic $n \times n$ matrix possesses a diagonal each of whose elements is greater than or equal to μ_n. Moreover, the phrase 'or equal to' cannot be omitted.*

Exercises 11.3

1. Deduce Theorem 11.1.3 from Frobenius's theorem (Corollary 11.2.6).

2. Let $\{x_1, \ldots, x_n\}$ and $\{y_1, \ldots, y_n\}$ be orthonormal sets of vectors in unitary n-dimensional space with inner product (,). Writing $a_{ij} = |(x_i, y_j)|^2$, show that $\|a_{ij}\|$ is a d.s. matrix. Deduce that, for each k with $1 \leqslant k \leqslant n$, the x's and y's can be renumbered such that

$$|(x_i, y_i)| \geqslant \mu_{nk}^{\frac{1}{2}} \qquad (1 \leqslant i \leqslant n - k + 1).$$

3. Determine the largest number λ_n such that at least half the elements on some diagonal of every d.s. $n \times n$ matrix are greater than or equal to λ_n. In particular, verify that $\lambda_n > 8/9n$ $(n \geqslant 1)$.

11.4 Doubly-stochastic patterns

The result of Theorem 11.1.3 prompts a more comprehensive inquiry into the distribution of positive elements in d.s. matrices. More precisely, we ask whether it is possible to prescribe in advance the position of positive elements in a d.s. matrix. This question is now to be discussed both for finite and for infinite matrices.

Two matrices of the same type are said to have *the same pattern* if their non-zero elements occupy the same places. A square matrix is said to have a *doubly-stochastic pattern* if it has the same pattern as some d.s. matrix. Again, if A, B are matrices of the same type, then A is said to be *contained* in B if $a_{ij} \neq 0$ implies $b_{ij} \neq 0$.

The solution of the problem for the finite case is entirely straightforward.

THEOREM 11.4.1. *Let M be a finite, non-zero, square matrix. The following statements are then equivalent.* (i) *M has a doubly-stochastic pattern.* (ii) *M cannot be reduced by means of permutations of rows and of columns to the form*

$$\left\| \begin{matrix} X & O \\ Y & Z \end{matrix} \right\|, \tag{1}$$

where X is a square matrix (of order less than that of M) and $Y \neq O$. (iii) *Every non-zero element of M belongs to a non-zero diagonal.*

Suppose, in the first place, that (i) holds and let D be a d.s. matrix with the same pattern as M. If (ii) were false, then D (as well as M) could be reduced to the form (1). In that case, denoting the type of X by $k \times k$, we see that the sum of all elements in X would be k when computed by rows and less than k when computed by columns. Thus (i) implies (ii).

Next, suppose that (iii) does not hold. Then there is a non-zero element x of M which does not belong to any non-zero diagonal. By permutations of rows and columns, we can ensure that M assumes the form

$$\left\| \begin{matrix} p & N \\ x & q \end{matrix} \right\|,$$

where the matrices of M, p, q, N are of type $n \times n$, $(n-1) \times 1$, $1 \times (n-1)$, $(n-1) \times (n-1)$ respectively and N has no non-zero diagonal. Hence, by Corollary 11.1.2, there exists an integer k in the range $1 \leqslant k \leqslant n-1$ and a set of k rows in N whose non-zero elements lie in (at most) $k-1$ columns. If these k rows and $k-1$ columns are moved into initial positions in N, then M assumes the form (1) with $Y \neq O$ since x is an element of Y. Thus (ii) is violated, and we conclude that (ii) implies (iii).

Finally, let m_{ij} be any non-zero element of M. If (iii) holds, then there exists a permutation matrix $P^{(ij)}$ which is contained in M and has a 1 in the (i, j)-th place. Denoting by t the number of non-zero elements in M, we see that

$$t^{-1} \sum_{m_{ij} \neq 0} P^{(ij)}$$

is a d.s. matrix with the same pattern as M. Thus (iii) implies (i), and the proof is complete.

Next, we turn to the case of infinite matrices. An infinite matrix is said to be *line-finite* if none of its lines contains infinitely many non-zero elements.

The following criterion is analogous to Corollary 11.1.2.

THEOREM 11.4.2. *An infinite, line-finite matrix possesses a non-zero diagonal if and only if, for each natural number k, any k rows resp. columns contain between them non-zero elements from at least k columns resp. rows.*

The necessity of the stated condition is obvious. To prove its sufficiency, we denote by I the set of all natural numbers and by $M = \|m_{ij}\|$ the given matrix. We now define

$$\mathrm{A}_i = \{(i, j) : j \in \mathrm{I},\ m_{ij} \neq 0\} \qquad (i \in \mathrm{I}),$$
$$\mathrm{B}_j = \{(i, j) : i \in \mathrm{I},\ m_{ij} \neq 0\} \qquad (j \in \mathrm{I}).$$

Since M is line-finite, it follows that each A and each B is finite. Since, moreover, the A's and also the B's are pairwise disjoint, it is clear that the families $(\mathrm{A}_i : i \in \mathrm{I})$, $(\mathrm{B}_i : i \in \mathrm{I})$ are relatively finite. Lastly, by our hypothesis, for every natural number k, the union of any k A's resp. B's intersects at least k B's resp. A's. Hence, by Theorem 10.1.5, the two families possess a CSR, and the desired conclusion follows.

In our next result, the assumption of line-finiteness is dropped.

LEMMA 11.4.3. *Let M be an infinite non-zero matrix. If every non-zero element of M belongs to a non-zero diagonal, then M has a doubly-stochastic pattern.*

Let m_{ij} be a non-zero element of M. Then, by hypothesis, M contains at least one permutation matrix which has a 1 in the (i, j)-th place. Choose one such permutation matrix and denote it by $P^{(ij)}$. Since the elements of M are denumerable, so are the permutation matrices $P^{(ij)}$ associated with the non-zero elements of M. Arranging them in a sequence P_k $(k = 1, 2, \ldots)$, we see that the matrix

$$\sum_{k=1}^{\infty} 2^{-k} P_k$$

is d.s. and has the same pattern as M.

We are now in a position to establish the analogue of Theorem 11.4.1 for line-finite matrices.

THEOREM 11.4.4. *The following statements relating to an infinite, line-finite, non-zero matrix M are equivalent.*

(i) *M has a doubly-stochastic pattern.*

(ii) *M cannot be reduced by permutations of rows and columns to either of the two forms*

$$\text{(a)} \left\| \begin{matrix} X_1 & O \\ Y_1 & Z_1 \end{matrix} \right\|, \qquad \text{(b)} \left\| \begin{matrix} X_2 & Y_2 \\ O & Z_2 \end{matrix} \right\|,$$

where X_1, X_2 are finite square matrices and $Y_1 \neq O$, $Y_2 \neq O$.

(iii) *Every non-zero element of M belongs to some non-zero diagonal.*

It should be noted that (in contrast to the finite case) the introduction of both forms (a) and (b) is essential. Thus, for example, the matrix

$$\left\| \begin{matrix} 0 & 1 & 0 & 0 & \ldots \\ 0 & 0 & 1 & 0 & \ldots \\ 0 & 0 & 0 & 1 & \ldots \\ \ldots & \ldots & \ldots & \ldots & \ldots \end{matrix} \right\|$$

cannot be reduced to the form (a); for every set of k rows in this matrix has non-zero elements belonging to exactly k columns, and the non-zero elements in these k columns lie entirely in the k rows in question. But plainly the matrix does not have a d.s. pattern and, of course, it is of the form (b).

We now come to the proof of the theorem. The fact that (i) implies (ii) is demonstrated in precisely the same way as the corresponding step in the proof of Theorem 11.4.1, except that now both the forms (a) and (b) need to be considered. Again, Lemma 11.4.3 shows that (iii) implies (i). It remains to show that (ii) implies (iii).

Assume that (iii) does not hold. Then there is a non-zero element x in M which does not belong to any non-zero diagonal. Hence the matrix N, obtained from M by the deletion of the row and column through x, has no non-zero diagonal. Hence, by Theorem 11.4.2, there is a natural number k and a set of k rows (or columns) in N whose non-zero elements lie in at most $k - 1$ columns (or rows). To fix our ideas, consider the first alternative and take the k rows and $k - 1$ columns in question as occupying initial positions in N, so that

$$N = \begin{Vmatrix} Q & O \\ R & S \end{Vmatrix},$$

where Q is of type $k \times (k - 1)$. By permutations of rows and columns, we can move x into the $(k + 1, k)$-th place in M. Then M assumes the form

$$\begin{Vmatrix} Q & p & O \\ q & x & r \\ R & s & S \end{Vmatrix},$$

where $X_1 = \|Q \quad p\|$ is of type $k \times k$, while the submatrix

$$Y_1 = \begin{Vmatrix} q & x \\ R & s \end{Vmatrix}$$

is non-zero since x is one of its elements. Thus M has been reduced to the form (a). Similarly, by considering the second alternative, we find that M can be reduced to the form (b). In either case, then, condition (ii) is violated; and we conclude that (ii) implies (iii).

Let us next consider the infinite matrix

$$M = \begin{Vmatrix} 1 & 1 & 1 & 1 & 1 & \dots \\ 1 & 1 & 1 & 1 & 1 & \dots \\ 1 & 1 & 0 & 0 & 0 & \dots \\ 1 & 0 & 1 & 0 & 0 & \dots \\ 1 & 0 & 0 & 1 & 0 & \dots \\ \dots & \dots & \dots & \dots & \dots & \dots \end{Vmatrix}.$$

Now any k rows resp. columns of M contain non-zero elements which belong to at least $k + 1$ columns resp. rows, and therefore M satisfies statement (ii) of Theorem 11.4.4. On the other hand, the matrix obtained by the deletion of the first row and first column in M has no non-zero diagonal, and so M fails to satisfy (iii) in Theorem 11.4.4. It follows therefore that Theorem 11.4.4 ceases to be valid for unrestricted infinite matrices. Nevertheless, something can be salvaged: our next theorem shows that (i) and (iii) remain equivalent.

THEOREM 11.4.5. *An infinite non-zero matrix M has a doubly-stochastic pattern if and only if every non-zero element of M belongs to some non-zero diagonal.*

In view of Lemma 11.4.3, it suffices to show that, if M has a d.s. pattern, then every one of its non-zero elements belongs to some non-zero diagonal. We shall prove this by reducing the case of a general infinite matrix to that of a line-finite matrix.

Let, then, M have a d.s. pattern and denote by D a d.s. matrix with the same pattern as M. Assume that some positive element d of D (which we may take as occupying the leading position) does not belong to a positive diagonal of D. Then, writing

$$D = \left\| \begin{matrix} d & a \\ b & E \end{matrix} \right\|,$$

we see that E has no positive diagonal.

It is plain that the sum of all elements in any k rows or any k columns of E is greater than or equal to $k-1+d$. Denote by ρ_m the sum of all elements in the m-th row of E. We now replace by zeros all but a finite number of positive elements in the m-th row of E, in such a way that the new row-sum exceeds $\rho_m - 3^{-m}d$; and we perform this operation for every natural number m. We then obtain a row-finite matrix F; and since the sum of all elements in E which have been replaced by zeros is less than

$$\sum_{m=1}^{\infty} 3^{-m}d = \tfrac{1}{2}d,$$

it follows that the sum of elements in any k rows or k columns of F exceeds $k-1+\frac{1}{2}d$.

Next, denote by σ_m the sum of all elements in the m-th column of F. We replace by zeros all but a finite number of positive elements in the m-th column of F, in such a way that the new column-sum exceeds $\sigma_m - 3^{-m}d$; and we perform this operation for every natural number m. As a result, we obtain a line-finite matrix G in which the sum of elements in any k rows or k columns exceeds $k-1$.

Now the positive elements of G in any k rows (columns) cannot be contained in fewer than k columns (rows) since the sum of elements in any $k-1$ columns or $k-1$ rows of G is at most $k-1$. Hence, by Theorem 11.4.2, G possesses a positive diagonal and so, therefore, does E. We thus arrive at a contradiction and conclude that the element d is, in fact, part of a positive diagonal of D. The proof is therefore complete.

Finally, we note the following immediate consequence of Theorem 11.4.5.

COROLLARY 11.4.6. *Every infinite doubly-stochastic matrix possesses a positive diagonal.*

The infinite analogue of Theorem 11.1.3 is therefore valid.

Exercises 11.4

1. Show that the omission of the phrase 'line-finite' invalidates Theorem 11.4.2.

2. An infinite, line-finite matrix is such that, for each natural number k, any k rows contain between them non-zero elements from at least k columns. Show that the matrix need not possess a non-zero diagonal.

3. Theorem 11.4.5 guarantees the existence of an infinite d.s. matrix all of whose elements are positive and also of one whose only zero elements are precisely the elements on the main diagonal. Give actual examples of such matrices.

4. Show that there exists an infinite d.s. matrix $\|d_{kj}\|$ such that, for every permutation π of $\{1, 2, 3, \ldots\}$,

$$\inf_{k \geqslant 1} d_{k,\pi(k)} = 0.$$

5. (i) Let M be infinite matrix; let ρ, σ be permutations of the set of natural numbers; and suppose that, for every $n \geqslant 1$, the n-th row of M contains at least $\rho(n)$ non-zero elements and the n-th column at least $\sigma(n)$ non-zero elements. By means of Corollary 1.3.5, show that M possesses a non-zero diagonal; and also verify that it need not possess more than one non-zero diagonal.

(ii) Let M be an infinite matrix and suppose that every line of M contains infinitely many non-zero elements. Prove that M has a non-zero diagonal.

6. Let A be an $n \times n$ matrix and, for $1 \leqslant k \leqslant n$, denote by r_k resp. s_k the number of non-zero elements in the k-th row resp. k-th column of A. Show that the sequences of integers $(r_1, \ldots, r_n)$, $(s_1, \ldots, s_n)$ do not determine whether A possesses a d.s. pattern.

11.5 Existence theorems for integral matrices

For a (rectangular) matrix Q, we shall denote by $R_i(Q)$ and $C_j(Q)$ the sum of its elements in the i-th row and j-th column respectively. We shall determine necessary and sufficient conditions for the existence of an integral matrix of given type whose elements, row-sums, and column-sums all lie between prescribed bounds.

THEOREM 11.5.1. *Let* $0 \leqslant r_i' \leqslant r_i$, $0 \leqslant s_j' \leqslant s_j$, $c_{ij} \geqslant 0$ $(1 \leqslant i \leqslant m,$ $1 \leqslant j \leqslant n)$ *be integers. Then there exists an* $m \times n$ *matrix* $Q = \|q_{ij}\|$ *with integral elements such that*

$$r_i' \leqslant R_i(Q) \leqslant r_i \qquad (1 \leqslant i \leqslant m), \tag{1}$$

$$s_j' \leqslant C_j(Q) \leqslant s_j \qquad (1 \leqslant j \leqslant n), \tag{2}$$

$$0 \leqslant q_{ij} \leqslant c_{ij} \qquad (1 \leqslant i \leqslant m,\ 1 \leqslant j \leqslant n) \tag{3}$$

if and only if, for all $\mathrm{I} \subseteq \{1, \ldots, m\}$, $\mathrm{J} \subseteq \{1, \ldots, n\}$,

$$\sum_{\substack{i \in \mathrm{I} \\ j \in \mathrm{J}}} c_{ij} \geqslant \max \left\{ \sum_{i \in \mathrm{I}} r_i' - \sum_{j \notin \mathrm{J}} s_j,\ \sum_{j \in \mathrm{J}} s_j' - \sum_{i \notin \mathrm{I}} r_i \right\}. \tag{4}$$

To prove this result, we put

$$\mathrm{E} = \{(i, j, k) : 1 \leqslant i \leqslant m,\ 1 \leqslant j \leqslant n,\ 1 \leqslant k \leqslant c_{ij}\},$$
$$\mathrm{E}_i = \{(i, j, k) : 1 \leqslant j \leqslant n,\ 1 \leqslant k \leqslant c_{ij}\} \qquad (1 \leqslant i \leqslant m),$$
$$\mathrm{F}_j = \{(i, j, k) : 1 \leqslant i \leqslant m,\ 1 \leqslant k \leqslant c_{ij}\} \qquad (1 \leqslant j \leqslant n).$$

Then $(\mathrm{E}_1, \ldots, \mathrm{E}_m)$ and $(\mathrm{F}_1, \ldots, \mathrm{F}_n)$ are partitions of E. Given a set $\mathrm{X} \subseteq \mathrm{E}$, we define the $m \times n$ matrix $Q = \|q_{ij}\|$ by the formula

$$q_{ij} = |\mathrm{X} \cap \mathrm{E}_i \cap \mathrm{F}_j|.$$

Then (3) is satisfied. Again, given $Q = \|q_{ij}\|$ subject to (3), we define $\mathrm{X} \subseteq \mathrm{E}$ by the formula

$$\mathrm{X} = \{(i, j, k) : 1 \leqslant i \leqslant m,\ 1 \leqslant j \leqslant n,\ 1 \leqslant k \leqslant q_{ij}\}.$$

In either case, we have

$$R_i(Q) = |\mathrm{X} \cap \mathrm{E}_i| \quad (1 \leqslant i \leqslant m), \qquad C_j(Q) = |\mathrm{X} \cap \mathrm{F}_j| \quad (1 \leqslant j \leqslant n).$$

Hence there exists a matrix Q with the requisite properties if and only if there exists a set $\mathrm{X} \subseteq \mathrm{E}$ such that

$$r_i' \leqslant |\mathrm{X} \cap \mathrm{E}_i| \leqslant r_i \quad (1 \leqslant i \leqslant m), \qquad s_j' \leqslant |\mathrm{X} \cap \mathrm{F}_j| \leqslant s_j \quad (1 \leqslant j \leqslant n).$$

Now, by Theorem 9.6.5, this is the case if and only if

$$\sum_{\substack{i \in \mathrm{I} \\ j \in \mathrm{J}}} |\mathrm{E}_i \cap \mathrm{F}_j| \geqslant \max \left\{ \sum_{i \in \mathrm{I}} r_i' - \sum_{j \notin \mathrm{J}} s_j,\ \sum_{j \in \mathrm{J}} s_j' - \sum_{i \notin \mathrm{I}} r_i \right\}$$

whenever $I \subseteq \{1, ..., m\}$, $J \subseteq \{1, ..., n\}$. But $|E_i \cap F_j| = c_{ij}$, and the assertion follows.

COROLLARY 11.5.2. (Fulkerson's 'linking principle') *Let* r_i, r_i', s_j, s_j', c_{ij} *be as in Theorem 11.5.1. If there exists an integral matrix* Q^* *such that*

$$r_i' \leqslant R_i(Q^*), \quad C_j(Q^*) \leqslant s_j, \quad 0 \leqslant q_{ij}^* \leqslant c_{ij} \quad (1 \leqslant i \leqslant m,\ 1 \leqslant j \leqslant n) \tag{5}$$

and an integral matrix Q^{**} *such that*

$$R_i(Q^{**}) \leqslant r_i, \ s_j' \leqslant C_j(Q^{**}), \ 0 \leqslant q_{ij}^{**} \leqslant c_{ij} \quad (1 \leqslant i \leqslant m,\ 1 \leqslant j \leqslant n), \tag{6}$$

then there exists an integral matrix Q *which satisfies* (1), (2), *and* (3).

Taking $r_i = \infty$ $(1 \leqslant i \leqslant m)$, $s_j' = 0$ $(1 \leqslant j \leqslant n)$ in Theorem 11.5.1, we see that Q^* subject to (5) exists if and only if

$$\sum_{\substack{i \in I \\ j \in J}} c_{ij} \geqslant \sum_{i \in I} r_i' - \sum_{j \notin J} s_j$$

for all I, J. Again, taking $r_i' = 0$ $(1 \leqslant i \leqslant m)$, $s_j = \infty$ $(1 \leqslant j \leqslant n)$, we see that Q^{**} subject to (6) exists if and only if

$$\sum_{\substack{i \in I \\ j \in J}} c_{ij} \geqslant \sum_{j \in J} s_j' - \sum_{i \notin I} r_i$$

for all I, J. It follows that, if there exist matrices Q^*, Q^{**} satisfying (5) and (6) respectively, then (4) holds for all I, J. Hence, by Theorem 11.5.1, there exists an integral matrix Q which satisfies (1), (2), and (3).

We next turn to the study of incidence matrices (i.e. matrices all of whose elements are equal to 0 or 1). In particular, we shall exhibit another, and in some ways more natural, proof of the Gale–Ryser criterion (Theorem 5.1.3).

Throughout, we shall write $0 \leqslant r_i' \leqslant r_i$ $(1 \leqslant i \leqslant m)$, $0 \leqslant s_j' \leqslant s_j$ $(1 \leqslant j \leqslant n)$: all these symbols will denote given integers. If x_i, y_i $(1 \leqslant i \leqslant n)$ are real numbers, let $\bar{x}_1, ..., \bar{x}_n$ be the numbers $x_1, ..., x_n$ arranged in non-ascending order of magnitude; and let $\bar{y}_1, ..., \bar{y}_n$ be defined analogously. If

$$\bar{x}_1 + ... + \bar{x}_k \leqslant \bar{y}_1 + ... + \bar{y}_k \tag{7}$$

for $1 \leqslant k \leqslant n$, we shall write

$$(x_1, ..., x_n) \ll (y_1, ..., y_n).$$

It will be recalled (cf. §5.1) that if (7) holds for $1 \leqslant k \leqslant n$ and if, in addition, there is equality for $k = n$, then we write

$$(x_1, ..., x_n) \prec (y_1, ..., y_n).$$

Again, for given numbers $x_1, \ldots, x_n$, we shall denote by x_k^* the number of x's greater than or equal to k.

LEMMA 11.5.3. *Let a_i $(1 \leqslant i \leqslant m)$, b_j $(1 \leqslant j \leqslant n)$ be non-negative integers. Then the inequality*

$$|\mathrm{I}|\,|\mathrm{J}| \geqslant \sum_{i \in \mathrm{I}} a_i - \sum_{j \notin \mathrm{J}} b_j \tag{8}$$

holds for all $\mathrm{I} \subseteq \{1, \ldots, m\}$, $\mathrm{J} \subseteq \{1, \ldots, n\}$ *if and only if*

$$(a_1, \ldots, a_m) \ll (b_1^*, \ldots, b_m^*). \tag{9}$$

We begin by observing that

$$\sum_{i=1}^{k} b_i^* = \sum_{j=1}^{n} \min(b_j, k) \qquad (\text{all } k). \tag{10}$$

Suppose that (8) holds for all I, J. Take $\mathrm{I} \subseteq \{1, \ldots, m\}$ and put

$$\mathrm{J} = \{j : 1 \leqslant j \leqslant n,\ b_j \geqslant |\mathrm{I}|\}.$$

Then, in view of (10), we have

$$\sum_{i \in \mathrm{I}} a_i \leqslant |\mathrm{I}|\,|\mathrm{J}| + \sum_{j \notin \mathrm{J}} b_j = \sum_{j \in \mathrm{J}} |\mathrm{I}| + \sum_{j \notin \mathrm{J}} b_j$$

$$= \sum_{j=1}^{n} \min(b_j, |\mathrm{I}|) = \sum_{i=1}^{|\mathrm{I}|} b_i^*.$$

Thus (8) implies (9).

Next, let (9) be satisfied so that

$$\sum_{i \in \mathrm{I}} a_i \leqslant \sum_{i=1}^{|\mathrm{I}|} b_i^*$$

whenever $\mathrm{I} \subseteq \{1, \ldots, m\}$. Taking any sets $\mathrm{I} \subseteq \{1, \ldots, m\}$, $\mathrm{J} \subseteq \{1, \ldots, n\}$, we have, by virtue of (10),

$$\sum_{i \in I} a_i \leqslant \sum_{i=1}^{|\mathrm{I}|} b_i^* = \sum_{j=1}^{n} \min(b_j, |\mathrm{I}|)$$

$$= \sum_{j \in \mathrm{J}} \min(b_j, |\mathrm{I}|) + \sum_{j \notin \mathrm{J}} \min(b_j, |\mathrm{I}|)$$

$$\leqslant \sum_{j \in \mathrm{J}} |\mathrm{I}| + \sum_{j \notin \mathrm{J}} b_j = |\mathrm{I}|\,|\mathrm{J}| + \sum_{j \notin \mathrm{J}} b_j.$$

The proof is therefore complete.

THEOREM 11.5.4. *There exists an* $m \times n$ *incidence matrix* Q *which satisfies the relations*

$$r_i' \leqslant R_i(Q) \leqslant r_i \quad (1 \leqslant i \leqslant m), \qquad s_j' \leqslant C_j(Q) \leqslant s_j \quad (1 \leqslant j \leqslant n)$$

if and only if

$$(r_1', \dots, r_m') \ll (s_1^*, \dots, s_m^*), \qquad (s_1', \dots, s_n') \ll (r_1^*, \dots, r_n^*).$$

Taking $c_{ij} = 1$ (all i, j) in Theorem 11.5.1, we infer that an incidence matrix Q subject to the given constraints exists if and only if, for all $\mathrm{I} \subseteq \{1, \dots, m\}$, $\mathrm{J} \subseteq \{1, \dots, n\}$, both the inequalities

$$|\mathrm{I}|\,|\mathrm{J}| \geqslant \sum_{i \in \mathrm{I}} r_i' - \sum_{j \notin \mathrm{J}} s_j,$$

$$|\mathrm{I}|\,|\mathrm{J}| \geqslant \sum_{j \in \mathrm{J}} s_j' - \sum_{i \notin \mathrm{I}} r_i$$

are valid. Now, by Lemma 11.5.3, the first of these requirements is satisfied if and only if

$$(r_1', \dots, r_m') \ll (s_1^*, \dots, s_m^*)$$

and the second if and only if

$$(s_1', \dots, s_n') \ll (r_1^*, \dots, r_n^*).$$

COROLLARY 11.5.5. *There exists an* $m \times n$ *incidence matrix* Q *such that* $R_i(Q) \leqslant r_i$ $(1 \leqslant i \leqslant m)$ *and* $s_j \leqslant C_j(Q)$ $(1 \leqslant j \leqslant n)$ *if and only if*

$$(s_1, \dots, s_n) \ll (r_1^*, \dots, r_n^*). \tag{11}$$

To prove this result, we simply take $r_i' = 0$ $(1 \leqslant i \leqslant m)$, $s_j = \infty$ $(1 \leqslant j \leqslant n)$ in Theorem 11.5.4 and write s_j in place of s_j'.

We are now able to derive afresh the Gale–Ryser criterion (Theorem 5.1.3). We first note that, in view of (10) and the given relations $r_1, \dots, r_m \leqslant n$, we have

$$\sum_1^m r_i = \sum_1^m \min(r_i, n) = \sum_1^n r_i^*. \tag{12}$$

Assume now that a matrix of the required kind exists. Then, by Corollary 11.5.5, relation (11) is valid. Further, by (12),

$$\sum_1^n s_j = \sum_1^m r_i = \sum_1^n r_i^*$$

and we therefore have

$$(s_1, \ldots, s_n) \prec (r_1^*, \ldots, r_n^*). \tag{13}$$

Conversely, if (13) holds, then so does (11). Hence, again by Corollary 11.5.5, there exists an incidence matrix Q with

$$R_i(Q) \leqslant r_i \quad (1 \leqslant i \leqslant m), \qquad s_j \leqslant C_j(Q) \quad (1 \leqslant j \leqslant n).$$

Therefore, in view of (12),

$$\sum_1^n s_j \leqslant \sum_1^n C_j(Q) = \sum_1^m R_i(Q) \leqslant \sum_1^m r_i = \sum_1^n r_i^* = \sum_1^n s_j.$$

The sign of equality therefore holds throughout and the matrix Q thus has the desired properties.

We shall conclude the section with a result of a somewhat different type.

THEOREM 11.5.6. *Let a_{ij}, r_i, s_j $(1 \leqslant i \leqslant m,\ 1 \leqslant j \leqslant n)$ be non-negative integers. Then*

$$\max_{\|x_{ij}\|} \sum_{\substack{1 \leqslant i \leqslant m \\ 1 \leqslant j \leqslant n}} x_{ij} = \min_{\mathrm{I},\mathrm{J}} \left\{ \sum_{i \in \mathrm{I}} r_i + \sum_{j \in \mathrm{J}} s_j + \sum_{\substack{i \notin \mathrm{I} \\ j \notin \mathrm{J}}} a_{ij} \right\}, \tag{14}$$

where the maximum is taken with respect to all $m \times n$ integral matrices $\|x_{ij}\|$ subject to the conditions

$$0 \leqslant x_{ij} \leqslant a_{ij} \qquad (1 \leqslant i \leqslant m, \quad 1 \leqslant j \leqslant n), \tag{15}$$

$$\sum_{j=1}^n x_{ij} \leqslant r_i \qquad (1 \leqslant i \leqslant m), \tag{16}$$

$$\sum_{i=1}^m x_{ij} \leqslant s_j \qquad (1 \leqslant j \leqslant n), \tag{17}$$

while the minimum is taken with respect to all subsets I *of* $\{1, \ldots, m\}$ *and* J *of* $\{1, \ldots, n\}$.

Write

$$\mathrm{R}_i = \{(i, j, k) : 1 \leqslant j \leqslant n,\ 1 \leqslant k \leqslant a_{ij}\} \qquad (1 \leqslant i \leqslant m),$$
$$\mathrm{S}_j = \{(i, j, k) : 1 \leqslant i \leqslant m,\ 1 \leqslant k \leqslant a_{ij}\} \qquad (1 \leqslant j \leqslant n).$$

Denote by $\mathfrak{R}$ the family consisting of r_i copies of R_i, $1 \leqslant i \leqslant m$, and by $\mathfrak{S}$

the family consisting of s_j copies of S_j, $1 \leqslant j \leqslant n$. By Corollary 9.3.3, the maximum cardinal (say t) of CPTs of $\mathfrak{R}$ and $\mathfrak{S}$ is given by

$$t = \sum_1^m r_i + \sum_1^n s_j + \min \left\{ \left| \bigcup_1^m \rho_i R_i \cap \bigcup_1^n \sigma_j S_j \right| - \sum_1^m \rho_i - \sum_1^n \sigma_j \right\},$$

where the minimum is taken with respect to all integers ρ_i, σ_j subject to the conditions

$$0 \leqslant \rho_i \leqslant r_i \quad (1 \leqslant i \leqslant m), \qquad 0 \leqslant \sigma_j \leqslant s_j \quad (1 \leqslant j \leqslant n).$$

It follows that

$$t = \sum_1^m r_i + \sum_1^n s_j + \min_{I, J} \left\{ \left| \bigcup_{i \in I} R_i \cap \bigcup_{j \in J} S_j \right| - \sum_{i \in I} r_i - \sum_{j \in J} s_j \right\}$$

$$= \min_{I, J} \left\{ \sum_{\substack{i \in I \\ j \in J}} a_{ij} + \sum_{i \notin I} r_i + \sum_{j \notin J} s_j \right\},$$

so that t is, in fact, equal to the right-hand side of (14).

Now denote by X a CPT of cardinal t of the families $\mathfrak{R}$ and $\mathfrak{S}$, and define ε_{ijk} as 1 or 0 according as $(i, j, k) \in X$ or $(i, j, k) \notin X$. Put

$$x_{ij} = \sum_{1 \leqslant k \leqslant a_{ij}} \varepsilon_{ijk}.$$

The integers x_{ij} then obviously satisfy (15), and

$$\sum_{\substack{1 \leqslant i \leqslant m \\ 1 \leqslant j \leqslant n}} x_{ij} = |X| = t. \tag{18}$$

Further, it is clear that X links x_{ij} copies of R_i to x_{ij} copies of S_j and that consequently (16) and (17) are satisfied. We have therefore demonstrated the existence of an integral $m \times n$ matrix $\|x_{ij}\|$ which satisfies (15), (16), (17), and (18). Thus, we have

$$\max_{\|x_{ij}\|} \sum_{\substack{1 \leqslant i \leqslant m \\ 1 \leqslant j \leqslant n}} x_{ij} \geqslant t, \tag{19}$$

where the maximum is taken with respect to matrices subject to (15), (16), and (17). On the other hand, if the integers x_{ij} satisfy (15), (16), and (17), then, for any $I \subseteq \{1, \ldots, m\}$, $J \subseteq \{1, \ldots, n\}$,

$$\sum_{\substack{1 \leqslant i \leqslant m \\ 1 \leqslant j \leqslant n}} x_{ij} \leqslant \sum_{\substack{i \in I \\ 1 \leqslant j \leqslant n}} x_{ij} + \sum_{\substack{1 \leqslant i \leqslant m \\ j \in J}} x_{ij} + \sum_{\substack{i \notin I \\ j \notin J}} x_{ij}$$

$$\leqslant \sum_{i \in I} r_i + \sum_{j \in J} s_j + \sum_{\substack{i \notin I \\ j \notin J}} a_{ij};$$

and therefore (19) continues to hold when the sign of inequality is reversed. The assertion is now proved.

Exercises 11.5

1. Deduce Theorems 9.6.5 and 9.6.6 from Theorem 11.5.1.

2. Verify directly the necessity of condition (4) in Theorem 11.5.1.

3. Let $m, n \geqslant 1$ and let r_i $(1 \leqslant i \leqslant m)$, s_j $(1 \leqslant j \leqslant n)$ be non-negative integers resp. real numbers. Show that the condition

$$\sum_{i=1}^{m} r_i = \sum_{j=1}^{n} s_j$$

is necessary and sufficient for the existence of an $m \times n$ matrix with non-negative integral resp. real elements and with row-sums r_i and column-sums s_j.

4. Deduce from Theorem 11.5.1 that there exists an integral matrix Q subject to conditions (1) and (2) if and only if

$$\sum_{1}^{m} r_i' \leqslant \sum_{1}^{n} s_j, \qquad \sum_{1}^{n} s_j' \leqslant \sum_{1}^{m} r_i.$$

Also give an ad hoc proof of this result.

5. Let $e \geqslant 0$, $0 \leqslant r' \leqslant r$, $0 \leqslant c' \leqslant c$ be integers, Show that there exists an integral $m \times n$ matrix whose elements, row-sums, and column-sums lie in the intervals $[0, e]$, $[r', r]$, $[c', c]$ respectively if and only if

$$r'm \leqslant \min(emn, cn), \qquad c'n \leqslant \min(emn, rm).$$

6. Let $0 \leqslant r_i' \leqslant r_i$, $0 \leqslant s_j' \leqslant s_j$, $c_{ij} \geqslant 0$ $(1 \leqslant i \leqslant m, 1 \leqslant j \leqslant n)$ be *real* numbers. Use Theorem 11.5.1 to prove that there exists a real $m \times n$ matrix $Q = \|q_{ij}\|$ which satisfies (1), (2), and (3) if and only if (4) holds for all I, J.

7. Let $r_1, \ldots, r_n, s_1, \ldots, s_n$ be non-negative integers. Show that there exists an $n \times n$ incidence matrix Q with zero trace and such that $r_i \leqslant R_i(Q)$ and $C_i(Q) \leqslant s_i$ $(1 \leqslant i \leqslant n)$ if and only if, for all $\mathrm{I}, \mathrm{J} \subseteq \{1, \ldots, n\}$,

$$\sum_{i \in \mathrm{I}} r_i \leqslant |\mathrm{I}|\,|\mathrm{J}| - |\mathrm{I} \cap \mathrm{J}| + \sum_{j \notin \mathrm{J}} s_j.$$

8. Exhibit König's theorem 11.2.1 as a special case of Theorem 11.5.6.

Notes on Chapter 11

§ 11.1 Although we have it on Frobenius's authority (**2**) that Theorem 11.1.3 is 'ein ganz spezieller Satz von geringem Werte', few mathematicians would now endorse this contemptuous verdict. In my opinion, this theorem is a particularly beautiful and suggestive result in combinatorial matrix theory. It was proved over half a century ago by König (**1**; cf. also **7**, 238) by means of the theory of graphs. Frobenius (**2**) exhibited it as a simple consequence of Corollary 11.2.6, and Egerváry (**1**) gave a generalization. For a direct proof, which is independent of combinatorial theory, see Koopmans & Beckmann (**1**).

Theorem 11.1.4 is the matrix version of an important result of Mendelsohn & Dulmage (**1**). It will be recalled that a transfinite extension of this result was discussed in § 10.1. Theorem 11.1.5 is due to König (**1**). I do not know to whom to attribute Theorem 11.1.6; but it appears to be a well-known result.

§ 11.2. The majority of results proved in this section form, together with those in § 3.2, a closely-knit group in which, roughly speaking, any one result can serve as a starting point of the discussion. A 'linear' presentation of the material cannot therefore do justice to the intricacy of relations between the various theorems. The reader is invited to work through the exercises at the end of the section so as to gain a first-hand impression of the interlocking pattern of mutual deductions. See also M. Hall (**4**, 66–67) for the relation between Hall's theorem and Theorem 11.2.1.

Theorem 11.2.1 is a celebrated result due to D. König (**5**; **6**; **7**, 240), which has stimulated a good deal of further research. We refer in passing to a recent self-contained proof of Entringer and Jackson (**1**), to an older constructive proof of M. M. Flood (**1**), and to generalizations of E. Egerváry (**1**), A. J. Hoffman (**1**), W. Vogel (**1**, **2**), and J. Edmonds (**2**).

Theorem 11.2.4 is also due to König (references as above). The special case of the theorem contained in Corollary 11.2.6 was discovered by Frobenius in the course of his well-known investigations on non-negative matrices (**1**) and was afterwards proved by him directly (**2**). Other direct proofs have since been given by Rado (**1**) and by Dulmage & Halperin (**1**). The derivation of Theorem 9.2.1 from Corollary 11.2.6 is taken from Maak's paper (**2**). Frobenius used the result of Corollary 11.2.6 to deduce the following striking consequence. Let A be an $n \times n$ matrix whose elements are independent indeterminates and, possibly, zeros; and suppose that the determinant of A does not vanish identically. Then the determinant of A is a reducible polynomial if and only if A contains a zero submatrix of width n. It may be of interest to note that Corollary 11.2.6 was the cause of acidic exchanges between Frobenius and König. For the tortuous history of this controversy, see the polemical footnote on page 240 of König's book (**7**).

Theorem 11.2.7 is due to R. Rado (**1**). It is one of the very few known results on representing sets.

The maximum number of scattered non-zero elements in a matrix Q is usually referred to in the literature as the 'term rank' of Q. (It is, of course, equal to the transversal index of the family, say $\mathfrak{A}$, associated with Q; and also to the rank of the formal incidence matrix of $\mathfrak{A}$.) Properties of the term rank have been the subject of intensive study, especially at the hands of H. J. Ryser. For further discussion and bibliographical references, see Ryser's book (**4**, 55–66) and his survey articles (**3**, **5**).

§ 11.3. G. Birkhoff's theorem 11.3.1 (proved in (**2**)) is one of the two fundamental results on finite d.s. matrices: the other is due to Hardy, Littlewood, and Pólya (**1**; **2**, Theorem 46). Birkhoff's own proof rests on Hall's theorem, though the induction step used in the present treatment is taken from a paper by Dulmage & Halperin (**1**). Subsequently, a number of other proofs were discovered; for information, see Mirsky's survey (**1**). Broadly speaking, all proofs fall into one or other of two categories; they depend either on combinatorial ideas or on properties of convex polytopes, and both modes of argument seem equally natural. The precise relation between the two methods is, however, still obscure: possibly future progress in applications of linear programming to combinatorial theory will bring some clarification.

Birkhoff's theorem has been extended to infinite d.s. matrices, with the notion of convex closure in a suitable topological vector space replacing that of convex hull. For references to work in this field, see Mirsky (**1**).

Van der Waerden's conjecture (see (**1**)) was enunciated in 1926. In the last decade a very determined assault on it has been mounted (by M. Marcus and others) which, if not yet sucessful, has resulted in the conquest of a good deal of ground. An account of the work up to about 1965 is given in the expository paper (**3**) of Marcus & Minc.

Theorem 11.3.3 was proved by Marcus & Ree (**1**). Indeed, it is a special case of the following result established by these authors. If $A \in \mathscr{D}_n$, $B = AA^T$, and if $\phi(x_1, \ldots, x_n)$ is convex in the sense that, for α, β with $0 < \alpha, \beta < 1, \alpha + \beta = 1$,

$$\phi(\alpha x_1 + \beta y_1, \ldots, \alpha x_n + \beta y_n) \leqslant \alpha\,\phi(x_1, \ldots, x_n) + \beta\phi(y_1, \ldots, y_n),$$

then there exists a permutation $\pi \in \mathfrak{S}_n$ such that

$$\phi(b_{11}, \ldots, b_{nn}) \leqslant \phi(a_{1,\,\pi(1)}, \ldots, a_{n,\,\pi(n)})$$

and $a_{k,\,\pi(k)} > 0\ (1 \leqslant k \leqslant n)$. The special choice $\phi(x_1, \ldots, x_n) = x_1 + \ldots + x_n$ leads easily to Theorem 11.3.3.

Theorems 11.3.4 and 11.3.5 are due to Marcus & Minc (**1**). The results of Theorems 11.3.3 and 11.3.4 admit of further development in the context of the spectral theory of d.s. matrices; see Marcus & Minc (**2**).

§ 11.4. The account offered here follows closely the argument of Perfect & Mirsky (**1**). Theorem 11.4.2 and Corollary 11.4.6 are due to J. R. Isbell (**1**). Moreover, the proof of Theorem 11.4.5 invokes Isbell's procedure for reducing the general case to the case of a line-finite matrix.

§ 11.5. This section is based on Mirsky's paper (**5**). However, Theorem 11.5.1 (and, indeed, a slightly more general result) is to be found in the work of H. G. Kellerer (**1**, **2**), where it is obtained in the context of a measure-theoretic study. A transfinite extension of Theorem 11.5.1 has been found by R. A.Brualdi (**10**).

Corollary 11.5.2 ('linking principle') is due to D. R. Fulkerson (**2**, Theorem 3) and is a special case of a very comprehensive result (**2**, Theorem 2) which asserts that, in suitable circumstances, if there exists an object X_1 satisfying condition $\mathscr{C}_1$ and an object X_2 satisfying condition $\mathscr{C}_2$, then there also exists an object satisfying both $\mathscr{C}_1$ and $\mathscr{C}_2$. Fulkerson's argument depends on consideration of 'flows in networks'—a subject which may be regarded as a quantification of the theory of graphs. A systematic account of this powerful and adaptable theory is given in Ford & Fulkerson's book (**2**). In particular, Chapter 2 of this work is largely devoted to the study of systems of representatives, transversals, and matrices.

Lemma 11.5.3 is implicit in the work of W. Vogel (**2**).

We may mention a variant of the Gale–Ryser criterion (Theorem 5.1.3) due to Fulkerson (**3**) which, too, can be deduced from Theorem 11.5.1. Let $r_1 \geqslant \ldots \geqslant r_n \geqslant 0$, $s_1 \geqslant \ldots \geqslant s_n \geqslant 0$ be integers. Then there exists an $n \times n$ incidence matrix Q with zero trace and such that $r_i \leqslant R_i(Q)$ and $C_i(Q) \leqslant s_i\ (1 \leqslant i \leqslant n)$ if and only if, for $1 \leqslant k \leqslant n$,

$$\sum_{i=1}^{k} r_i \leqslant \sum_{i=1}^{k} s_i^* - \min(k, s_k^*).$$

For the case of an incidence matrix $\|a_{ij}\|$, Theorem 11.5.6 was discovered by W. Vogel (**1**, **2**). I owe the proof given above to Professor R. A. Brualdi, who has also found a still more general result (**11**).

12 Conclusion

12.1 Current trends in transversal theory

Until very recently, transversal theory was little more than a patchwork of isolated results. If it has now attained the status of a distinctive discipline, the discipline is still very young. A few results can be traced back half a century, but a majority of theorems treated here were discovered during the last ten or fifteen years, and many during the last five. The current rate of progress is exceptionally rapid and the subject is still in a state of flux. Indeed, even a cursory comparison between a survey article written in 1965 (Mirsky & Perfect (**1**)) and the contents of the present work reveals the decisive changes that transversal theory has undergone within the last few years. It is, to be sure, possible to discern the general shape of the subject, but there cannot be any certainty that the shape will remain unaltered, even in broadest outline. It is therefore plain that the attempt, made in the preceding chapters, to codify transversal theory must necessarily have a provisional character.

At present, it is even difficult to speak with any degree of confidence of the standing of individual results within the theory. Of the fundamental nature of Philip Hall's original criterion (Theorem 2.2.1) there can admittedly be no doubt. It is not the most powerful or the most comprehensive result; indeed, by comparison with many subsequent discoveries, it deals with a very restricted situation. Nevertheless, its position is secure, for it stands at the beginning of the road and points the way to further progress. Transversal theory, at least as it appears now, is predominantly a study concerned with the existence of transversals and transversal-like objects. (Enumerative questions are no less interesting but are generally much less tractable.) Hall's theorem settles the first significant problem in the field; and the solution is framed in a form that has served as a pattern for more general investigations undertaken subsequently. It should, of course, be borne in mind that there is a whole series of results equivalent or very closely related to Hall's theorem and therefore of comparable importance. We recall, in particular, Theorem 2.2.4, Theorem 3.2.1 (Hall–Ore), Theorem 3.2.4, Theorem 3.2.6, Theorem 11.2.1 (König), Corollary 11.2.6 (Frobenius).

There is a second and more profound reason why Hall's theorem must be accorded a distinguished place. It is, in fact, not merely the prototype but the actual source of a large part of transversal theory, and especially the part that is concerned with finite families. We know that Hall's theorem is 'self-refining': by applying it to more recondite families constructed from the original family, we are in many cases led to significant generalizations of the theorem itself. This procedure—the method of 'elementary constructions' introduced in Chapter 3—turns out to be extremely effective so that it is very nearly true to describe the entire body of finite transversal theory as a chain of corollaries of Hall's theorem. It must be added that to deal with transfinite theory we require, naturally enough, a stock of fresh ideas and that, even in the finite case, arguments based on 'elementary constructions' are not uniformly illuminating.

A very rich source of new insight is found in the study of abstract independence. This field of inquiry was created in its own right with the object of placing on an axiomatic footing the theory of linear independence in vector spaces. Nevertheless, as it has turned out, results and methods in this field can be harnessed effectively to the development of transversal theory. Certain key results enable us to link the two disciplines. The theorem of Edmonds and Fulkerson (Theorem 6.5.2) states that a transversal structure is always a pre-independence structure. Rado's theorem on independent transversals (Theorem 6.2.1) achieves for general pre-independence structures what Hall's theorem had done for universal structures. Rado's theorem and its transfinite analogue (Theorem 6.2.4) must, in the present state of our knowledge, be regarded as the central results in the whole of transversal theory. In any case, the power of the method based on considerations of independence would be difficult to overstate: it seems to add a new dimension to our inquiry. The reader will have noted that this method enables us to derive in an easy and natural manner results which seem inaccessible, or only accessible with difficulty, by alternative methods. In particular, applications of the notion of independence yield a very transparent treatment of problems involving common transversals. We recall, as an instance, the proof of the Ford–Fulkerson criterion (Theorem 9.3.2) and of Theorem 9.4.3. It seems highly likely that the continued study of abstract independence, and especially a more searching analysis of 'induced structures', will lead to decisive advances in transversal theory. The prognosis, at any rate, is favourable.

Another line of research which may well emerge in the next year or two is the attempt to superimpose on independence structures some additional structure, say metrical, topological, or ordinal. The resulting objects will naturally have a much richer texture than pure independence structures, and an understanding of their constitution may well lead to progress in transversal theory. The general idea of studying mixed structures is, of course, rather obvious; but so far very little research in this field seems to have been undertaken. However, the work of D. Gale (**3**) and of D. J. A. Welsh (**2**) should be mentioned.

It may also be of interest to refer to a simple but surprisingly fruitful idea, that of duality. Nearly everyone working in transversal theory must have observed the duality between sets and elements. From a purely formal point of view, this observation is trite but it performs nevertheless a very useful service in practice since, by interchanging the roles played by sets and elements figuring in a theorem, we are led at once to another theorem which may bear little superficial resemblance to the original one. Alternatively, given two known theorems, we are sometimes able to recognize their dual relation and so their essential identity. The notion of duality has been implicit in transversal theory from the beginning, but in the last year or two an attempt has been made to formalize and exploit it systematically.

When we examine the actual content of theorems in transversal theory, three pervasive features force themselves on our attention. In the first place, many theorems have the form of 'linking principles', i.e. statements which may be described in the following terms. Let A, B be given systems (such as families of sets) and suppose that there exist two objects satisfying the asymmetric conditions $\Omega(\mathrm{A}, B)$ and $\Omega(B, A)$ respectively. A linking principle then asserts the existence of an object satisfying both the conditions $\Omega(A, B)$, $\Omega(B, A)$ and so related symmetrically to the systems A and B. Possibly the simplest result of this kind is the Schröder–Bernstein theorem 1.3.6. Other instances are provided by Ore's theorem 1.3.4 (and its deltoid form embodied in Theorem 2.3.1), by Theorem 10.1.7, and by Fulkerson's principle (Corollary 11.5.2). We may, in passing, also refer to more sophisticated results of J. S. Pym (**3**, **4**) and of R. A. Brualdi (**3**). It is a common feature of linking principles that they do not involve postulates of local finiteness.

Again, a number of theorems relate the local character of a mathematical system to its global character. A simple and representative example is M. Hall's theorem 4.2.1, (b) and (c), which asserts that an infinite family of finite sets possesses a transversal if every finite subfamily has this property. Other examples of this type of statement are to be found in Theorem 10.1.5 (about common systems of representatives) and in Theorem 10.4.5, (i) and (ii) (about common transversals). Proofs of such results are normally based on Rado's selection principle and consequently involve some postulate of local finiteness. (Thus, in M. Hall's theorem, we require the sets of the given family to be finite.)

Another aspect of transversal theory is the distinction between 'qualitative' and 'quantitative' results.† Let A be some system and S an 'existential' statement, i.e. a statement of the type 'there exists an object X related in such and such a manner to A'. Suppose that S implies some further existential statement S^*. It can happen that the converse inference is also valid, i.e. the truth of S^* implies that of S. We call such an inference a qualitative result. As a

† In the present context, the term 'quantitative' has no relation to enumerative problems.

typical instance, we may cite Theorem 6.6.2. It is trivial that, if a family $\mathfrak{A}$ possesses a transversal which contains a set M, then $\mathfrak{A}$ possesses a transversal and also a subfamily of which M is a transversal. Theorem 6.6.2 asserts that the converse inference is still valid.

In other theorems, the existence of an object X related in a prescribed manner to a system A may be secured by a set of inequalities usually resembling Hall's condition. In such a situation we speak of a quantitative result. Hall's theorem and its transfinite extension (Theorem 4.2.1, (a) and (c)) are obvious instances of this type of result; for other examples, we may refer to Theorem 3.3.6, (i) and (ii), Theorem 6.6.3, Theorem 10.1.5, (i) and (ii), and Theorem 10.4.5, (i) and (iii). A number of theorems in transversal theory exhibit more than one of the three features we have singled out, and in some of the most attractive results all three are present.

In the course of our exposition of transversal theory, we have had occasion from time to time to refer to results from the theory of graphs, though no attempt was made to explore systematically the links between these two areas of combinatorial analysis. In fact, the relation between transversal theory and the theory of graphs is close. Thus it will be recalled that the earliest formulation, due to D. König, of a result intimately related to Hall's theorem was couched in the language of graphs (Theorem 1.7.1). Indeed, numerous results in transversal theory can be rephrased in graph-theoretic terms (which usually involve statements about bipartite graphs). This feature of the theory is, of course, rather superficial as it is a matter of language only. However, the very interesting work of Ford & Fulkerson (**2**) on 'flows in networks', which may be regarded as a quantitative variant of the theory of graphs, provides an astonishingly diverse range of applications in transversal theory. Again, more recently, Hazel Perfect (**2**; **6**, chap. VIII) demonstrated that a large part of transversal theory can be derived from Menger's graph theorem 1.7.2. This development seems likely to have considerable repercussions on the future shape of the subject.

In our treatment, we have sought to present transversal theory not so much as a collection of arresting individual results as a systematic discipline, and to point to what appear to be its salient features. But, inevitably, part of the outline is still hidden from view. Thus, to take an isolated example, what precisely is the status of Dilworth's decomposition theorem? Its immediate and striking appeal will scarcely be denied, but it is much harder to assign to it a definite place within transversal theory. A number of results we have discussed (such as König's theorem 11.2.1) can be readily deduced from Dilworth's theorem; yet it does not seem that in any of these instances its power is exploited to the full. If we incline to the belief that its importance in transversal theory is likely to grow, our opinion is conjectural rather than based on tangible evidence.

Another feature of transversal theory that has perplexed a number of observers is the ubiquity of 'standard extremal theorems', i.e. theorems which assert the equality of a maximum and a minimum. A very rapid survey of the material in the preceding chapters will convince us of the abundance of such theorems. We may mention, for example, the following results (some of which need to be trivially modified to conform to the desired pattern): Theorem 1.7.2 (Menger), Corollary 3.2.3, Theorem 3.2.4, Corollary 4.4.2 (Dilworth), Theorem 6.2.2 (Tverberg), Theorem 8.1.2 (Nash-Williams), Theorem 8.4.2 (Brualdi), Theorem 8.4.3 (Brualdi), Corollary 9.3.3, Theorem 11.2.1 (König), Theorem 11.2.7 (Rado), Theorem 11.5.6 (Vogel, Brualdi). In areas of combinatorial analysis other than transversal theory, standard extremal theorems are also very thick on the ground (see e.g. Harper & Rota's report (**1**)). The pervasiveness of this type of theorem is hardly likely to be fortuitous, and we naturally look for an underlying pattern. However, all we can say at the moment is that the profusion of standard extremal results points forcibly in the direction of methods based on the duality theorem of linear programming. Such methods have, indeed, been applied successfully to transversal theory. A brief description of the basic idea will be found in the survey referred to earlier (Mirsky & Perfect (**1**)); for a more technical discussion, see Hoffman (**2**) and Kuhn & Tucker (**1**). But the last word on the subject has certainly not been spoken, and the precise relevance of linear programming in combinatorial theory is still eluding attempts at clarification.

12.2 Future research and open questions

We shall conclude our exposition by listing a series of open problems, many of which are current coin among mathematicians working in combinatorial analysis. This collection is inevitably very heterogeneous, a fact which reflects, perhaps, to some extent, the present fluid state of transversal theory. Some of the problems stated below are of a purely technical nature—possibly no more than exercises which can be solved completely within the context of the theory developed so far (e.g. by the use of suitable 'elementary constructions'.) By contrast, the solution of other questions would certainly extend the scope of the present theory. Many of the questions are entirely specific; in other cases, we indicate not so much individual problems as areas of research. Again, the questions are not independent, and the reader will recognize that the solution of some of them would readily lead to the solution of others. It seems, nevertheless, worthwhile to include the easier problems in case the harder ones should turn out to be intractable.

Before we enumerate the individual problems, we shall attempt to identify the main areas of research where progress is likely to make a significant impact on transversal theory as a whole.

(i) At present, transversal theory is largely concerned with the determination of criteria for the existence of transversal-like objects associated with a single family or with two families. In certain directions, work has gone a long way and many extremely sophisticated results are known. But one must cherish the hope that some, at any rate, of future efforts will be devoted to problems involving three or more families.

(ii) There has been a pronounced tendency in modern transversal theory not to rest content with results relating to finite sets but to seek to establish their transfinite analogues; and, up to a point, attempts to do so have been very successful. Nevertheless, many of the results that have been proved are still burdened with restrictions involving some kind of local finiteness. The removal or modification of these restrictions cannot be easy since, in some ways, they are entirely natural; but progress in this field would certainly be of very great interest.

(iii) The introduction of ideas derived from the study of abstract independence has changed transversal theory in a decisive manner, and has brought clarity and order into situations which seemed previously difficult and obscure. There can be little doubt that further work on independence will supply transversal theory with a new range of effective tools. In particular, the investigation of 'induced' independence structures and their rank functions is likely to prove rewarding.

(iv) Broadly speaking, transversal theory (and, more generally, combinatorial analysis) have so far developed as branches of the theory of sets. It may be profitable to seek to extend and enrich the existing theories by studying combinatorial problems for sets which carry some additional structure. Since this programme of work lies wholly in the future, it is difficult to speak in concrete terms, but certain questions put forward by Dr Pym (see No. 5 below) illustrate the approach we have in mind. In particular, it may be of interest to analyse independence structures defined on sets endowed with some algebraic, topological, or ordinal character.

(v) It may be of equal interest to go in precisely the opposite direction and instead of seeking to enrich the objects we study, to impoverish them by considering structures 'weaker' than independence structure. We are thinking, specifically, of the investigation of hereditary structures which satisfy some postulate (which?) less exacting than the replacement axiom: the particular model that may guide us here is the collection of all common partial transversals of two families of sets.

(vi) Finally, we mention briefly the topic of 'submodular functions', i.e. non-decreasing functions of sets which satisfy the modular inequality. The origin of the subject goes back to a paper of R. P. Dilworth (**1**) published as long ago as 1944. Very recently, there has been a fresh spate of investigations in this field, by Edmonds, Rota, Welsh (**6**), and Pym & Perfect (**1**).

We have not found it possible to include a discussion of submodular functions in our account, and so cannot now enter into details. It suffices to say that the recent development of the subject, especially at the hands of J. S. Pym and Hazel Perfect, seems extremely promising. The investigations of these authors yield, in particular, numerous models of independence structures and provide some striking applications to transversal theory. Further work in this field should be of the highest interest.

* * * *

Let us now look in greater detail at what appear to be gaps or inadequacies in current transversal theory.

1. Let $(A_1, \ldots, A_n)$ be a family of sets; and let $p \geqslant 1$, $1 \leqslant k \leqslant n$. Under what conditions do there exist pairwise disjoint sets $X_1, \ldots, X_k$ and integers $i_1, \ldots, i_k$ such that $1 \leqslant i_1 < \ldots < i_k \leqslant n$, $X_1 \subseteq A_{i_1}, \ldots, X_k \subseteq A_{i_k}$, $|X_1| = \ldots = |X_k| = p$? In view of Theorem 3.3.1, one necessary and sufficient condition is

$$\max_{J} \min_{I} \{|A(I)| - p|I \cap J|\} \geqslant 0,$$

where the minimum is taken with respect to all $I \subseteq \{1, \ldots, n\}$ and the maximum with respect to all $J \subseteq \{1, \ldots, n\}$ such that $|J| = k$. This formulation does not, however, seem to be entirely satisfactory and we seek a criterion nearer in type to Hall's theorem.

2. $(A_1, \ldots, A_n)$ is a finite family of subsets of a finite set E. What conditions are necessary and sufficient for the existence of a set $R \subseteq E$ such that $|R \cap A_i| = 1$ $(1 \leqslant i \leqslant n)$?

3. Let $\mathfrak{A} = (A_i : i \in I)$ be an infinite family of subsets of a set E. When does $\mathfrak{A}$ possess a system of representatives in which no element occurs infinitely often? It is known (cf. Ex. 4.2.6) and is easy to prove that, when I is denumerable, a necessary and sufficient condition is that, for every infinite subset J of I, A(J) should be an infinite set. When I is non-denumerable, this condition is no longer sufficient (as may be seen on taking I as the set of real numbers and, for each $i \in I$, A_i as the set of rational numbers), and the question remains open. An evidently necessary condition is that, for every infinite subset J of I, we should have $|A(J)| \geqslant |J|$.

A variant of the problem is concerned with conditions in order that $\mathfrak{A}$ should possess a system of representatives in which the frequency of occurrence of every element of E is less than $|I|$.

4. The transfinite form of Hall's theorem (Theorem 4.2.1) is valid when the individual sets in the given family are finite. When this postulate is dropped,

other conditions have to be introduced. Rado and Jung (Theorem 4.3.1) disposed of the case when just one set in the family is infinite. More recently, Brualdi & Scrimger (**1**) and Folkman (**1**) discussed the case of finitely many infinite sets; but both these papers are very difficult (and Folkman's also involves an 'existential' condition). It would be of great interest to settle even the case of *two* infinite sets in a relatively simple manner.

5. Let E, I be certain mathematical systems and let $\mathfrak{A} = (A_i : i \in I)$ be a family of subsets (or, possibly, subsystems) of E. What conditions are necessary and sufficient in order that $\mathfrak{A}$ should possess an injective choice function which, at the same time, is a structure-preserving mapping of I into E? For example, if E, I are partially ordered (or totally ordered) sets, when does $\mathfrak{A}$ possess an injective and order-preserving choice function? Or, again, if E, I are abelian groups (in which case the A_i may need to be taken as subgroups of E), when does $\mathfrak{A}$ possess a choice function which is a monomorphism of I into E? It is clearly easy to formulate further problems of this character.

[J. S. Pym]

6. A question easier than that indicated in No. 5 is as follows. Let $\mathfrak{A}$ be a finite family of subsets of a finite partially ordered set E. Under what circumstances does $\mathfrak{A}$ possess a transversal which is an antichain?

[J. S. Pym]

7. Let E be a finite set; let a non-negative 'weight' $w(x)$ be associated with each element $x \in E$; and, for $F \subseteq E$, write

$$w(F) = \sum_{x \in F} w(x).$$

If $\mathfrak{A} = (A_1, \ldots, A_n)$ is a family of subsets of E, what can be said about max $w(X)$, where the maximum is taken with respect to all partial transversals X of $\mathfrak{A}$? (When $w(x) = 1$ for all $x \in E$, the problem reduces to the determination of the transversal index of $\mathfrak{A}$.) Needless to say, countless problems in transversal theory can be generalized by the introduction of 'weights'.

8. Let $\mathscr{E}$ be an independence structure on an infinite set E and let $\mathfrak{F} = (F_i : i \in I)$ be a partition of E. When does $\mathfrak{F}$ possess an independent transversal?

[R. A. Brualdi]

9. At present, the relation between Dilworth's decomposition theorem 4.4.1 and transversal theory is rather tenuous; but further study may reveal unexpected connections. It would, in any case, be worth examining systematically which results in transversal theory can be exhibited as consequences of Dilworth's theorem. Equally, it would be interesting to know whether Menger's graph theorem can be deduced from Dilworth's theorem.

10. A sequence $(a_1, \ldots, a_n)$ of n real terms is said to be 'convex' if, for any three consecutive terms,

$$2a_{k+1} \leqslant a_k + a_{k+2}.$$

It is said to be 'concave' if this inequality is reversed. It is required to estimate $f(n)$, the largest integer such that *every* sequence of n real terms possesses either a convex or a concave subsequence of $f(n)$ terms. (Cf. Theorem 4.4.4.) Analogously, we can formulate other 'three-term problems'.

11. Gale and Ryser established necessary and sufficient conditions (Theorem 5.1.3) for the existence of a rectangular incidence matrix with prescribed line-sums. Ryser has proposed the further, probably very difficult, problem of determining the number of such matrices.

12. It is clear from the proof of Theorem 5.4.2 that the necessary and sufficient conditions appearing in that result are equivalent to a *finite* subset of conditions. It would be of interest to verify that the conditions in this finite set are independent. (Cf. the remarks at the end of § 2.1.)

[R. Rado]

13. Theorem 5.2.1 gives a lower estimate for the number of strict systems of distinct representatives of a given family. It would be equally interesting to obtain (upper and lower) estimates for the number of transversals of a given family. Questions such as these can also be given a statistical slant. For example, what is the 'expected' number of transversals of a family of n sets none of which contains more than m elements? Further, how many families among those mentioned possess a transversal? There are obvious variants of all these questions when 'transversal' is replaced by 'independent transversal'.

14. Under what circumstances does a finite family of sets possess a unique independent transversal? (Cf. Ex. 6.2.5.) More particularly, under what circumstances do two finite families of sets possess a unique common transversal?

15. Most investigations so far have been concerned with individual independence structures, but it may be rewarding to examine the collection of *all* independence structures on a given set. Can we impose any recognizable structure on this collection, say by defining suitable operations?

16. Let $\mathfrak{X}$ be a collection of subsets of a set E. Under what circumstances is $\mathfrak{X}$ precisely the collection of sets common to two independence structures on E?

17. Let $\mathscr{E}$ be an independence structure on a finite set E; let $\mathfrak{A} = (A_1, \ldots, A_n)$ be a family of subsets of E; and let $1 \leqslant k \leqslant n$. When does $\mathfrak{A}$ possess a transversal whose rank is equal to k? (The question is almost trivial if we require the rank of the transversal to be greater than or equal to k.)

18. Let $\mathfrak{A} = (A_1, \ldots, A_n)$, $\mathfrak{B} = (B_1, \ldots, B_n)$ be families of subsets of a finite set E; let $M \subseteq E$; and let $k \leqslant |M|$. What conditions are necessary and sufficient for the existence of a common transversal X of $\mathfrak{A}$ and $\mathfrak{B}$ such that $|X \cap M| \geqslant k$? A harder variant of the problem is concerned with an *independent* transversal X of $\mathfrak{A}$, and we can also seek to bound $|X \cap M|$ above.

19. Let $\mathscr{E}$ be an independence structure on a finite set E, let $\mathfrak{A} = (A_1, \ldots, A_n)$ be a family of subsets of E, and let k be a positive integer. When is it possible to partition E into k independent partial transversals of $\mathfrak{A}$?

20. For the collections $\mathscr{E}_1, \mathscr{E}_2$ of subsets of a finite set, let the sum $\mathscr{E}_1 + \mathscr{E}_2$ be defined by the equation

$$\mathscr{E}_1 + \mathscr{E}_2 = \{X_1 \cup X_2 : X_1 \in \mathscr{E}_1, X_2 \in \mathscr{E}_2\}.$$

What structural properties are preserved under the operation '+'? We know, by Theorem 8.1.1, that the sum of two independence structures is an independence structure; and it is easy to verify that the sum of two transversal structures is a transversal structure. Is the analogous conclusion valid for e.g. collections of sets common to two independence structures? It is, of course, easy to frame other questions of the same type.

21. Let E be a finite set, $\mathfrak{A}$ a family of subsets of E, and $\mathscr{E}$ an independence structure on E. We shall say that $\mathfrak{A}$ *conforms* to $\mathscr{E}$ if the set of all independent partial transversals of $\mathfrak{A}$ is again an independence structure. There are a number of natural questions we can ask about the relation of $\mathfrak{A}$ and $\mathscr{E}$, and we mention two of these. (i) Characterize $\mathscr{E}$ such that *every* family conforms to it. (ii) Characterize $\mathscr{E}$ such that a *given* family conforms to it. The answer to (i) is not difficult: $\mathscr{E}$ is a 'truncated restriction' of the universal structure on E (cf. Misc. Exs. No. 20). Question (ii) seems to be harder.

[H. Perfect (7)]

22. The sum of a number of independence structures defined on a finite set is, by Theorem 8.1.1, again an independence structure. Can we assert that every independence structure can be expressed as the sum of independence structures which are, in some sense or other, 'irreducible'? If such a decomposition exists, is it unique?

23. Two collections $\mathscr{E}, \mathscr{E}'$ of subsets of a set E are said to be 'isomorphic' if there exists a permutation σ of E such that $X\ (\subseteq E)$ is a member of $\mathscr{E}$ if and only if $\sigma(X)$ is a member of $\mathscr{E}'$.

Let $I(n)$ resp. $L(n)$ resp. $T(n)$ denote the number of non-isomorphic independence structures resp. linear structures resp. transversal structures on a set of n elements. Then, trivially,

$$T(n) \leqslant L(n) \leqslant I(n).$$

What can be said about the rate of growth of these functions? D. J. A. Welsh noted that $T(n) \leqslant n2^{n^2}$ and, in his paper (**3**), he showed that $T(n) \geqslant 2^n$. More recently, B. Bollobás (**1**) proved that

$$\liminf \frac{\log_2 L(n)}{n^2} \geqslant \frac{1}{12}$$

and Welsh observed that a slight modification of Bollobás's argument yields the stronger relation

$$\liminf \frac{\log_2 T(n)}{n^2} \geqslant \frac{1}{12}.$$

It is plain that these conclusions still leave many questions unanswered. In particular, is it true that almost all independence structures are non-transversal in the sense that $T(n) = o\big(\mathrm{I}(n)\big)$ as $n \to \infty$?

24. We recall that an independence structure is not necessarily a transversal structure; but possibly a 'small' adjustment is sufficient in every, or in almost every, case to convert an independence structure to a transversal structure. The interested reader may be able to follow up this vague suggestion by investigating what type of adjustment will suffice and how 'small' it can be kept.

25. What conditions are necessary and sufficient in order that a given collection of sets should be a transversal structure?

26. D. J. A. Welsh's characterization of transversal structures (Theorem 8.3.1) does not provide us with a systematic method for deciding whether a given finite independence structure is transversal. It would be of interest to devise a simple procedure fulfilling this requirement.

27. Let $\mathscr{E}_1$, $\mathscr{E}_2$ be collections of subsets of a finite set E. We denote by $\mathscr{E}_1 \cap \mathscr{E}_2$ the collection of all sets common to $\mathscr{E}_1$ and $\mathscr{E}_2$, and by $\mathscr{E}_1 + \mathscr{E}_2$ the collection of all sets expressible in the form $X_1 \cup X_2$, where $X_1 \in \mathscr{E}_1$, $X_2 \in \mathscr{E}_2$.
It is easy to show that, for any two independence structures $\mathscr{E}_1$, $\mathscr{E}_2$, we have

$$(\mathscr{E}_1 \cap \mathscr{E}_2) + (\mathscr{E}_1 \cap \mathscr{E}_2) \subseteq (\mathscr{E}_1 + \mathscr{E}_1) \cap (\mathscr{E}_2 + \mathscr{E}_2).$$

Can the sign of inclusion be replaced by that of equality? For transversal structures this is certainly the case, as is implicit in the work of R. A. Brualdi (**12**); but it seems unlikely that this is so generally. We may, then, seek to characterize those independence structures for which the above relation holds with equality.

[H. Perfect]

28. The theory of independence structures leans heavily on the axiom of finite character. R. Rado (**9**) raised the question of the development of a non-

trivial theory free from this restriction (i.e. of a theory of pre-independence structures). For some recent work in this field, see D. A. Higgs (**1**).

29. Let $\mathscr{E}$ be a (finite or infinite) independence structure. Under what conditions is $\mathscr{E}$ linear (over a given division ring or over some unspecified division ring)? This fundamental problem, implicit in the work of H. Whitney (**1**), has been solved for the case of fields by P. Vámos (**1**); but the general case remains open.

There are other questions in this area, and we mention one of them. Given a set $\mathscr{D}$ of division rings, it is required to find the greatest integer $k(\mathscr{D})$ such that every independence structure on a set of $k(\mathscr{D})$ elements is linear over some division ring in $\mathscr{D}$. (For example, it is known that, if $\mathscr{D}$ consists of just the field of rational numbers, then $k(\mathscr{D}) = 6$.)

30. Let $\mathscr{E}$ be a finite independence structure, and let p be 0 or a prime number. We say that p is a 'characteristic number' of $\mathscr{E}$ if $\mathscr{E}$ is linear over some field of characteristic p. The set of all characteristic numbers of $\mathscr{E}$ is called the 'characteristic set' of $\mathscr{E}$. A set of numbers is called a *characteristic set* if it is the characteristic set of some finite independence structure.

The determination of all characteristic sets is an interesting and, no doubt, formidable problem. A few facts are known. Thus Rado (**7**) proved that, if a characteristic set contains 0, then it also contains all but a finite number of primes. Again, Vámos showed that, if a characteristic set contains infinitely many primes, then it also contains 0. Further, any single prime constitutes a characteristic set; and it is conjectured that the only finite characteristic sets are single primes.

[P. Vámos]

31. Let $\mathscr{E}$ be an independence structure on a finite set, $\mathfrak{B}$ the family of bases of $\mathscr{E}$, and $\mathscr{E}'$ the transversal structure of $\mathfrak{B}$. What can be said about the relation between $\mathscr{E}$ and $\mathscr{E}'$? In particular, how are their rank functions related?

32. When is a linear structure a transversal structure? (Cf. Theorems 7.1.3 and 7.1.5(i).)

33. Let $\mathscr{E}$ be a collection of subsets of an infinite set E, and let P be a property which such a collection may possess. We shall say that P is *finitary* if, whenever every finite restriction of $\mathscr{E}$ (i.e. the restriction of $\mathscr{E}$ to every finite subset of E) has property P, then so has $\mathscr{E}$ itself. What can be said about the class of finitary properties? For example, it has been shown by Vámos that the property of being linear (but not the property of being linear over a *specified* field) is finitary; so is the property of being transversal for independence structures, as has been discovered by J. H. Mason (**1**) and by others. The investigation is presumably much easier (and correspondingly less interesting)

if we confine ourselves to the consideration of structures in which the cardinals of members are bounded.

34. Certain set-theoretic models of independence structures are described in §§ 6.5 and 7.4, and many other models of this type are exhibited in the work of Hazel Perfect (**3**). However, the interest of the topic is far from exhausted, and it would be well worth while to identify further models.

35. An easy transfinite analogue of Corollary 8.2.2 is provided by Corollary 8.2.3. It is less obvious how to extend Corollary 8.2.5. If, then, $\mathscr{E}$ is an independence structure on an infinite set E and k is a natural number, under what circumstances can E be partitioned into k spanning sets?

36. What conditions must a finite family of finite sets satisfy in order that the complementary structure of its transversal structure should again be a transversal structure?

[D. J. A. Welsh (**8**)]

37. At present, effectively nothing is known about links of more than two families. The determination of criteria for the existence of a common system of representatives or a common transversal of three (or more) families would probably constitute the most decisive advance in transversal theory that can result from the solution of an individual problem.

38. Let $\mathscr{E}_1, \ldots, \mathscr{E}_k$ be independence structures on a finite set E. What conditions are necessary and sufficient for the existence of sets $X_i \in \mathscr{E}_i$, $1 \leqslant i \leqslant k$, such that $(X_1, \ldots, X_k)$ is combinatorially equivalent to a given family of sets? (Special cases of this problem are dealt with by Theorems 5.4.2 and 8.2.4.)

The question is certainly difficult for it contains, in particular, the problem of common transversals of k families of sets. (Cf. No. 37 above.) Let us formulate a very special case in which the answer is still not obvious. When does a family of sets possess two transversals whose intersection has a prescribed cardinal?

39. Let $\mathscr{E}$ be an independence structure on a set E, and let $\mathfrak{A}$ be a finite family of subsets of E. What conditions are necessary and sufficient for the existence of m pairwise disjoint independent partial transversals of $\mathfrak{A}$ with prescribed cardinals? (When $\mathscr{E}$ is the universal structure, the answer to this question is provided by Theorem 5.1.1.) Even the following special case of the above problem seems difficult: when do two families $\mathfrak{A}$, $\mathfrak{B}$ of sets possess pairwise disjoint CPTs of prescribed cardinals? (When the prescribed cardinals are all equal, the problem has been solved recently; see Brualdi (**12**) and Fulkerson (**5**).)

40. Let $\mathfrak{A} = (A_1, \ldots, A_n)$, $\mathfrak{B} = (B_1, \ldots, B_n)$ be two families of sets, and let $(t_1, \ldots, t_n)$ be a sequence of non-negative integers. Under what circumstances

do there exist permutations α, β of $\{1, \ldots, n\}$ such that

$$|A_{\alpha(k)} \cap B_{\beta(k)}| \geqslant t_k \qquad (1 \leqslant k \leqslant n)?$$

Again, under what circumstances do there exist permutations α, β and pairwise disjoint sets $X_1, \ldots, X_n$ such that $|X_k| = t_k$ $(1 \leqslant k \leqslant n)$ and

$$X_k \subseteq A_{\alpha(k)} \cap B_{\beta(k)} \qquad (1 \leqslant k \leqslant n)?$$

(For $t_1 = \ldots = t_n = 1$, the first problem reduces to the problem of common systems of representatives, cf. Theorem 9.2.1; the second problem reduces to the problem of common transversals, cf. Corollary 9.3.4.)

41. Let $(E_1, \ldots, E_p)$ and $(F_1, \ldots, F_q)$ be two partitions of a set E; let $0 \leqslant e_i' \leqslant e_i$ $(1 \leqslant i \leqslant p)$, $0 \leqslant f_j' \leqslant f_j$ $(1 \leqslant j \leqslant q)$ be integers; and let $\mathfrak{A} = (A_1, \ldots, A_n)$, $\mathfrak{B} = (B_1, \ldots, B_n)$ be two families of subsets of E. When do $\mathfrak{A}$ and $\mathfrak{B}$ possess a common transversal X such that

$$e_i' \leqslant |X \cap E_i| \leqslant e_i \qquad (1 \leqslant i \leqslant p)? \qquad (*)$$

When does $\mathfrak{A}$ possess a transversal X which satisfies (*) and

$$f_j' \leqslant |X \cap F_j| \leqslant f_j \qquad (1 \leqslant j \leqslant q)?$$

(If the answer to the first question were known, the problem of common systems of representatives of three families could be solved.)

42. Let $\mathscr{E}_1$, $\mathscr{E}_2$ be independence structures, with rank functions ρ_1, ρ_2, on a finite set E. When is $\mathscr{E}_1 \cap \mathscr{E}_2$ an independence structure? In view of Corollary 6.7.3 and Theorem 8.4.2, a necessary and sufficient condition is that the function σ, given by

$$\sigma(A) = \min_{X \subseteq A} \{\rho_1(X) + \rho_2(A \setminus X)\} \qquad (A \subseteq E),$$

should satisfy the modular inequality. However, this form of answer is not easy to apply in practice. For instance, it does not seem to yield a reasonable criterion for deciding under what circumstances the set of all common partial transversals of two given families of sets is an independence structure.

43. Let an independence structure be defined on a set E, and let $\mathfrak{A}$, $\mathfrak{B}$ be families, each consisting of n subsets of E. What conditions are necessary and sufficient for the existence of an independent common transversal of $\mathfrak{A}$ and $\mathfrak{B}$?

A more general question runs as follows. Let $\mathscr{E}_1$, $\mathscr{E}_2$ be independence structures on E, and let $\mathfrak{A}$ be a family of subsets of E. When does $\mathfrak{A}$ possess a transversal T with $T \in \mathscr{E}_1$, $T \in \mathscr{E}_2$?

It is easy to formulate a still more comprehensive problem. Let $\mathscr{E}_1$, $\mathscr{E}_2$, $\mathscr{E}_3$

be independence structures on a finite set E. Is there a simple formula for the maximum cardinal of sets common to all three structures? (Cf. Theorem 8.4.2 and Ex. 8.4.3.)

44. Does there exist an analogue of the Hoffman–Kuhn theorem (Corollary 9.6.2) in which 'transversal' is replaced by 'independent transversal'?

45. Let $\mathscr{E}$ be an independence structure on a set E and let $\mathfrak{A}$ be a family of subsets of E. What conditions are necessary and sufficient for the existence of a system $\mathfrak{X}$ of representatives of $\mathfrak{A}$ such that the frequency of occurrence of each element of E in $\mathfrak{X}$ lies between prescribed bounds while the range of $\mathfrak{X}$ is an independent set? An analogous problem can, of course, be formulated for common systems of representatives of two given families.

46. Let $\mathfrak{A} = (A_1, ..., A_n)$ be a family of subsets of $\{x_1, ..., x_m\}_{\neq}$; let k be a natural number; and let c_i, d_i, a_{ij} $(1 \leqslant i \leqslant k;\ 1 \leqslant j \leqslant m)$ be non-negative integers. Under what conditions does $\mathfrak{A}$ possess a system $\mathfrak{X}$ of representatives such that

$$c_i \leqslant \sum_{j=1}^{m} a_{ij} f(\mathfrak{X}; x_j) \leqslant d_i \qquad (1 \leqslant i \leqslant k)?$$

(The case when $k = m$ and $\|a_{ij}\|$ is a diagonal $m \times m$ matrix with positive diagonal elements is settled by Theorem 10.2.2.)

47. Let $\mathfrak{A}$ be a finite family of finite, non-empty sets. Denote by M^* the maximum number of pairwise disjoint sets in $\mathfrak{A}$ and by M_* the minimum number of elements in a representing set of $\mathfrak{A}$. What conditions are necessary and sufficient for the validity of the relation $M^* = M_*$? (Cf. Theorem 11.2.7.)

[R. Rado]

48. Several results in §11.3 are concerned with quantitative refinements of Theorem 11.1.3. Are there significant quantitative refinements of the infinite analogue of Theorem 11.1.3, namely Corollary 11.4.6? For example, let Δ denote a diagonal of an infinite matrix M, and let $\Delta(n)$ be the sum of the elements of Δ contained in the first n rows of M. Is it possible to estimate a function $f(n)$ with the property that every infinite d.s.matrix M possesses a positive diagonal Δ with $\Delta(n) \geqslant f(n)$ for all $n \geqslant 1$?

49. Let condition (4) in Theorem 11.5.1 be satisfied. What additional requirements must be met if the matrix Q, subject to (1), (2), and (3), is to be unique.

50. What can be said about the number, say $P(n)$, of $n \times n$ incidence matrices which possess a positive diagonal? In particular, is the relation $P(n) = o(2^{n^2})$ valid? We can also seek to estimate the number of $n \times n$ incidence matrices with a d.s. pattern.

Miscellaneous Exercises

1. Let $(s_k: 1 \leqslant k \leqslant n)$ be a family of n integers. Show that there exists a set K with $\varnothing \subset K \subseteq \{1, \ldots, n\}$ such that

$$\sum_{k \in K} s_k$$

is divisible by n.

2. Deduce Theorem 3.3.1 (for a finite E) and Theorem 5.1.1 (for finite E and I) from Theorem 8.2.4.

3. Let E be a finite set, $\mathfrak{A}$ a finite family of subsets of E, $\mathscr{E}$ the transversal structure of $\mathfrak{A}$, and k a natural number. Show that the collection of sets $\{X \in \mathscr{E}: |X| \leqslant k\}$ is not necessarily a transversal structure.

4. The set N^n of n-tuples of non-negative integers is partially ordered by the requirement that $(x_1, \ldots, x_n) \leqslant (y_1, \ldots, y_n)$ if and only if $x_k \leqslant y_k$ $(1 \leqslant k \leqslant n)$. Show that all antichains in N^n are finite.

5. Let $\mathscr{E}$ be an independence structure, with rank function ρ, on a finite set E. Show that (i) if $A \subseteq E$, $x \in E$, then

$$\rho(A) \leqslant \rho(A \cup \{x\}) \leqslant \rho(A) + 1;$$

(ii) if $x, y \in E$, $A \subseteq E$, and

$$\rho(A) = \rho(A \cup \{x\}) = \rho(A \cup \{y\}),$$

then $\rho(A) = \rho(A \cup \{x, y\})$.

6. Let E be a finite set and let ρ be a mapping of $\mathscr{P}(E)$ into the set of non-negative integers. Suppose that $\rho(\varnothing) = 0$ and that ρ has the properties (i) and (ii) listed in the preceding question. Show that the collection of sets

$$\mathscr{E} = \{X \subseteq E: \rho(X) = |X|\}$$

is an independence structure and that ρ is its rank function.

7. Show that the complementary structure of a finite transversal structure need not be a transversal structure.

[D. J. A. Welsh (**8**)]

8. Let N be the set of all natural numbers. Show that the set of all permutations of N has the same cardinal as the set of all real numbers.

9. Let $\mathscr{E}$ be an independence structure, with rank function ρ, on a finite set E; and let (E_1, E_2) be a partition of E. Show that the following statements are equivalent.

(i) $\mathscr{E}$ is the sum of two independence structures one of which is trivial on E_1 while the other is trivial on E_2.

(ii) $\rho(E) = \rho(E_1) + \rho(E_2)$.

(iii) $\rho(X) = \rho(X \cap E_1) + \rho(X \cap E_2)$ for all $X \subseteq E$.

10. Let $\mathscr{E}$ be an independence structure, with rank function ρ, on a finite set E; and let A be a subset of E. Show that

$$\rho(X) + \rho(A) = \rho(X \cup A) + \rho(X \cap A)$$

for all $X \subseteq E$ if and only if $\mathscr{E}$ is the sum of two independence structures one of which is trivial on A while the other is trivial on $E \setminus A$.

11. Let $\mathscr{E}$ be an independence structure on a finite set E; let B_1, B_2 be bases of $\mathscr{E}$; and let $x \in B_1 \setminus B_2$. Show that there exists an element $y \in B_2 \setminus B_1$ such that $(B_2 \setminus \{y\}) \cup \{x\}$ is a basis of $\mathscr{E}$.

12. Let $\mathfrak{A} = (A_i : i \in I)$ be a fiinte family of subsets of a finite set E, and suppose that

$$E = \bigcup_{i \in I} A_i.$$

Further, let k be a natural number. Show that E can be partitioned into k pairwise disjoint partial transversals of $\mathfrak{A}$ if and only if

$$k \geqslant \max_{J \subset I} \frac{|E \setminus A(J)|}{|I \setminus J|}.$$

13. Let $(r_k : k \geqslant 1)$ and $(s_k : k \geqslant 1)$ be arbitrary sequences of positive integers. Show that there exists an infinite incidence matrix with row-sums r_k $(k \geqslant 1)$ and column-sums s_k $(k \geqslant 1)$.

14. Show that the number of $m \times n$ incidence matrices in which no line consists entirely of zeros is equal to

$$(-1)^m \sum_{i=0}^{m} (-1)^i \binom{m}{i} (2^i - 1)^n.$$

[D. A. Klarner]

15. Let E be a finite set, $F \subseteq E$, and $\mathscr{E}$ an independence structure on E. Let the collection $\mathscr{E}^*$ of subsets of E be specified by the requirement that $X \in \mathscr{E}^*$ if and only if there exists a maximal independent subset H of F such that $X \cup H \in \mathscr{E}$. Show that $\mathscr{E}^*$ is an independence structure, and that the rank functions ρ, ρ^* of $\mathscr{E}$, $\mathscr{E}^*$ respectively are connected by the formula

$$\rho^*(X) = \rho(X \cup F) + \rho(X \cap F) - \rho(F) \qquad (X \subseteq E).$$

[H. Perfect (7)]

16. Let E, I be sets and, for each $i \in I$, let $\mathscr{F}_i$ be a finite collection of arbitrary subsets of E. Suppose that, for each $J \subset\subset I$, there exists a family $(F_i : i \in J)$ of pairwise disjoint sets with $F_i \in \mathscr{F}_i$ $(i \in J)$. Show that there exists a family $(E_i : i \in I)$ of pairwise disjoint sets with $E_i \in \mathscr{F}_i$ $(i \in I)$.

17. Show that the sum of two transversal structures (defined on an arbitrary ground set) is again a transversal structure.

18. Let $(E, \mathscr{E})$ be a finite independence space and let $X, Y \in \mathscr{E}$, $|X| = |Y|$, $x \in X \setminus Y$. Show that there exists an element $y \in Y \setminus X$ such that

$$(X \setminus \{x\}) \cup \{y\} \in \mathscr{E}, \qquad (Y \setminus \{y\}) \cup \{x\} \in \mathscr{E}.$$

[R. A. Brualdi]

19. Let $\mathfrak{A} = (A_1, ..., A_m)$, $\mathfrak{B} = (B_1, ..., B_n)$ be two families of subsets of E, and let $M \subseteq E$, $k \leqslant \min(m, n)$. Use Theorem 9.4.3 to show that $\mathfrak{A}$ and $\mathfrak{B}$ possess a CPT of cardinal k which contains M if and only if the inequalities

$$|A(I) \cap M| \geqslant |I| + |M| - m, \qquad |B(J) \cap M| \geqslant |J| + |M| - n,$$
$$|A(I) \cap B(J)| + |\{A(I) \cup B(J)\} \cap M| \geqslant |I| + |J| + |M| - m - n + k$$

are satisfied for all $I \subseteq \{1, ..., m\}$, $J \subseteq \{1, ..., n\}$.

Hence obtain necessary and sufficient conditions in order that one of $\mathfrak{A}$, $\mathfrak{B}$ and a subfamily of the other should possess a CT which contains M.

20. Let $\mathscr{E}$ be an independence structure on a finite set E. Establish the equivalence of the following statments.

(i) There exists an integer m and a subset F of E such that $\mathscr{E} = \{X \subseteq F: |X| \leqslant m\}$.

(ii) $\mathscr{E}$ has the 'universal replacement property': if $X, Y \in \mathscr{E}$ and $|Y| = |X| + 1$, then $X \cup \{y\} \in \mathscr{E}$ for *every* element $y \in Y \setminus X$.

(iii) The independent partial transversals of every finite family of subsets of E constitute an independence structure.

(iv) The sets common to $\mathscr{E}$ and any independence structure on E form again an independence structure.

[H. Perfect]

21. Let $(A_1, A_2, A_3, ...)$ be a sequence of finite non-empty sets and suppose that $E = \bigcup \{A_n: n \geqslant 1\}$ is an infinite set. Let Δ be a subset of $E \times E$ such that, for each $n \geqslant 1$ and each $x \in A_{n+1}$, there exists some $y \in A_n$ with $(y, x) \in \Delta$. Prove that there exists a sequence $(x_n: n \geqslant 1)$ of elements of E such that $x_n \in A_n$, $(x_n, x_{n+1}) \in \Delta$ for all $n \geqslant 1$. (This result is known as König's 'infinity lemma'.)

[D. König (4)]

22. Let $x_1, ..., x_n$ be real numbers; put $I_0 = \{1, ..., n\}$; and, for $\emptyset \subset I \subseteq I_0$, write

$$M(I) = \max\{x_i: i \in I\}, \qquad m(I) = \min\{x_i: i \in I\}.$$

Show that

$$M(I_0) = \sum_{\emptyset \subset I \subseteq I_0} (-1)^{|I|-1} m(I),$$

$$m(I_0) = \sum_{\emptyset \subset I \subseteq I_0} (-1)^{|I|-1} M(I).$$

23. Let E be a finite set; let ϕ be a mapping of $\mathscr{P}(E) \setminus \{\varnothing\}$ into the set of real numbers; and let the mapping ψ be defined by the equation

$$\psi(A) = \sum_{\varnothing \subset X \subseteq A} (-1)^{|X|-1} \phi(X) \qquad (\varnothing \subset A \subseteq E).$$

Verify that

$$\phi(A) = \sum_{\varnothing \subset X \subseteq A} (-1)^{|X|-1} \psi(X) \qquad (\varnothing \subset A \subseteq E).$$

Hence show that either identity in the preceding question implies the other and that the identity in Ex. 1.3.1 remains valid when the symbols $\cup$ and $\cap$ are interchanged.

24. Let M be a rectangular matrix with elements in an arbitrary field. Show that, if the first r rows and also the first s columns of M are linearly independent, then M contains a scattered set of non-zero elements which is incident with both the first r rows and the first s columus.

Use this result to give an alternative proof of Theorem 11.1.4.

[H. Perfect (**1**)]

25. Let $\mathscr{E}$ be a collection of subsets of an arbitrary set E, let m be a natural number, and let the collection $\mathscr{E}^*$ be specified by the requirement that $X \in \mathscr{E}^*$ if and only if there exists a set $Y \in \mathscr{E}$ such that $Y \subseteq X$, $|X \setminus Y| \leqslant m$. Show that, if $\mathscr{E}$ is a pre-independence structure (independence structure), then so is $\mathscr{E}^*$.

[H. Perfect]

26. Show that the two families of sets $\mathfrak{A} = (A_1, \ldots, A_n)$ and $\mathfrak{B} = (B_1, \ldots, B_n)$ possess k pairwise disjoint common transversals if and only if, for all $I, J \subseteq \{1\ldots, ,n\}$,

$$|A(I) \cap B(J)| \geqslant k\{|I| + |J| - n\}.$$

[R. A. Brualdi (**12**), D. R. Fulkerson (**5**)]

27. Let $\mathfrak{A} = (A_i : i \in I)$ and $\mathfrak{B} = (B_j : j \in J)$ be finite families of subsets of a finite set E, and let k be a natural number. Denote by $\mathfrak{A}^{(k)}$ the family consisting of k copies of each of the sets A_i, $i \in I$; and let $\mathfrak{B}^{(k)}$ be defined analogously.

Establish the equivalence of the following statements. (i) E can be partitioned into k CPTs of $\mathfrak{A}$ and $\mathfrak{B}$. (ii) E is a CPT of $\mathfrak{A}^{(k)}$ and $\mathfrak{B}^{(k)}$.

Hence establish the equivalance of the following statements. (a) E can be partitioned into k CPTs of $\mathfrak{A}$ and $\mathfrak{B}$; (b) E can be partitioned into k PTs of $\mathfrak{A}$ and also into k PTs of $\mathfrak{B}$; (c) for all $I^* \subseteq I$, $J^* \subseteq J$,

$$|A(I^*) \cap B(J^*)| + k|I \setminus I^*| + k|J \setminus J^*| \geqslant |E|.$$

Exhibit Theorem 3.3.5 as a special case of this result.

[R. A. Brualdi]

28. Let $\mathfrak{A} = (A_i : i \in I)$, $\mathfrak{B} = (B_j : j \in J)$ be finite families of finite sets, and let k be a natural number.

(i) Show that $\mathfrak{A}$ and $\mathfrak{B}$ possess transversals the cardinal of whose intersection is *at least* k if and only if

$$|A(I^*) \cap B(J^*)| + |I \setminus I^*| + |J \setminus J^*| \geqslant k,$$
$$|A(I^*)| \geqslant |I^*|, \qquad |B(J^*)| \geqslant |J^*|$$

whenever $I^* \subseteq I$, $J^* \subseteq J$.

(ii) Show that $\mathfrak{A}$ and $\mathfrak{B}$ possess transversals the cardinal of whose intersection is *at most* k if and only if

$$|A(I^*) \cup B(J^*)| \geqslant |I^*| + |J^*| - k,$$
$$|A(I^*)| \geqslant |I^*|, \qquad |B(J^*)| \geqslant |J^*|$$

whenever $I^* \subseteq I$, $J^* \subseteq J$.

[H. Perfect]

29. Let $\mathscr{D}_n$ denote the set of all d.s. $n \times n$ matrices and, for $A = ||a_{kj}|| \in \mathscr{D}_n$, $\pi \in \mathfrak{S}_n$, write

$$d_\pi(A) = a_{1,\pi(1)} + \ldots + a_{n,\pi(n)}.$$

Show (without invoking the deeper Theorem 11.3.3) that, given $A \in \mathscr{D}_n$, there exists $\pi \in \mathfrak{S}_n$ such that $d_\pi(A) \geqslant 1$. Deduce that, for all $n \geqslant 1$,

$$\min_{A \in \mathscr{D}_n} \max_{\pi \in \mathfrak{S}_n} d_\pi(A) = 1.$$

30. Show that the sum of two linear structures need not be a linear structure.

31 Let $\mathfrak{A} = (A_1, \ldots, A_{n+1})$ be a family of non-empty subsets of a set of cardinal n. Show that there exist non-empty, disjoint subsets I, J of $\{1, 2, \ldots, n + 1\}$ such that $A(I) = A(J)$.

[B. Lindström]

32. Let E be an arbitrary set and let $\mathscr{E}$ denote the collection of all subsets of E whose cardinal does not exceed the natural number m. Show that $\mathscr{E}$ is linear over any field F such that $|E| \leqslant |F|$.

33. Deduce the Ford–Fulkerson criterion (Theorem 9.3.2) from Brualdi's theorem 8.4.2.

34. Let $E = \{1, 2, \ldots, 6\}$ and denote by $\mathscr{E}$ the collection of all subsets of E of cardinal at most 2, with the exception of the sets $\{1, 2\}$, $\{3, 4\}$, $\{5, 6\}$. Show that $\mathscr{E}$ is linear over every field except the field of 2 elements.

35. Let $\mathscr{E}_1$, $\mathscr{E}_2$ be two independence structures on a finite set E, and suppose that E has the same rank in $\mathscr{E}_1$ as in $\mathscr{E}_2$. Show that, in general, the collection of subsets of common bases of $\mathscr{E}_1$ and $\mathscr{E}_2$ is not an independence structure.

36. Let ρ be a mapping of the power set of the finite set E into the set of non-negative integers. Show that ρ satisfies the relations

$$\rho(A) \leqslant \rho(B) \qquad (A \subseteq B \subseteq E),$$
$$\rho(A \cup B) + \rho(A \cap B) \leqslant \rho(A) + \rho(B) \qquad (A, B \subseteq E)$$

if and only if it satisfies

$$\rho(A) + \rho(A \cup B \cup C) \leqslant \rho(A \cup B) + \rho(A \cup C) \qquad (A, B, C \subseteq E).$$

[A. W. Ingleton]

37. Let S_n denote the collection, partially ordered by inclusion, of the $2^n - 1$ non-empty subsets of a set of n elements. Show that the number of chains of

cardinal $k \geqslant 2$ in S_n is equal to

$$\frac{n!}{(n-k+2)!}\left\{(k+1)^{n-k+2} - 2k^{n-k+2} + (k-1)^{n-k+2}\right\}.$$

38. Let E be an arbitrary set. Show that an independence structure $\mathscr{E}$ on E possesses a unique basis if and only if $\mathscr{E}$ is a restriction of the universal structure on E.

39. Let $\mathfrak{A}, \mathfrak{B}, \mathfrak{A}', \mathfrak{B}'$ be families of sets such that $\mathfrak{A}' \subseteq \mathfrak{A}$, $\mathfrak{B}' \subseteq \mathfrak{B}$, and let t be a natural number. Suppose that (i) $\mathfrak{A}$ and $\mathfrak{B}$ have a CPT of cardinal at least t; (ii) $\mathfrak{A}'$ and a subfamily of $\mathfrak{B}$ have a CT; (iii) $\mathfrak{B}'$ and a subfamily of $\mathfrak{A}$ have a CT. Show that there exist families $\mathfrak{A}_0$, $\mathfrak{B}_0$ with $\mathfrak{A}' \subseteq \mathfrak{A}_0 \subseteq \mathfrak{A}; \mathfrak{B}' \subseteq \mathfrak{B}_0 \subseteq \mathfrak{B}$ which have a CT of cardinal at least t.

[R. A. Brualdi (**11**)]

40. Let E be a finite set, M a subset of E, n a natural number, and $|M| \leqslant n$. Show that the collection of sets

$$\mathscr{E} = \{X \subseteq E: |X \cup M| \leqslant n\}$$

is an independence structure, and that its rank function ρ is given by the equation

$$\rho(X) = |X| - \{|X \cup M| - n\}^+ \qquad (X \subseteq E).$$

Hence obtain necessary and sufficient conditions (given in Theorem 3.3.6) for the family $(A_1, \ldots, A_n)$ of subsets of E to possess a transversal which contains M.

[D. J. A. Welsh]

41. Let $\mathfrak{A} = (A_i: i \in I)$ be an arbitrary family of subsets of an arbitrary set E, and let k be a natural number. Suppose that each A_i contains at least k elements and that each element of E belongs to at most k sets of $\mathfrak{A}$. Show that $\mathfrak{A}$ possesses a transversal, but that the converse inference is false.

42. Let $\mathscr{E}$ be an independence structure on a finite set E; let B be a basis of $\mathscr{E}$; and let $x \in E \setminus B$. Show that there exists a unique subset D of B such that $D \cup \{x\}$ is a minimal dependent set.

43. Let $G = (N, E)$ be a finite graph. A set of edges of the form

$$\{\{e_1, e_2\}, \{e_2, e_3\}, \ldots, \{e_{m-1}, e_m\}, \{e_m, e_1\}\},$$

where $m \geqslant 3$ and $\{e_1, e_2, \ldots, e_m\}_{\neq} \subseteq N$, is called a *circuit*. Let $\mathscr{E}$ be the collection of subsets of E such that X ($\subseteq$ E) belongs to $\mathscr{E}$ if and only if it contains no circuit. Show that $\mathscr{E}$ is an independence structure and, indeed, that it is linear over the field of 2 elements.

44. (i) Let $\mathscr{E}$ be independence structure on a set E, and let $M \subset\subset E$. Show that the set of integers

$$\{|B \cap M|: B \text{ a basis of } \mathscr{E}\}$$

is an interval.

(ii) Let $\mathfrak{A}$ be a family of subsets of a set E which possesses a transversal, and let $M \subset\subset E$. Show that the set of integers

$$\{|T \cap M|: T \text{ a transversal of } \mathfrak{A}\}$$

is an interval.

(iii) Let $\mathscr{E}$ be an independence structure on a finite set E, let $\mathfrak{A}$ be a family of subsets of E which possesses an independent transversal, and let $M \in \mathscr{E}$. Show that the set of integers

$$\{|T \cap M| : T \text{ an independent transversal of } \mathfrak{A}\}$$

need not be an interval.

45. Show that the complementary structure of an independence structure on an infinite set need not have finite character.

46. Let $E = \{1, 2, ..., 6\}$ and denote by $\mathscr{E}$ the collection of all subsets X of E such that $|X| \leqslant 3$ with the exception of $\{1, 2, 6\}$, $\{1, 4, 5\}$, $\{2, 3, 5\}$, $\{3, 4, 6\}$. Show that $\mathscr{E}$ is linear over *every* field.

47. Let E be a finite set; $\mathfrak{A} = (A_1, ..., A_n)$ a family of subsets of E; $M \subseteq E$; and $1 \leqslant k \leqslant n$. Use the theorem of Hoffman & Kuhn (Corollary 9.6.2) to establish necessary and sufficient conditions for $\mathfrak{A}$ to possess a transversal X such that (i) $|X \cap M| \geqslant k$; (ii) $|X \cap M| \leqslant k$; (iii) $|X \cap M| = k$. (Cf. Ex. 6.2.8.)

Suppose that $\mathfrak{A}$ possesses transversals X_1, X_2 with $|X_1 \cap M| \geqslant k$, $|X_2 \cap M| \leqslant k$. Show that $\mathfrak{A}$ possesses a transversal X with $|X \cap M| = k$. (Cf. No. 44.)

48. Let $\mathscr{E}$ be an independence structure on a set E; let A, B be independent sets; and suppose that $|A| \leqslant |B|$. Show that there exists an independent set C with $A \subseteq C \subseteq A \cup B$ and $|C| = |B|$.

[R. Rado (9)]

49. Let $\mathscr{E}$ be an independence structure, with rank function ρ, on a finite set E; let $(E_1, ..., E_p)$ be a partition of E; and let $r_1, ..., r_p$ be non-negative integers. Show that $\mathscr{E}$ possesses a basis B with $|B \cap E_j| \geqslant r_j$ $(1 \leqslant j \leqslant p)$ if and only if, for all $J \subseteq \{1, ..., p\}$,

$$\rho\left(\bigcup_{j \in J} E_j\right) \geqslant \sum_{j \in J} r_j.$$

Let $\mathfrak{A} = (A_1, ..., A_n)$ be a family of subsets of E. Deduce necessary and sufficient conditions for $\mathfrak{A}$ to possess a transversal T with $|T \cap E_j| \geqslant r_j$ $(1 \leqslant j \leqslant p)$.

50. Let $\mathscr{E}$ be a pre-independence structure, with rank function ρ, on an arbitrary set E; let $\mathfrak{A} = (A_1, ..., A_n)$ be a family of subsets of E; let $M \subset\subset E$; and suppose that $\rho(M) \leqslant m \leqslant n$. Show that $\mathfrak{A}$ possesses an independent partial transversal of cardinal m which contains a maximal independent subset of M if and only if, for all $I \subseteq \{1, ..., n\}$, the inequalities

$$\rho(A(I) \cup M) + \rho(A(I) \cap M) \geqslant \rho(M) + |I| - n + m,$$
$$\rho(A(I) \cap M) \geqslant \rho(M) + |I| - n$$

are valid.

[H. Perfect (7)]

Bibliography

AIGNER, M. & DOWLING, T. A.

1. Matching theorems for combinatorial geometries. *Bull. Amer. Math. Soc.* **76** (1970), 57–60.

ASHE, D. S.

1. Minimal dependent sets. *J. Australian Math. Soc.* **6** (1966), 259–262.

BANACH, S.

1. Un théorème sur les transformations biunivoques. *Fundamenta Math.* **6** (1924), 236–239.

BELL, J. L. & SLOMSON, A. B.

1. *Models and Ultraproducts: An Introduction.* (North-Holland Publishing Company, Amsterdam, 1969).

BERGE, C.

1. *Théorie des Graphes et ses Applications* (Dunod, Paris, 1958). Translated as *The Theory of Graphs and its Applications* (Methuen, London, 1962).
2. The rank of a family of sets and some applications to graph theory. *Recent Progress in Combinatorics* (ed. W. T. Tutte; Academic Press, New York, 1969), 49–57.

BIRKHOFF, G.

1. Abstract linear dependence and lattices. *Amer. J. Math.* **57** (1935), 800–804.
2. Tres observaciones sobre el algebra lineal. *Univ. Nac. Tucumán, Rev.* Ser. A, **5** (1946), 147–151.

BLEICHER, M. N. & PRESTON, G. B.

1. Abstract linear dependence relations. *Publ. Math. Debrecen* **8** (1961), 55–63.

BOLLOBÁS, B.

1. A lower bound for the number of non-isomorphic matroids. *J. Combinatorial Theory* **7** (1969), 366–368.

BRUALDI, R. A.

1. Matroid analogues of some distinct representatives theorems. Not published, 1967.
2. Symmetrized form of R. Rado's theorem on independent representatives. Not published, 1967.
3. An extension of Banach's mapping theorem. *Proc. Amer. Math. Soc.* **20** (1969), 520–526.
4. Common transversals and strong exchange systems. *J. Combinatorial Theory* **8** (1970), 307–329.
5. A very general theorem on systems of distinct representatives. *Trans. Amer. Math. Soc.* **140** (1969), 149–160.
6. Comments on bases in dependence structures. *Bull. Australian Math. Soc.* **1** (1969), 161–167.
7. A general theorem concerning common transversals. *Proceedings of the Oxford Conference on Combinatorial Mathematics and its Applications, 1969*. To appear.

8. Admissible mappings between dependence structures. *Proc. London Math. Soc.* (3) **21** (1970), 296–312.

9. On families of finite independence structures. *Proc. London Math. Soc.* To appear.

10. Existence theorem for infinite integral matrices. *Mathematika.* **17** (1970), 91–97.

11. An extremal problem for a class of integral matrices. Not yet published.

12. Common partial transversals and integral matrices. Not yet published.

BRUALDI, R. A. & PYM, J. S.

1. A general linking theorem in directed graphs. *Studies in Pure Mathematics* (Academic Press, London, 1971.) To appear.

BRUALDI, R. A. & SCRIMGER, E. B.

1. Exchange systems, matchings and transversals. *J. Combinatorial Theory* **5** (1968), 244–257.

DE BRUIJN, N. G.

1. Gemeenschappelijke representantensystemen van twee klassenindeelingen van een verzameling. *Nieuw Arch. Wisk.* (2) **22** (1943), 48–52.

DE BRUIJN, N. G. & ERDÖS, P.

1. A colour problem for infinite graphs and a problem in the theory of relations. *Kon. Ned. Akad. Wetensch. Proc.* Ser. A, **54** (1951), 371–373.

BUSHAW, D.

1. *Elements of General Topology* (John Wiley & Sons, New York, 1963).

CHAPMAN, H. W.

1. A note on the elementary theory of groups of finite order. *Messenger Math.* **42** (1913), 132–134.

CHASE, P. J.

1. Generalized matroids. Not yet published.

COHN, P. M.

1. *Universal Algebra* (Harper & Row, New York, 1965).

DANTZIG, G. B. & HOFFMAN, A. J.

1. Dilworth's theorem on partially ordered sets. *Linear Inequalities and Related Systems* (Annals of Mathematics Studies No. 38; Princeton University Press, 1956), 207–214.

DILWORTH, R. P.

1. Dependence relations in a semi-modular lattice. *Duke Math. J.* **11** (1944), 575–587.

2. A decomposition theorem for partially ordered sets. *Annals of Math.* (2) **51** (1950), 161–166.

3. Some combinatorial problems on partially ordered sets. *Proceedings of Symposia in Applied Mathematics. Vol. 10. Combinatorial Analysis* (Providence, 1960), 85–90.

DLAB, V.

1. Axiomatic treatment of bases in arbitrary sets. *Czech. Math. J.* **15** (1965), 554–563.

DULMAGE, L. & HALPERIN, I.

1. On a theorem of Frobenius–König and J. von Neumann's game of hide and seek. *Trans. Roy. Soc. Canada,* Sect. III, (3) **49** (1955), 23–29.

EDMONDS, J.

1. Minimum partition of a matroid into independent subsets. *J. Res. Nat. Bur. Standards* **69B** (1965), 67–72.

2. Paths, trees, and flowers. *Canad. J. Math.* **17** (1965), 449–467.
3. Systems of distinct representatives and linear algebra. *J. Res. Nat. Bur. Standards* **71B** (1967), 241–245.

EDMONDS, J. & FULKERSON, D. R.
1. Transversals and matroid partition. *J. Res. Nat. Bur. Standards* **69B** (1965), 147–153.

EGERVÁRY, E.
1. Matrixok kombinatorius tulajdonságairól. *Mat. Fiz. Lapok* **38** (1931), 16–28. [Hungarian with German summary.]

ENTRINGER, R. C. & JACKSON, D. E.
1. A direct proof of the duality theorem of König. *Archiv der Math.* **20** (1969), 666–667.

ERDÖS, P. & KAPLANSKY, I.
1. The asymptotic number of Latin rectangles. *Amer. J. Math.* **68** (1946), 230–236.

ERDÖS, P. & SZEKERES, G.
1. A combinatorial problem in geometry. *Compositio Math.* **2** (1935), 463–470.

EVERETT, C. J. & WHAPLES, G.
1. Representations of sequences of sets. *Amer. J. Math.* **71** (1949), 287–293.

FLOOD, M. M.
1. An alternative proof of a theorem of König as an algorithm for the Hitchcock distribution problem. *Proceedings of Symposia in Applied Mathematics. Vol. 10. Combinatorial Analysis* (Providence, 1960), 299–307.

FOLKMAN, J.
1. Transversals of infinite families with finitely many infinite members. RAND Corporation Memorandum RM–5676–PR, 1968.

FOLKMAN, J. & FULKERSON, D. R.
1. Flows in infinite graphs. *J. Combinatorial Theory* **8** (1970), 30–44.

FORD, L. R. JR. & FULKERSON, D. R.
1. Network flow and systems of representatives. *Canad. J. Math.* **10** (1958), 78–84.
2. *Flows in Networks* (Princeton University Press, 1962).

FOSTER, B. L.
1. Short proof of a theorem of Rado on graphs. *Proc. Amer. Math. Soc.* **15** (1964), 865–866.

FOURNIER, J.-C.
1. Sur la représentation sur un corps des matroides à sept et huit éléments. *C. R. Acad. Sci.* (Paris), Ser. A, **270** (1970), 810–813.

FROBENIUS, G.
1. Über Matrizen aus nicht negativen Elementen. *Sitzber. Preuss. Akad. Wiss.* 1912, 456–477.
2. Über zerlegbare Determinanten. *Sitzber. Preuss. Akad. Wiss.* 1917, 274–277.

FUCHS, L.
1. *Partially Ordered Algebraic Systems* (Pergamon Press, Oxford, 1963).

FULKERSON, D. R.
1. Note on Dilworth's decomposition theorem for partially ordered sets. *Proc. Amer. Math. Soc.* **7** (1956), 701–702.
2. A network-flow feasibility theorem and combinatorial applications. *Canad. J. Math.* **11** (1959), 440–451.
3. Zero-one matrices with zero trace. *Pacific J. Math.* **10** (1960), 831–836.

4. The maximum number of disjoint permutations contained in a matrix of zeros and ones. *Canad. J. Math.* **16** (1964), 729–735.
5. Disjoint common partial transversals of two families of sets. *Studies in Pure Mathematics* (Academic Press, London, 1971). To appear.

GALE, D.
1. A theorem on flows in networks. *Pacific J. Math.* **7** (1957), 1073–1082.
2. *The Theory of Linear Economic Models* (McGraw–Hill, New York, 1960).
3. Optimal assignments in an ordered set: an application of matroid theory. *J. Combinatorial Theory* **4** (1968), 176–180.

GALLAI, T.
1. Maximum-Minimum Sätze über Graphen. *Acta Math. Acad. Sci. Hung.* **9** (1958), 395–434.

GALLAI, T. & MILGRAM, A. N.
1. Verallgemeinerung eines graphentheoretischen Satzes von Rédei. *Acta Sci. Math.* (Szeged) **21** (1960), 181–186.

GOTTSCHALK, W. H.
1. Choice functions and Tychonoff's theorem. *Proc. Amer. Math. Soc.* **2** (1951), 172.

HALL, M. JR.
1. An existence theorem for Latin squares. *Bull. Amer. Math. Soc.* **51** (1945), 387–388.
2. Distinct representatives of subsets. *Bull. Amer. Math. Soc.* **54** (1948), 922–926.
3. An algorithm for distinct representatives. *Amer. Math. Monthly* **63** (1956), 716–717.
4. A survey of combinatorial analysis. *Surveys in Applied Mathematics. Vol. 4. Some Aspects of Analysis and Probability* (John Wiley & Sons, New York, 1958), 37–104.

HALL, P.
1. On representatives of subsets. *J. London Math. Soc.* **10** (1935), 26–30.

HALMOS, P. R.
1. *Naive Set Theory* (Van Nostrand, Princeton, 1960).
2. *Lectures on Boolean Algebras* (Van Nostrand, Princeton, 1963).

HALMOS, P. R. & VAUGHAN, H. E.
1. The marriage problem. *Amer. J. Math.* **72** (1950), 214–215.

HAMMER, P. L. & RUDEANU, S.
1. *Boolean Methods in Operations Research and Related Areas* (Springer, Berlin, 1968).

HARARY, F.
1. *Graph Theory* (Addison–Wesley, Reading, Mass., 1969).

HARARY, F. & WELSH, D. J. A.
1. Matroids versus graphs. *Lecture Notes in Mathematics* (Springer, Berlin) **110** (1969), 155–170.

HARDY, G. H., LITTLEWOOD, J. E., & PÓLYA, G.
1. Some simple inequalities satisfied by convex functions. *Messenger Math.* **58** (1929), 145–152.
2. *Inequalities* (Cambridge University Press, 1934).

HARPER, L. H. & ROTA, G.-C.
1. *Matching Theory, an Introduction.* Not yet published.

HENKIN, L.

1. Some interconnections between modern algebra and mathematical logic. *Trans. Amer. Math. Soc.* **74** (1953), 410–427.

HIGGINS, P. J.

1. Disjoint transversals of subsets. *Canad. J. Math.* **11** (1959), 280–285.

HIGGS, D. A.

1. Matroids and duality. *Colloq. Math.* **20** (1969), 215–220.

HOFFMAN, A. J.

1. Generalizations of a theorem of König. *J. Washington Acad. Sci.* **46** (1956), 211–212.
2. Some recent applications of the theory of linear inequalities to extremal combinatorial analysis. *Proceedings of Symposia in Applied Mathematics. Vol. 10. Combinatorial Analysis* (Providence, 1960), 113–127.

HOFFMAN, A. J. & KUHN, H. W.

1. Systems of distinct representatives and linear programming. *Amer. Math. Monthly* **63** (1956), 455–460.
2. On systems of distinct representatives. *Linear Inequalities and Related Systems* (Annals of Mathematics Studies No. 38, Princeton University Press, 1956), 199–206.

HORN, A.

1. A characterization of unions of linearly independent sets. *J. London Math. Soc.* **30** (1955), 494–496.

INGLETON, A. W.

1. A note on independence functions and rank. *J. London Math. Soc.* **34** (1959), 49–56.
2. Representation of matroids. Not yet published.

ISBELL, J. R.

1. Birkhoff's problem 111. *Proc. Amer. Math. Soc.* **6** (1955), 217–218.

JACOBSON, N.

1. *Lectures in Abstract Algebra,* 3 vols. (Van Nostrand, Princeton & New York, 1951–64).

KELLERER, H. G.

1. Funktionen auf Produkträumen mit vorgegebenen Marginal-Funktionen. *Math. Annalen* **144** (1961), 323–344.
2. Allgemeine Systeme von Repräsentanten. *Z Wahrscheinlichkeitstheorie* **2** (1964), 306–309.

KELLEY, J. L.

1. *General Topology* (Van Nostrand, Princeton, 1955).

KERTÉSZ, A.

1. On independent sets of elements in algebra. *Acta Sci. Math.* (Szeged) **21** (1960), 260–269.

KNASTER, B.

1. Un théorème sur les fonctions d'ensembles. *Annales Soc. Polonaise Math.* **6** (1927), 133.

KNESER, H.

1. Das Auswahlaxiom und das Lemma von Zorn. *Math. Zeitschrift* **96** (1967), 62–63.

KÖNIG, D.

1. Über Graphen und ihre Anwendung auf Determinantentheorie und Mengenlehre. *Math. Annalen* **77** (1916), 453–465.

2. Sur un problème de la théorie générale des ensembles et la théorie des graphes. *Rev. Métaphys. Morale* **30** (1923), 443–449.
3. Sur les correspondances multivoques des ensembles. *Fundamenta Math.* **8** (1926), 114–134.
4. Über eine Schlussweise aus dem Endlichen ins Unendliche. *Acta Lit. Sci.* Sect. Sci. Math. (Szeged) **3** (1927), 121–130.
5. Graphok és matrixok. *Mat. Fiz. Lapok* **38** (1931), 116–119. [Hungarian with German summary.]
6. Über trennende Knotenpunkte in Graphen (nebst Anwendungen auf Determinanten und Matrizen). *Acta Lit. Sci.* Sect. Sci. Math. (Szeged) **6** (1932–4), 155–179.
7. *Theorie der endlichen und unendlichen Graphen* (Leipzig, 1936).

KÖNIG, D. & VALKÓ, S.
1. Über mehrdeutige Abbildungen von Mengen. *Math. Annalen* **95** (1926), 135–138.

KOOPMANS, T. C. & BECKMANN, M. J.
1. Assignment problems and the location of economic activities. *Econometrica* **25** (1957), 53–76.

KREWERAS, G.
1. Extension d'un théorème sur les répartitions en classes. *C. R. Acad. Sci.* (Paris) **222** (1946), 431–432.

KUHN, H. W. & TUCKER, A. W. (editors)
1. *Linear Inequalities and Related Systems* (Annals of Mathematics Studies No. 38, Princeton University Press, 1956).

LAZARSON, T.
1. Independence functions in algebra (PhD Dissertation, University of London, 1957).
2. The representation problem for independence functions. *J. London Math Soc.* **33** (1958), 21–25.

LÖWIG, H.
1. Über die Dimension linearer Räume. *Studia Math.* **5** (1934), 18–23.

LUBEL, D.
1. A short proof of Sperner's lemma. *J. Combinatorial Theory* **1** (1966), 299.

LUNDINA, D. Š.
1. On a refinement of a theorem of Rado. *Sibirsk. Mat. Ž.* **7** (1966), 718–719 [Russian].

LUXEMBURG, W. A. J.
1. A remark on a paper by N. G. de Bruijn and P. Erdös. *Kon. Ned. Akad. Wetensch. Proc.* Ser. A, **65** (1962), 343–345.

MAAK, W.
1. Eine neue Definition der fastperiodischen Funktionen. *Abh. math. Sem. Hamburg. Univ.* **11** (1936), 240–244.
2. Ein Problem der Kombinatorik in seiner Formulierung von H. Weyl. *Math.-phys. Semesterber.* (Göttingen) **2** (1952), 251–256.

MACLANE, S.
1. Some interpretations of abstract linear dependence in terms of projective geometry. *Amer. J. Math.* **58** (1936), 236–240.
2. A lattice formulation for transcendence degrees and p-bases. *Duke Math. J.* **4** (1938), 455–468.

MANN, H. B. & RYSER, H. J.

1. Systems of distinct representatives. *Amer. Math. Monthly* **60** (1953), 397–401.

MARCUS, M. & MINC, H.

1. Some results on doubly stochastic matrices. *Proc. Amer. Math. Soc.* **13** (1962), 571–579.
2. Diagonal products in doubly stochastic matrices. *Quart. J. Math.* (Oxford) (2) **16** (1965), 32–34.
3. Permanents. *Amer. Math. Monthly* **72** (1965), 577–591.

MARCUS, M. & REE, R.

1. Diagonals of doubly stochastic matrices. *Quart. J. Math.* (Oxford) (2) **10** (1959), 296–302.

MASON, J. H.

1. Representations of independence spaces (PhD Dissertation, University of Wisconsin, 1969).

MENDELSOHN, N. S. & DULMAGE, A. L.

1. Some generalizations of the problem of distinct representatives. *Canad. J. Math.* **10** (1958), 230–241.

MENGER, K.

1. Zur allgemeinen Kurventheorie. *Fundamenta Math.* **10** (1927), 96–115.

MILLER, G. A.

1. On a method due to Galois. *Quart. J. Pure Appl. Math.* **41** (1910), 382–384.

MINC, H.

1. Upper bounds for permanents of (0, 1)-matrices. *Bull. Amer. Math. Soc.* **69** (1963), 789–791.

MIRSKY, L.

1. Results and problems in the theory of doubly-stochastic matrices. *Z. Wahrscheinlichkeitstheorie* **1** (1962–3), 319–334.
2. Transversals of subsets. *Quart. J. Math.* (Oxford) (2) **17** (1966), 58–60.
3. Systems of representatives with repetition. *Proc. Cambridge Phil. Soc.* **63** (1967), 1135–1140.
4. A theorem on common transversals. *Math. Annalen* **177** (1968), 49–53.
5. Combinatorial theorems and integral matrices. *J. Combinatorial Theory* **5** (1968), 30–44.
6. Transversal theory and the study of abstract independence. *J. Math. Analysis Appl.* **25** (1969), 209–217.
7. Hall's criterion as a 'self-refining' result. *Monatsh. für Math.* **73** (1969), 139–146.
8. Pure and applied combinatorics. *Bull. Inst. Math. Appl.* **5** (1969), 2–4.
9. A proof of Rado's theorem on independent transversals. *Studies in Pure Mathematics* (Academic Press, London, 1971). To appear.

MIRSKY, L. & PERFECT, HAZEL

1. Systems of representatives. *J. Math. Analysis Appl.* **15** (1966), 520–568.
2. Applications of the notion of independence to problems of combinatorial analysis. *J. Combinatorial Theory* **2** (1967), 327–357.
3. Comments on certain combinatorial theorems of Ford and Fulkerson. *Archiv der Math.* **19** (1968), 413–416.

NASH-WILLIAMS, C. ST.J. A.

1. An application of matroids to graph theory. *Théorie des Graphes, Journées Internationales d'Études, Rome, 1966* (Dunod, Paris, 1967), 263–265.

NEUMANN, B. H.

1. An embedding theorem for algebraic systems. *Proc. London Math. Soc.* (3) **4** (1954), 138–153.

ORE, O.

1. Graphs and matching theorems. *Duke Math. J.* **22** (1955), 625–639.
2. On coset representatives in groups. *Proc. Amer. Math. Soc.* **9** (1958), 665–670.
3. *Theory of Graphs* (Amer. Math. Soc. Colloquium Publications No. 38, Providence, 1962).
4. *Graphs and their Uses* (Random House, New York, 1963).

OSTRAND, P. A.

1. Systems of distinct representatives, II. *J. Math. Analysis Appl.* **32** (1970) 1–4.

PERFECT, HAZEL

1. Symmetrized form of P. Hall's theorem on distinct representatives. *Quart. J. Math.* (Oxford) (2) **17** (1966), 303–306.
2. Applications of Menger's graph theorem. *J. Math. Analysis Appl.* **22** (1968), 96–111.
3. Independence spaces and combinatorial problems. *Proc. London Math. Soc.* (3) **19** (1969), 17–30.
4. Remark on a criterion for common transversals. *Glasgow Math. J.* **10** (1969), 66–67.
5. A generalization of Rado's theorem on independent transversals. *Proc. Cambridge Phil. Soc.* **66** (1969), 513–515.
6. Studies in transversal theory (PhD Dissertation, University of London, 1969).
7. Marginal elements in transversal theory. *Studies in Pure Mathematics* (Academic Press, London, 1971). To appear.

PERFECT, HAZEL & MIRSKY, L.

1. The distribution of positive elements in doubly-stochastic matrices. *J. London Math. Soc.* **40** (1965), 689–698.

PERFECT, HAZEL & PYM, J. S.

1. An extension of Banach's mapping theorem, with applications to problems concerning common representatives. *Proc. Cambridge Phil. Soc.* **62** (1966), 187–192.

PERLES, M. A.

1. A proof of Dilworth's decomposition theorem for partially ordered sets. *Israel J. Math.* **1** (1963), 105–107.
2. On Dilworth's theorem in the infinite case. *Israel J. Math.* **1** (1963), 108–109.

PIFF, M. J. & WELSH, D. J. A.

1. On the vector representation of matroids. *J. London Math. Soc.* (2) **2** (1970), 284–288.

PYM, J. S.

1. Some results on transversals of families of subsets. *Monatsh. für Math.* **71** (1967), 446–451.
2. A proof of Menger's theorem. *Monatsh. für Math.* **73** (1969), 81–83.
3. The linking of sets in graphs. *J. London Math. Soc.* **44** (1969), 542–550.
4. A proof of the linkage theorem. *J. Math. Analysis Appl.* **27** (1969), 636–638.

PYM, J. S. & PERFECT, HAZEL

1. Submodular functions and independence structures. *J. Math. Analysis Appl.* **30** (1970), 1–31.

RADO, R.

1. Bemerkungen zur Kombinatorik im Anschluss an Untersuchungen von Herrn D. König. *Sitzber. Berliner Math. Ges.* **32** (1933), 60–75.
2. A theorem on general measure functions. *Proc. London Math. Soc.* (2) **44** (1938), 61–91.
3. A theorem on independence relations. *Quart. J. Math.* (Oxford) **13** (1942), 83–89.
4. Theorems on linear combinatorial topology and general measure. *Annals of Math.* (2) **44** (1943), 228–276.
5. Axiomatic treatment of rank in infinite sets. *Canad. J. Math.* **1** (1949), 337–343.
6. Factorization of even graphs. *Quart. J. Math.* (Oxford) **20** (1949), 95–104.
7. Note on independence functions. *Proc. London Math. Soc.* (3) **7** (1957), 300–320.
8. A combinatorial theorem on vector spaces. *J. London Math. Soc.* **37** (1962), 351–353.
9. Abstract linear dependence. *Colloq. Math.* **14** (1966), 257–264.
10. On the number of systems of distinct representatives of sets. *J. London Math. Soc.* **42** (1967), 107–109.
11. Note on the transfinite case of Hall's theorem on representatives. *J. London Math. Soc.* **42** (1967), 321–324.

RANKIN, R. A.

1. Common transversals. *Proc. Edinburgh Math. Soc.* (2) **15** (1966–7), 147–154.

RICE, N. M.

1. Stone's representation theorem for boolean algebra. *Amer. Math. Monthly* **75** (1968), 503–504.
2. A general selection principle, with applications in analysis and algebra. *Canad. Math. Bull.* **11** (1968), 573–584.

ROBERTSON, A. P. & WESTON, J. D.

1. A general basis theorem. *Proc. Edinburgh Math. Soc.* (2) **11** (1958–9), 139–141.

ROTMAN, B. & KNEEBONE, G. T.

1. *The Theory of Sets and Transfinite Numbers* (Oldbourne, London, 1966).

RYSER, H. J.

1. A combinatorial theorem with an application to Latin rectangles. *Proc. Amer. Math. Soc.* **2** (1951), 550–552.
2. Combinatorial properties of matrices of zeros and ones. *Canad. J. Math.* **9** (1957), 371–377.
3. Matrices of zeros and ones. *Bull. Amer. Math. Soc.* **66** (1960), 442–464.
4. *Combinatorial Mathematics* (Carus Mathematical Monographs No. 14; 1963).
5. Matrices of zeros and ones in combinatorial mathematics. *Recent Advances in Matrix Theory* (The University of Wisconsin Press, Madison and Milwaukee, 1964), 103–124.

SCHERK. P.

1. On sets of vectors. *Math. Scand.* **16** (1965), 38–40.

SCORZA, G.

1. A proposito di un teorema del Chapman. *Boll. Unione Mat. Ital.* **6** (1927), 1–6.

SEIDENBERG, A.
1. A simple proof of a theorem of Erdös and Szekeres. *J. London Math. Soc.* **34** (1959), 352.

SHMUSHKOVITCH, V.
1. On a combinatorial theorem of the theory of sets. *Mat. Sbornik* N.S. **6** (1939), 138–147 [Russian with English summary].

SHÜ, S.
1. On the common representative system of residue classes of infinite groups. *J. London Math. Soc.* **16** (1941), 101–104.

SIERPIŃSKI, W.
1. *Algèbre des Ensembles* (Monografie Matematyczne No. 23, Warsaw, 1951).

SIMMONS, G. F.
1. *Introduction to Topology and Modern Analysis* (McGraw–Hill, New York, 1963).

SPERNER, E.
1. Note zu der Arbeit von Herrn B. L. van der Waerden: 'Ein Satz über Klasseneinteilungen von endlichen Mengen'. *Abh. math. Sem. Hamburg. Univ.* **5** (1927), 232.
2. Ein Satz über Untermengen einer endlichen Menge. *Math. Zeitschrift* **27** (1928), 544–548.

STONE, M. H.
1. The theory of representations for Boolean algebras. *Trans. Amer. Math. Soc.* **40** (1936), 37–111.

SZPILRAJN, E.
1. Sur l'extension de l'ordre partiel. *Fundamenta Math.* **16** (1930), 385–389.

TITCHMARSH, E. C.
1. *The Theory of Functions* (Second edition, Oxford University Press, 1939).

TUTTE, W. T.
1. The factorization of linear graphs. *J. London Math. Soc.* **22** (1947), 107–111.
2. Matroids and graphs. *Trans. Amer. Math. Soc.* **90** (1959), 527–552.
3. Lectures on matroids. *J. Res. Nat. Bur. Standards* **69B** (1965), 1–47.
4. Menger's theorem for matroids. *J. Res. Nat. Bur. Standards* **69B** (1965), 49–53.
5. Introduction to the theory of matroids. RAND Corporation Report R–448–PR, 1966.

TVERBERG, H.
1. On Dilworth's decomposition theorem for partially ordered sets. *J. Combinatorial Theory* **3** (1967), 305–306.

VÁMOS, P.
1. On the representation of independence structures. Not yet published.

VOGEL, W.
1. Lineare Programme und allgemeine Vertretersysteme. *Math. Zeitschrift* **76** (1961), 103–115.
2. Bemerkungen zur Theorie der Matrizen aus Nullen und Einsen. *Archiv der Math.* **14** (1963), 139–144.

VAN DER WAERDEN, B. L.
1. Aufgabe 45. *Jber. Deutsch. Math. Ver.* **35** (1926), 117.
2. Ein Satz über Klasseneinteilungen von endlichen Mengen. *Abh. math. Sem. Hamburg. Univ.* **5** (1927), 185–188.
3. *Moderne Algebra* (Second edition; Springer, Berlin, 1937).

WELSH, D. J. A.

1. Some applications of a theorem of Rado. *Mathematika* **15** (1968), 199–203.
2. Kruskal's theorem for matroids. *Proc. Cambridge Phil. Soc.* **64** (1968), 3–4.
3. A bound for the number of matroids. *J. Combinatorial Theory* **6** (1969), 313–316.
4. Transversal theory and matroids. *Canad. J. Math.* **21** (1969), 1323–1330.
5. On matroid theorems of Edmonds and Rado. *J. London Math. Soc.* (2) **2** (1970), 251–256.
6. Submodular functions, transversals and matroids. Not published, 1969.
7. Generalized versions of Hall's theorem. *J. Combinatorial Theory.* To appear.
8. On the relation between matroids and transversal theory. Not yet published.
9. Related classes of set functions. *Studies in Pure Mathematics* (Academic Press, London, 1971). To appear.

WESTON, J. D.

1. A short proof of Zorn's lemma. *Archiv der Math.* **8** (1957), 279.

WEYL, H.

1. Almost periodic invariant vector sets in a metric vector space. *Amer. J. Math.* **71** (1949), 178–205.

WHITNEY, H.

1. On the abstract properties of linear dependence. *Amer. J. Math.* **57** (1935), 509–533.

WOLK, E. S.

1. The comparability graph of a tree. *Proc. Amer. Math. Soc.* **13** (1962), 789–795.
2. A note on 'the comparability graph of a tree'. *Proc. Amer. Math. Soc.* **16** (1965), 17–20.
3. On theorems of Tychonoff, Alexander and R. Rado. *Proc. Amer. Math. Soc.* **18** (1967), 113–115.

YAMAMOTO, K.

1. On the asymptotic number of Latin rectangles. *Japan. J. Math.* **21** (1951), 113–119.

ZORN, M.

1. A remark on method in transfinite algebra. *Bull. Amer. Math. Soc.* **41** (1935), 667–670.

Index of Symbols

The numbers refer to the pages on which the symbols are introduced.

Index of Authors

General Index

Where an entry has more than one page reference, the page printed in bold type indicates the place of the relevant definition or statement of theorem.

Mathematics in Science and Engineering

A Series of Monographs and Textbooks

Edited by RICHARD BELLMAN, *University of Southern California*

1. T. Y. Thomas. Concepts from Tensor Analysis and Differential Geometry. Second Edition. 1965

2. T. Y. Thomas. Plastic Flow and Fracture in Solids. 1961

3. R. Aris. The Optimal Design of Chemical Reactors: A Study in Dynamic Programming. 1961

4. J. LaSalle and S. Lefschetz. Stability by by Liapunov's Direct Method with Applications. 1961

5. G. Leitmann (ed.). Optimization Techniques: With Applications to Aerospace Systems. 1962

6. R. Bellman and K. L. Cooke. Differential-Difference Equations. 1963

7. F. A. Haight. Mathematical Theories of Traffic Flow. 1963

8. F. V. Atkinson. Discrete and Continuous Boundary Problems. 1964

9. A. Jeffrey and T. Taniuti. Non-Linear Wave Propagation: With Applications to Physics and Magnetohydrodynamics. 1964

10. J. T. Tou. Optimum Design of Digital Control Systems. 1963.

11. H. Flanders. Differential Forms: With Applications to the Physical Sciences. 1963

12. S. M. Roberts. Dynamic Programming in Chemical Engineering and Process Control. 1964

13. S. Lefschetz. Stability of Nonlinear Control Systems. 1965

14. D. N. Chorafas. Systems and Simulation. 1965

15. A. A. Pervozvanskii. Random Processes in Nonlinear Control Systems. 1965

16. M. C. Pease, III. Methods of Matrix Algebra. 1965

17. V. E. Benes. Mathematical Theory of Connecting Networks and Telephone Traffic. 1965

18. W. F. Ames. Nonlinear Partial Differential Equations in Engineering. 1965

19. J. Aczel. Lectures on Functional Equations and Their Applications. 1966

20. R. E. Murphy. Adaptive Processes in Economic Systems. 1965

21. S. E. Dreyfus. Dynamic Programming and the Calculus of Variations. 1965

22. A. A. Fel'dbaum. Optimal Control Systems. 1965

23. A. Halanay. Differential Equations: Stability, Oscillations, Time Lags. 1966

24. M. N. Oguztoreli. Time-Lag Control Systems. 1966

25. D. Sworder. Optimal Adaptive Control Systems. 1966

26. M. Ash. Optimal Shutdown Control of Nuclear Reactors. 1966

27. D. N. Chorafas. Control System Functions and Programming Approaches (In Two Volumes). 1966

28. N. P. Erugin. Linear Systems of Ordinary Differential Equations. 1966

29. S. Marcus. Algebraic Linguistics; Analytical Models. 1967

30. A. M. Liapunov. Stability of Motion. 1966

31. G. Leitmann (ed.). Topics in Optimization. 1967

32. M. Aoki. Optimization of Stochastic Systems. 1967

33. H. J. Kushner. Stochastic Stability and control. 1967

34. M. Urabe. Nonlinear Autonomous Oscillations. 1967

35. F. Calogero. Variable Phase Approach to Potential Scattering. 1967

36. A. Kaufmann. Graphs, Dynamic Programming, and Finite Games. 1967

37. A. Kaufmann and R. Cruon. Dynamic Programming: Sequential Scientific Management. 1967

38. J. H. Ahlberg, E. N. Nilson, and J. L. Walsh. The Theory of Splines and Their Applications. 1967

39. Y. Sawaragi, Y. Sunahara, and T. Nakamizo. Statistical Decision Theory in Adaptive Control Systems. 1967

40. R. Bellman. Introduction to the Mathematical Theory of Control Processes Volume I. 1967 (Volumes II and III in preparation)

41. E. S. Lee. Quasilinearization and Invariant Imbedding. 1968

42. W. Ames. Nonlinear Ordinary Differential Equations in Transport Processes. 1968

43. W. Miller, Jr. Lie Theory and Special Functions. 1968

44. P. B. Bailey, L. F. Shampine, and P. E. Waltman. Nonlinear Two Point Boundary Value Problems. 1968

45. Iu. P. Petrov. Variational Methods in Optimum Control Theory. 1968

46. O. A. Ladyzhenskaya and N. N. Ural'tseva. Linear and Quasilinear Elliptic Equations. 1968

47. A. Kaufmann and R. Faure. Introduction to Operations Research. 1968

48. C. A. Swanson. Comparison and Oscillation Theory of Linear Differential Equations. 1968

49. R. Hermann. Differential Geometry and the Calculus of Variations. 1968

50. N. K. Jaiswal. Priority Queues. 1968

51. H. Nikaido. Convex Structures and Economic Theory. 1968

52. K. S. Fu. Sequential Methods in Pattern Recognition and Machine Learning. 1968

53. Y. L. Luke. The Special Functions and Their Approximations (In Two Volumes). 1969

54. R. P. Gilbert. Function Theoretic Methods in Partial Differential Equations. 1969

55. V. Lakshmikantham and S. Leela. Differential and Integral Inequalities (In Two Volumes). 1969

56. S. H. Hermes and J. P. LaSalle. Functional Analysis and Time Optimal Control. 1969

57. M. Iri. Network Flow, Transportation, and Scheduling: Theory and Algorithms. 1969

58. A. Blaquiere, F. Gerard, and G. Leitmann. Quantitative and Qualitative Games. 1969

59. P. L. Falb and J. L. de Jong. Successive Approximation Methods in Control and Oscillation Theory. 1969

60. G. Rosen. Formulations of Classical and Quantum Dynamical Theory. 1969

61. R. Bellman. Methods of Nonlinear Analysis, Volume I. 1970

62. R. Bellman, K. L. Cooke, and J. A. Lockett. Algorithms, Graphs, and Computers. 1970

63. E. J. Beltrami. An Algorithmic Approach to Nonlinear Analysis and Optimization. 1970

64. A. H. Jazwinski. Stochastic Processes and Filtering Theory. 1970

65. P. Dyer and S. R. McReynolds. The Computation and Theory of Optimal Control. 1970

66. J. M. Mendel and K. S. Fu (eds.). Adaptive, Learning, and Pattern Recognition Systems: Theory and Applications. 1970

67. C. Derman. Finite State Markovian Decision Processes. 1970

68. M. Mesarovic, D. Macko, and Y. Takahara. Theory of Hierarchial Multilevel Systems. 1970

69. H. H. Happ. Diakoptics and Networks. 1971

70. Karl Astrom. Introduction to Stochastic Control Theory. 1970

71. G. A. Baker, Jr. and J. L. Gammel (eds.). The Padé Approximant in Theoretical Physics. 1970

72. C. Berge. Principles of Combinatorics. 1971

73. Ya. Z. Tsypkin. Adaptation and Learning in Automatic Systems. 1971

74. Leon Lapidus and John H. Seinfeld. Numerical Solution of Ordinary Differential Equations. 1971

75. L. Mirsky. Transversal Theory. 1971

In preparation

Harold Greenberg. Integer Programming

E. Polak. Computational Methods in Optimization: A Unified Approach

Thomas G. Windeknecht. A Mathematical Introduction to General Dynamical Processes

Andrew P. Sage and James L. Melsa. System Identification

R. Boudarel, J. Delmas, and P. Guichet. Dynamic Programming and Its Application to Optimal Control

William Stenger and Alexander Weinstein. Methods of Intermediate Problems for Eigenvalues Theory and Ramifications